BIODIVERSITY

This book, *Biodiversity: Law, Policy and Governance*, which is a compilation of articles and papers, fittingly discusses, critiques, analyses as well as examines the widening gap in biodiversity governance. It deliberates on the problem, discerns the various forms of innovative governance adopted and provides an in-depth analysis of the judicial discourses as well as the management system. It scrutinizes the main issues that plague a country that hopes to implement a method for securing a positive upshot by discussing the widening gap between the theoretical aspirations and the practical execution, where the obligations undertaken seldom transform from mere ink and paper.

Justice A. K. Sikri, Judge, Supreme Court of India, New Delhi, India

Conservation of biodiversity is a fundamental concern towards securing a sustainable future. This volume argues that despite various domestic and international policies and legal frameworks on biodiversity conservation – be it forest, wildlife, marine, coastal, etc. – their implementation suffers from many deficiencies. It explores the factors that hinder effective implementation of these policies and frameworks. It also analyses existing laws, both international and domestic, to identify inherent problems in the existing legal system. The book maintains that careful adherence to established procedures and protocols, public awareness, filling the lacuna in legal framework, and a strong political will are sine qua non for effective conservation of biodiversity and sustainable development. The volume defends the protection of traditional knowledge and participation of indigenous communities along with reinforcements of intellectual property in this regard. It also commends the role played by the Indian judiciary, especially the Supreme Court of India and India's National Green Tribunal, for the preservation and enhancement of natural resources by applying established as well as evolving principles of environmental law.

This book will be useful to scholars and researchers of environmental studies, development studies, policy studies and law related to biodiversity and conservation.

Usha Tandon is Professor and Head at Campus Law Centre, University of Delhi, India.

Mohan Parasaran is a senior advocate at the Supreme Court of India and was previously the Solicitor General of India and Additional Solicitor General of India.

Sidharth Luthra is a senior advocate at the Supreme Court of India and was earlier the Additional Solicitor General of India.

BIODIVERSITY

Law, Policy and Governance

Edited by
Usha Tandon, Mohan Parasaran
and Sidharth Luthra

Routledge
Taylor & Francis Group

LONDON AND NEW YORK

First published 2018 by Routledge

2 Park Square, Milton Park, Abingdon, Oxfordshire OX14 4RN
52 Vanderbilt Avenue, New York, NY 10017

Routledge is an imprint of the Taylor & Francis Group, an informa business

First issued in paperback 2019

British Library Cataloguing-in-Publication Data
A catalogue record for this book is available from the British Library

Library of Congress Cataloging-in-Publication Data
A catalog record for this book has been requested

ISBN: 978-1-138-28819-5 (hbk)
ISBN: 978-0-367-88677-6 (pbk)

Typeset in Sabon
by Apex CoVantage, LLC

To,
Our Childhood Feathered Friend

Passer Domesticus (House Sparrow)

Loss of Biodiversity has taken a severe toll on many aspects of Life in Delhi.
In growing up days, a familiar sight was the House Sparrow.
Despite being declared the State Bird in 2012, the sparrow is now nearly extinct in the Capital.
This book is dedicated to the House Sparrow.

CONTENTS

CONTENTS

CONTENTS

CONTRIBUTORS

Niharika Bahl is an advocate and independent law researcher based in New Delhi, India. She is currently a guest faculty member at Campus Law Centre, University of Delhi. She is a recipient of the L. R. Sivasubramaniam Gold Medal, 2011. Her areas of interest include gender relations, indigenous law and constitutional law. She has also been engaged with civil society organizations in the field of democratic and transparency reforms.

Stellina Jolly is Assistant Professor at South Asian University (SAU), New Delhi, India. Her teaching interests include international environmental law, conflict of laws and bioethics. She has published articles in *European Asian Journal of Law and Governance*, *UNESCO Journal of Bioethics*, *International Journal of Public Law and Policy*, *International Journal of Juridical Sciences*, etc. She has been part of editorial boards of peer-reviewed journals including Springer Books, *Indian Journal of Human Rights*, *International Journal of Bioethics* and *NUJS International Journal of Legal Studies and Research*.

Amrendra Kumar is Doctoral Fellow at the Centre for International Legal Studies, Jawaharlal Nehru University, New Delhi, India. He is currently pursuing his doctorate on the topic 'International Legal Regime on Access and Benefit Sharing Over Genetic Resources: A Critical Study'. His research interests lie in the areas of Access and Benefit Sharing relating to genetic resources, traditional knowledge and Intellectual Property Rights under biodiversity conservation law.

Niraj Kumar is Assistant Professor at National Law University (NLU), New Delhi, India. He is also a member of Center for Comparative Law at NLU. Before joining NLU, he taught at the Faculty of Law,

University of Delhi and has been a member of the editorial committee of *NCLJ*, published by Law Center II, Faculty of Law, University of Delhi. He has presented papers in public law in national and
international conferences. He has also prepared modules on judicial
review for UGC e-pathshala.

Pushpa Kumar Lakshmanan is an Associate Professor at Nalanda
University, India. He teaches Environmental Law and Policy. As
a Fulbright scholar he pursued Post-Doctoral research at Harvard
Law School, Harvard University. He is a Research Professor (Hon),
at World Institute of Scientific Exploration, Baltimore. Dr. Lakshmanan is an interdisciplinary scholar specializing in International
Environmental Law primarily on the Convention on Biological
Diversity, climate change, global environmental governance and
sustainable development. His current focus lies in comparative
study of environmental and biodiversity laws of South Asian and
ASEAN nations. He is the Founding Convenor of Green Fulbright-
ers Forum, India.

Shreeyash Uday Lalit is pursuing an LL.B. (Bachelor of Law) from
the Campus Law Center, University of Delhi and graduated with
a B.Tech. (Bachelor of Technology) from the Indian Institute
of Technology Guwahati, India. He has interned with several
renowned legal personalities, including a Supreme Court judge and
a Union Government Law Officer, and intends to litigate in future.

Sidharth Luthra is a senior advocate at the Supreme Court of India. He
has been the Additional Solicitor General of India and represented
the Union and State Governments in a variety of matters including
those relating to fundamental rights, environmental law, criminal
law, juvenile rights, education and policy issues. Mr. Luthra is a
fellow of the Cambridge Commonwealth Trust Society and visiting
Professor at Northumbria University at Newcastle, United Kingdom and an Honorary Professor at the Amity University, Noida,
Uttar Pradesh. He was recently conferred the Doctor of Laws (Hon-
oris Causa) by Amity University, Noida, Uttar Pradesh, India.

Vandana Mahalwar is Assistant Professor at the Indian Law Institute,
New Delhi, India. She holds a doctorate in law from National Law
University, New Delhi, India. Her doctoral work is on 'Character
Merchandising Under Intellectual Property Regime: International
Practice & Indian Perspective'. With various publications in reputed
journals, she has examined issues pertaining to the Right of Public-
ity, Character Merchandising, *vis-à-vis* Intellectual Property Laws.

She has participated and presented papers in many conferences and workshops. Before joining the Indian Law Institute, she was Assistant Professor at the Campus Law Centre, University of Delhi.

Erimma Gloria Orie teaches environmental law, oil and gas law and commercial law at the National Open University of Nigeria. She is the Head of the Department, Private and Property Law at the university and also represents the faculty of law on the board of the postgraduate school at the university. She has published several peer-reviewed articles and is a member of several professional bodies, such as the Nigerian Bar Association, Maritime Law Association, Chartered Institute of Arbitrators of Nigeria and the World Commission on Environmental Law (IUCN).

Mohan Parasaran is a senior advocate at the Supreme Court of India. Mr. Parasaran has been the Solicitor General of India and Additional Solicitor General of India for a long period of time. As the top law officer of the country, he is well versed with Government policies and schemes including those on biodiversity and mitigation of climate change. He was the Chairman of the Organizing Committee for the National Consultation for the Second Generation Reforms on Legal Education, wherein the Government of India unveiled its Vision Document road maps for the reforms in legal education in 2010. He represented before the UN in connection with changes to be made to uniform the International Commercial Arbitration.

Parikshet Sirohi is Assistant Professor, Campus Law Centre, University of Delhi, India and teaches Intellectual Property Rights and Business Regulation. He has delivered lectures at several prestigious institutions in the areas of Intellectual Property Law, Constitution Law, Competition Law, Consumer Protection Law and Criminal Law. He has authored a book entitled *Interface of Design Law with Copyright Law and Trademark Law* (2015) and co-edited *A Critical Approach to the Right to Information Legislation in India* (2015). He is pursuing a doctorate from Jamia Millia Islamia, New Delhi on the topic 'Food Security Law: An Endeavour to Tackle the Problem of Hunger through Legislation'.

Marie Valerie Uppiah is Lecturer in International Trade and Business Law at the University of Mauritius. She graduated in Law and Management from the University of Mauritius and has an LL.M. in International Business Law from Birmingham City University, UK. She has undertaken research in the fields of international trade law and maritime law. She is also a member of the West Indian Ocean Marine Science Association (WIOMSA).

Usha Tandon is Professor and Head at Campus Law Centre, University of Delhi. With 28 years of teaching experience, she is a world-renowned scholar for her work on human development focusing on environmental protection and women empowerment. She has numerous research publications to her credit and has presented her research work at various national and international platforms in various countries including the US, the UK, France, Germany, the Netherlands, China, Singapore, South Korea, Malaysia, Bhutan and Nepal. Her most recent research work includes her contribution as an expert, for the recent World Bank's Project, 'Enabling the Business of Agriculture- 2017 Report'.

Moses S-N Watulo is studying Law at the Campus Law Centre, University of Delhi, India. He is an ardent environmental enthusiast and conservation activist. He served as the Chairman, Kyambogo Environmental Education and Management Association (KEEMA). He is a member of the Institute of Internal Auditors (IIA, USA) and the Association of Chartered Certified Accountants (ACCA, UK). As a member of the ACCA, he deals with current issues of environmental and social reporting as required under integrated financial reporting by Chartered Accountants.

PREFACE

The discourse on law, policy and governance in the context of biodiversity requires a deep understanding of changing relations between human beings and natural environment. From 'development v. environment' to 'sustainable development', the scholars, world over, are now debating over 'anthropocentric' and 'eco-centric' approaches to environmental protection. It is, however, consistently and universally accepted that human beings have the responsibility to all biological life on earth because, apart from being the most consuming species of all, they are capable of thinking and perceiving Earth as a whole.

Thus, the conservation and enhancement of biological diversity is fundamental to ensure a healthy environment for present and future generations of humans as well as other living species. However, despite the UN's recognition of this problem and warnings, threats to biodiversity and environmental degradation have continued unabated, resulting in failure of meeting the original target of reducing the loss of biodiversity by the year 2010 and to meet Millennium Development Goals by 2015. The efforts to preserve biodiversity require a broad ecosystem approach rather than on merely preventing the extinction of species or parks, as the interdependencies of living organisms in ecosystems are quite complex. As knowledge of the causes and consequences of environmental problems continues to improve, it is imperative that legal concepts be adapted to take into account new information and to improve the capacity of law to respond to environmental imperatives. Further, environmental law must develop the legal tools that will turn established and evolving scientific and policy recommendations into enforceable norms. At a global level, this must be built on the principle of Common But Differentiated Responsibilities.

The genesis of this book lies in a three-day international conference organized by the Campus Law Centre, University of Delhi from February 12th to 14th, 2016. The book contains a few selected papers presented at the conference that have been thoroughly revised by the contributors and scrupulously edited by the editors. We are thankful to all contributors for seriously attending to editorial comments and gracefully meeting the timeline. Various persons at different stages have helped for the onerous task of reviewing and editing this book. We would like to place on record the hard work put in by Neeraj Gupta, Akash Anand, Priti Rana, Nancy Dhunna, Anju Sinha, Harleen Kaur, Shourie Anand, Varun Bansal, Niharika Bahl and Vishwadeep. Mr Neeraj Gupta, Research Scholar, Faculty of Law, University of Delhi, India deserves special appreciation for religiously going through several versions of chapters, in the process of several rounds of editing the manuscript.

We express our thanks to Professor Klaus Bosselmann from the University of Auckland for writing the Foreword to the book. We are grateful to Routledge, Taylor and Francis Group, and particularly Dr Shashank S. Sinha, for undertaking the task of publishing this book.

Usha Tandon, Mohan Parasaran
and Sidharth Luthra

FOREWORD

For too long, threats to biological diversity have been ignored. Twenty-five years after the Convention on Biological Diversity (CBD) was adopted in 1992, life on the planet has lost much of its robustness and resilience. We are today facing the sixth mass extinction with dire consequences for people and the planet.

Biodiversity is generally understood as vital to the functioning of biological cycles and life itself, but also as a prerequisite to humanity's economic and social development. In theory, the 196 signatory countries to the CBD agree that conservation of biodiversity must take priority in national policies and laws. In practice, however, countries have acted in almost complete defiance – as if there is no real problem.

Since 1992, the planet has lost nearly half of its wildlife with an average annual decline of 2%. Each year, we are losing several thousand, perhaps up to 100,000 species depending on the assumed number of existing species. Some 1.7 million species have been identified, although the actual number is far higher. Estimates range from 3 to 30 million with some studies suggesting that there may be over 100 million species on Earth.

We do know, however, that one-third of all known species are threatened with extinction. We do know that biodiversity loss occurs at an ever-accelerating speed. And we also know that it is caused by a single species, i.e. human beings with their devastating impact on habitats, natural cycles, the atmosphere and the oceans.

How then is it possible that we know about the human-caused threats to biodiversity, yet continue to ignore them in law, policy and governance?

If twenty-five years of experience with the CBD and its related protocols, commitments and conferences have taught us anything, it is that the efforts of the international community have not reversed the trend. Something is missing, and it would be too simple to merely discern a lack of implementation. The implementation gap in international law is well known, but far more important, yet less known, is the systematic gap.

The systematic gap in law and governance points to the mismatch between the complexity of a problem and its representation in law, policy and governance. Problems of low complexity are more likely to be represented accurately. Take local air pollution, for example, or even the ozone layer hole. In both instances, laws and policies once enacted have proven to be successful. Problems of high complexity, on the other end, require far more sophisticated approaches than the 1992 UN Framework Convention on Climate Change and the 2015 Paris Agreement for climate change, for example, or the CBD-regime for biodiversity. The very fact that twenty-five years of operationalizing the international climate and biodiversity conventions have been so unsuccessful is proof of the need for a more systematic approach.

The good news is that States themselves have, in principle, long accepted the need for a more systematic approach. They called it 'sustainable development'.

But talk about 'sustainable development' is cheap as long as it remains undefined and uncodified. Until today, States have resisted ecologically sustainable development as a defined concept that must shape and inform all our laws and policies. There are many reasons for such resistance and most have to do with the prevailing growth paradigm and the unwillingness to change the relationship between human beings and the natural environment. It is the unwillingness – or shall we say stupidity? – to alter the short-sighted anthropocentric outlook on ecological systems that is at the heart of the systematic gap.

This book illustrates the systematic gap in biodiversity governance. Its editors and authors provide insights into various forms of ecosystem governance, management regimes, judicial responses and innovative approaches. Together, they can be seen as advocacy for systematic change.

As long as States and their institutions continue to favour unilateral policies over multilateral governance, short-term measures over long-term solutions and anthropocentric short-sightedness over eco-centric foresight, conservation of biodiversity will not succeed. Or positively speaking, we need to take a new, systematic approach to biodiversity as being presented in this volume.

Klaus Bosselmann
Professor of Law, Director, New Zealand Centre for
Environmental Law (NZCEL), The University of
Auckland, and Chair, IUCN Ethics Specialist Group,
World Commission on Environmental Law

INTRODUCTION

*Usha Tandon, Mohan Parasaran and
Sidharth Luthra*

The main thesis of this book is based on the argument that despite international and national law and policy on conservation and protection of biodiversity – be it forest, wildlife, marine, coastal, etc. – its implementation suffers from lots of deficiencies. The book explores various factors that hinder the effective implementation of conservation of biodiversity and critically examines the legal issues involved in the protection of biodiversity, such as fair and equitable benefit sharing of natural resources, violation of intellectual property rights and the use of genetically modified crops. It vehemently argues that though various international instruments such as the United Nations Convention on the Law of the Sea (UNCLOS) and the Convention on Biological Diversity (CBD) along with its Protocols provide that the sovereign States have exclusive rights over the natural resources found in their territory and they can exploit them for their benefit, yet, the exploitation of natural resources by individual Nations has to be sustainable and not indiscriminate. This is more true for India, Brazil and South Africa, as the physical features and climatic situations of these countries have helped to harbour and sustain immense biodiversity under diversified ecological habitats such as forest, wetlands, mangroves, marine, mountains and deserts. It calls for the application of the *Urgenda* principle in holding governments responsible for fulfilling their obligations to mitigate climate change to protect biodiversity. The book asserts that for effective implementation of laws, the participation of indigenous people cannot be undermined. It applauds the commendable role played by Indian judiciary, especially the Supreme Court of India and India's National Green Tribunal, for the preservation and enhancement of natural resources by applying established as well as evolving principles of environmental law. It argues that careful adherence to the established procedures and protocols, awareness among the masses, filling the lacuna in legal framework, a strong

1

political will and strict implementation of laws are *sine qua non* for effective conservation of biodiversity and sustainable development.

While critically examining the Convention on Biological Diversity (CBD) 1992, the book argues that in order to further the aims of the Convention, there is a need for incorporation of some compulsory provisions under the Agreement on Trade-Related Aspect of Intellectual Property (TRIPS Agreement) which all member States should be mandated to comply with, and the proposed amendment to TRIPS should incorporate three disclosure requirements as to the i) source and country of origin of biological material; ii) evidence of prior informed consent; and iii) evidence of a benefit-sharing agreement. The book discusses the Indian Biological Diversity Act (BDA) 2002, which has provided a solid base for implementation of the Nagoya Protocol, and suggests that considering the rich biodiversity and biological resources available in the country, India should make use of the Nagoya Protocol to convert this instrument as a tool of opportunity to help local communities and conservation efforts. Thus, the BDA needs to be fine-tuned to incorporate compliance measures, checkpoints and involvement of local communities in the access and benefit sharing (ABS) process in a big way. The book strongly argues for the protection of traditional knowledge (TK) as many indigenous communities depend on TK for their survival. The existing forms of intellectual property or a combination of various forms of intellectual property can be used to protect TK till the development of a comprehensive *sui generis* legislative regime. The case for and against Genetically Modified (GM) crops has also been dealt with to argue that the choice that confronts us today is not between conventional crops and GM crops, but between sustainable and unsustainable farming practices, especially to meet the challenge of feeding the hungry world.

Investigating the role of gender, the book critically examines the current legal regime that fails to appreciate the crucial role played by women in food subsistence and managing biodiversity. It builds up a case for the feminist dimension of biodiversity challenges as women and nature are intrinsically linked. It argues that law makers and policy planners have to ensure that women are not simply 'added' to the conservation programme, but rather biodiversity itself should be defined in broader, more inclusive and fluid terms, incorporating diverse gendered experiences of different groups. It brings out a case for the creation of a 'World Environment Organization' after having critically examined the effectiveness of two global associations and institutions which have taken up the task of conservation, namely the International Union for Conservation of Nature (IUCN) and the United Nations Environment Programme (UNEP).

Having introduced the main thesis advanced in the book, we now briefly introduce various chapters, chronologically, bringing out the key issues and arguments raised by the contributors.

Erimma Gloria Orie underlines Nigeria's wetlands in Lagos Lagoon that remain neglected, resulting in loss of its rich biodiversity, despite Nigeria being a signatory to various international conventions and having enacted various domestic laws and policies for the regulation of wetlands. She identifies various challenges for such loss which include absence of proper law, capacity building and lack of awareness, the top-down approach, invasive species, widespread poverty, urbanization, flooding and erosion, etc. Exploring the role of wetlands in the protection, improvement and conservation of biodiversity, the author laments that although Nigeria is a signatory to the Ramsar convention, there is no specific national law regulating the maintenance of wetlands in Nigeria and the institutional framework for the management of the wetlands is also not adequate. She concludes by recommending the establishment of a national law to achieve sustainable use of the wetlands and effective mitigation of the impact of climate change.

Marie Valerie Uppiah takes up the issue of marine governance and draws attention to the consequences of damage caused to the sea which include not only destruction to the marine ecosystem but also an adverse effect on the socio-economic life of many States as the destruction of marine ecosystem leads to a decrease in fishing activities for local fishermen in various small islands. She analyzes the concept of Marine Spatial Planning (MSP) to regulate sustainable use of marine resources to help States to sustainably explore and exploit the marine resources present in their Exclusive Economic Zone (EEZ) and other parts of their marine territory. The chapter studies the case of Mauritius which has recently devised a roadmap for the development of its ocean economy involving elements of MSP which are being implemented for an effective and sustainable use of the marine resources. Providing an overview of the United Nations Convention on the Law of the Sea (UNCLOS) 1982, she indulges in in-depth analysis of MSP with special reference to Mauritius and strongly argues for the use of MSP, with its ecosystem-based approach, as the most appropriate way to protect the marine environment and at the same time enhance economic development at sea.

Amrendra Kumar provides a comparative analysis of legal governance regarding biodiversity protection in India, South Africa and

Brazil. He discusses their laws, policies and diplomacy addressing bio-diversity conservation and management in the contemporary age, and examines the constitutional provisions, legislations, regulations, plans and policies of these countries contributing to the principles and practices in this regard. The author proposes that these countries should amend existing legislations and improve the plans and policies for better biodiversity conservation and management; take initiative and leadership through diplomacy in attaining the objectives of the CBD and other MEAs at the regional and national levels; and actively participate in events and negotiations formalizing the practice and procedure for the conservation, access, transfer and equitable sharing of the benefits of the biological resources integrating these States' interest as biodiversity-rich countries.

Moses S-N Watulo examines the biodiversity conservation and management in East African countries with special reference to Uganda. Geographically, Uganda, he explains, is a land-locked country, that is, it has no access to the sea, being bordered by five countries; and this geographical location, he argues, has repercussions as far as the conservation and management of biodiversity is concerned because of the interlinked nature of the environmental masses, which in many cases traverse the national boundaries. He cites the example of the great River Nile that has its source at Jinja in Uganda on Lake Victoria and flows all the way to the Red Sea through South Sudan, Sudan and Egypt on its way, and this forms the Lake Victoria Basin composed of a network of rivers and the freshwater inland lake shared between Uganda, Kenya and Tanzania. He writes that vast species of flora and fauna thrive, notwithstanding the glaring negative effects of human activity, within the diverse geographical environment marked by breathtaking physical features like mountains, lakes, rivers, tropical forests and vast swamps. While analyzing the nexus between environment and industrialization, he provides a comprehensive description of the status of biodiversity conservation in Uganda, biodiversity risks in Uganda, role of various agencies, judicial efforts and the scattered legal framework, and argues for the enactment of consolidated legislation for biodiversity conservation and management in Uganda.

Pushpa Kumar Lakshmanan critically examines the implementation of the Nagoya Protocol on the Convention of Biological Diversity (CBD) by the Indian Legal system. He explains that while recognizing the sovereign rights of the States over their resources and their authority to determine the access to genetic resources through domestic legislation, CBD pleaded for conservation of biodiversity; utilization of biological resources in a sustainable manner; and fair and equitable

4

sharing of benefits arising out of the utilization of genetic resources and traditional knowledge, including their appropriate access. India ratified the CBD in 1994 and rolled out a domestic regulatory mechanism by enacting the Biological Diversity Act, 2002 and the Protection of Plant Varieties and Farmers' Rights Act, 2001. The scholar argues that even before the coming into force of the Biological Diversity Act, India dealt with ABS pertaining to genetic resources and traditional knowledge and cites the Kani-Tropical Botanic Garden and Research Institute (TBGRI) as an example to demonstrate the first ABS model in India. While referring to the major provisions of the Nagoya Protocol that require a fair and non-arbitrary regulatory mechanism within the jurisdictions of the Contracting Parties with proper access rules, benefit sharing procedures and compliance mechanisms based on prior informed consent and mutually agreed terms, he investigates how far the Biological Diversity (BD) Act in India is meeting the requirements of the Nagoya Protocol and what needs to be done in giving full effect to the provisions of the Protocol. He identifies the major challenge for the National Biodiversity Authority (NBA) and the State Biodiversity Boards (SBBs), constituted under India's BD Act, which is to convince the local bodies and constituting Biodiversity Management Committees (BMCs). To ensure better implementation of the ABS laws in India, he argues for strengthening the institutional capacities with regard to ABS, focusing on capacity building, training, conducting awareness programs for the stakeholders, including indigenous communities, and developing education materials.

Vandana Mahalwar draws attention to the tension between the Trade-Related Aspects of Intellectual Property Rights (TRIPS) Agreement and the Convention on Biological Diversity (CBD), and critically analyzes the proposed amendments made by bio-rich countries such as Bolivia, Brazil, China, Columbia, Cuba, Ecuador, Indonesia, Kenya, Pakistan, Peru, Thailand, Venezuela, Zimbabwe and India relating to the problem of bio-piracy and responses to such proposal by some developed countries. She highlights that though CBD has been made to provide access and benefit sharing on biological diversity, other branches of international laws are not in consonance and coordination with the CBD. The author argues that the Agreement on the Trade-Related Aspect of Intellectual Property Law has not taken, till date, the development in the area of biological diversity law, and it does not provide any mandatory provisions regarding the disclosure of the source or origin of the biological resources used in any particular invention Similarly, it contains no provision as to consensual agreement on the sharing and access of biological resources. In this context, she stresses the need to bring harmony in these two conflicting international agreements and hails the proposed amendment incorporating a genetic resource disclosure requirement which, she believes, would be a significant step towards the protection of owners of genetic resources.

Parikshet Sirohi deals with the legal issues involved in commercially grown, Genetically Modified (GM) crops. The author categorizes the concerns about GM crops into three main categories *viz.* health, environment, and the manner in which GM seeds are brought to the market and argues that the last category, namely the commercialization of GM technology, is not so much a criticism of the technology, but more a concern as to who benefits from it. He explains that the primary criticism of the GM system all over the world is that seed companies make it illegal for the farmer to save the seed and use it for propagation. In pursuance of this goal, these companies have even sought to sue farmers in North America whose crops were cross-pollinated by GM pollen. The author studies the impact of GM crops upon plants developed by traditional breeding techniques, which has been recognized the world over as an accepted and integral part of agricultural practices. After having dealt with all the aspects of GMC in its favour as well as against the use of it, he argues that in an increasingly hungry world, GM technology cannot be completely ruled out and calls for the judicious use of it.

Mohan Parasaran and Sidharth Luthra critically examine the role played by Public Interest Litigation (PIL) in the protection and enhancement of biodiversity in India. They explain how the higher Indian judiciary, particularly the Supreme Court of India, has intervened through PIL to apply the principle of Sustainable Development to deal with a variety of issues related to conservation of forests, protection of wildlife, protection of the rights of tribal people and forest dwellers etc. While focusing on conservation of forests, they provide a critical appraisal of important cases like *T.N. Godaverman*, *Ganesh Wood*, *Lafarge*, *CEL* and *WWF* along with the judgments involving application of sustainable development, the precautionary principle and public trust, and argue for the due compliance of the Court's Orders and Directions regarding conservation along with the effective implementation of laws.

Usha Tandon assesses the role of India's National Green Tribunal (NGT), established in 2010 by the Central Government, in protecting and enhancing the biodiversity in India. Consisting of judicial and scientific experts, NGT is a quasi-judicial body aimed to provide quick disposal of cases. Providing a brief overview of the provisions of the NGT Act, she identifies and critically analyzes the major orders passed by the NGT since its inception on biodiversity, like the preservation of biodiversity of Kaziranga National Park, Kanha National Park, Ghana Jungle, Sarguja Forests, the Aravali Ridge Area, Yamuna Biodiversity, Ganga Biodiversity, Bombay Marine Ecology, etc. She describes how the NGT has evolved innovative techniques, case after case, to remove hindrances coming in its way to protect the natural environment, which earned it an envious reputation. After having examined the sharp reactions of powerful stakeholders for developmental projects, she concludes by stating that, till date, the NGT has been successful in surviving all odds against it and argues for its bright future for the protection of biodiversity, ecology and natural resources.

Stellina Jolly presents a comparative analysis between the International Court of Justice and the Indian judiciary in biodiversity enforcement and compliance. In this context, she maintains that the judiciary is uniquely positioned to bring about strong compliance to international norms and bring environmental justice to the critical stakeholders, and explores the central and pivotal role that the judiciary plays in enforcing biodiversity laws. Through a catena of cases, she dwells into the role played by the Indian judiciary in protecting and bringing compliance to India's international commitment for biodiversity protection, like the case of *Vedanta*. She argues that the judicial insistence on consultation of local communities is based on not just natural justice principles but a host of international law principles and conventions, including the Convention on Biodiversity and the Rio Declaration on Environment and Development, and she looks forward to a more effective role played by the Indian judiciary in this regard.

Niharika Bahl demonstrates how the degradation of abundance of biodiversity has resulted in worsening the economic conditions of women in India. She explains the linkage between the indiscriminate use of biodiversity in resource-rich belts and the consequent exploitation of poor sections of the society, largely women, and argues that women's contribution in the management and conservation of biodiversity has been ignored and undervalued at national as well as international levels. She claims that barring the exhortation in the Convention on Biological Diversity, 1992, there is hardly any mention elsewhere to recognize women's vital role in conservation and sustainable use of biological diversity. Bearing in mind the rapid shrinking of biodiversity across the globe, she advocates for an urgent need to bring to the fore the role of women in sustaining and safeguarding the biodiversity.

Shreeyash Uday Lalit, while establishing the link between loss of biodiversity and climate change, critically analyzes the recent decision in the *Urgenda Foundation case* of the Netherlands Hague District Court, wherein the Court compelled the Government of the Netherlands to fulfill its obligations by virtue of its duty of due care to achieve its objective of a 25% reduction in greenhouse gas emissions under various international obligations, including the United Nations Framework Convention on Climate Change (UNFCCC). The author analyzes *Urgenda* in light of the civil, constitutional and international law; various provisions of the European Convention on Human Rights (ECHR); as well as the international obligations of the Netherlands. He examines the impact of the decision on foreign courts, and on the future deliberation of international climate change policy. Comparing *Urgenda* to the legal scenario in India, he argues for the possibility of Indian courts importing the tort of negligence or the duty of care as against the States for its failure to mitigate climate change to protect biodiversity.

Niraj Kumar, after enquiring into the possibility of a Global Environmental Organization with special reference to the IUCN and UNEP, argues for the creation of such an organization. The major arguments in favour of a Global Environmental Organization, according to him, include a coherent emphasis on environmental issues both internally, i.e. between various environmental organizations like the IUCN, UNEP, etc., and externally *vis-à-vis* other issue-based international organizations like the WIPO, WTO and WHO, and maybe also with international agencies like the IAEA. The author believes that achieving the above goal will not be very problematic because of the fact that most of the other organizations and agencies have their own in-house environmental branches, and because it has been accepted by almost all players that there can't be even benign neglect of the environment. One of the important structural issues which he explores in the chapter is the role of non-governmental organizations in this kind of organization.

Part I

BIODIVERSITY CONSERVATION
Issues and challenges

1

SUSTAINABLE WETLANDS AND BIODIVERSITY CONSERVATION

Nigeria's Lagos Lagoon in focus

Erimma Gloria Orie

Wetlands are land areas covered with water or where water is present at or near the soil surface all year or varying periods of the year. Wetlands exist in every country and in every climatic zone, from the polar regions to the tropics. Wetlands are some of the planet's most diverse and productive ecosystems, representing around 6% of the Earth's land area – some 570 million hectares, of which 2% are lakes, 30% bogs, 26% fens, 20% swamps and 15% floodplains.[1] They are distributed around the world and cover an area that is 33% larger than the USA. Throughout history, they have been integral to human survival and development and are valued for their contribution to ecological balance and biodiversity.[2] Notable functions of wetlands include flood control, groundwater recharge, coastal protection, sediment traps, atmospheric equilibrium and waste treatments as well as biological productivity, which provide nurseries for aquatic life and habitat for upland mammals like deer.[3]

Although one of the world's most important environmental assets, wetlands have been progressively lost and degraded from human activities. The rate of their loss is known to be greater than for any other type of ecosystem.[4] A review of 189 reports of change in wetland areas finds that the reported long-term loss of natural wetlands averages between 54% and 57% but loss may have been as high as 87% since 1700 AD.[5] There has been a much (3.7 times) faster rate of wetland loss during the twentieth and early twenty-first centuries, with a loss of 64–71% of wetlands since 1900 AD.[6] Over 50% of the world's wetlands have disappeared in the last century[7] while about 60% of European wetlands were lost before the 1990s[8] due to increased anthropogenic interference.[9] Although the rate of wetland loss in Europe has slowed, and in North America has remained low since the 1980s, the rate has remained high in Asia where large-scale and rapid conversion of coastal and inland natural wetlands is continuing. The situation is the same to an extent in Africa and Nigeria in particular.

Nigeria is richly endowed with abundant wetlands ecosystems, most of which are located in the Niger, Benue and Chad basins. These wetlands represent about 2.6% of the country's area of about 923,768km[2].[10] Nigeria has about 14 identified major wetlands.[11] These include Sokoto–Rima, Komadugu Yobe, Makurdi, Cross River, Lower Niger, Upper Niger and Kainji Lake, Middle Niger–Lokoja–Jebba–Lower Kaduna, Niger Delta, Benin–Owena, Lake Chad, Badagry and Yewa creeks, and Lagos Lagoon.[12] The Lagos Lagoon is one of the most extensive wetlands in the southern region of Nigeria. Both Lagos and Lekki Lagoons have a combined size of 646km[2].[13] The only western outlet for the two lagoons is through the Commodore Channel which links Lagos Lagoon to Bight of Benin/Atlantic Ocean. With direct connection to the sea, salinity is generally higher in Lagos Lagoon and the waters are brackish while the Lekki Lagoon is of freshwater.[14] The Lagos Lagoon is found in 10 out of the 20 Local Government Areas (LGA) in the State. These include Epe, Etiosa, Ibeju/Lekki, Ikorodu, part of Amuwo–Odofin, Apapa, Kosofe, Lagos Island, Mainland and Shomolu LGAs.

In spite of the acclaimed benefits or 'ecosystem services' that the wetlands provide to the environment, the Lagos Lagoon has witnessed more than 96% loss due mainly to anthropogenic factors. In addition, like most of the Nigerian wetlands, it is not on gazette, neither is it well documented.[15] There is also no specific national law regulating the maintenance of wetlands. For example, in the entire country, only 11 wetland sites are recognized as Ramsar sites,[16] both inland and coastal.[17] Even the Niger Delta, Nigeria's largest and richest biodiversity region, is yet to be recognized and placed on gazette as a Ramsar site.[18]

The implication is that these wetlands are therefore at the risk of degradation and biodiversity loss resulting mainly from human activities which will be exacerbated by climate change. Degradation of wetlands through land development, poor water management, etc. reduces the capacity of wetlands to provide significant ecosystem services. The water regime of such a wetland can be so altered that the land can no longer support the wetland vegetation and maintain hydric soils. For example, if a wetland is lost, most if not all of its wetland functions are also lost.[19] Thus, the loss of the Lagos Lagoon will have dire consequences on the food chain[20] and health of the entire agrarian populace of Lagos State and beyond. There is therefore the urgent need for establishment and proper enforcement of regulation on the protection of wetlands to ensure sustainable use and conservation of the Lagos Lagoon wetlands.

I. Conceptual clarification

A. Wetlands

The most widely accepted definition of wetlands is as provided in the Convention on Wetlands of International Importance (Ramsar) Convention on Wetlands.[21] The convention defines wetlands as "areas of marsh, fen, peatland or water, whether natural or artificial, permanent or temporary, with water that is static or flowing, fresh, brackish or salt, including areas of marine water the depth of which at low tide does not exceed six metres." It "may incorporate riparian and coastal zones adjacent to the wetlands, and islands or bodies of marine water deeper than six metres at low tide lying within the wetlands."[22] Wetlands as broadly defined by the Ramsar convention include lakes and rivers, swamps and marshes, wet grasslands and peatlands, oases, estuaries, deltas and tidal flats, near-shore marine areas, mangroves and coral reefs, and human-made sites such as fish ponds, rice paddies, reservoirs and salt pans. They are not exclusively land or water environments. However, in general wetlands are those areas where water is the primary factor controlling the environment and the associated habitats.

The thrust of the Convention is for contracting parties to[23]:

i   work towards the wise use of all their wetlands;
ii designate suitable wetlands for the list of Wetlands of International Importance (the 'Ramsar List') and ensure their effective management; and
iii cooperate internationally on trans-boundary wetlands, shared wetland systems and shared species.

In line with the Convention, therefore, Nigeria is expected among other things to designate suitable wetlands for the Ramsar List, and ensure their effective management while also working towards the wise use of all their wetlands.

B. Biodiversity

Biodiversity denotes the variability of living organisms from all sources including terrestrial, marine and other aquatic ecosystems and the ecological complexes of which they are part. It encompasses the variety of all forms of life on earth, which provides the building blocks for human existence and our ability to adapt to environmental changes in the future.[24] Biological diversity involves genetic, species and ecosystem diversity. So many of these unique ecological characteristics and ecosystems, such as mountains, mangroves, wetlands, savannas, rainforests and transitory sites for migratory species, are vital for sustainable environment, hence the need for conservation.

C. Conservation

The word *conservation* means the preservation from destructive influences, natural decay, or waste; the preservation of the environment, especially of natural resources; and the maintenance of essential ecological processes and life-support systems, the preservation of genetic diversity, and the sustainable utilization of species and ecosystems.[25]

As used in this monograph, conservation is the protection, preservation, safeguarding, maintenance and management of a thing to secure that thing (in this case, the Lagos Lagoon and by extension Nigeria's biodiversity) for future generations.[26] The Convention on Biological Diversity (CBD) signed in June 1992 is probably the most all-encompassing international agreement on biodiversity ever adopted. Nigeria was among the 153 countries that signed. The essence of the convention is to ensure that there is a balance between use and conservation to achieve sustainability/sustainable use.

D. Sustainable/unsustainable use

The exploitation of natural resources in a manner that will ensure their availability for the future is what is referred to as sustainable use. On the other hand, unsustainable use is the use of a thing in a manner that cannot be sustained, that can render it extinct or unavailable for use for the next generation. This may be due to factors like over-extraction of natural resources, over-harvesting, wrong techniques like peeling off bark, or cutting down or uprooting a tree, pollution of wetlands, large-scale poaching and random hunting of all species. In the context of biodiversity in Nigeria, unsustainable use of biodiversity and forest products is the norm. Most people in rural areas depend in part on the extraction of resources like seafoods from the swamps and riverine areas. In the urban areas, the tendency is to convert and/or develop every available land for monetary gain. As population increases, extraction increases until there are no resources left to harvest. Even more troubling is the widespread over-extraction of seafood products purely for profit.

II. Role of wetlands in conserving biodiversity

Wetlands are indispensable for the countless benefits or 'ecosystem services' that they provide humanity. The goods and services that wetlands provide to humanity are numerous and fundamental to meeting some of our most basic needs namely:

A. GDP booster

It has been estimated that the ratio of conservation costs to Nigeria was about 3.8% of gross domestic product (GDP) while the aggregate contribution of biodiversity to the GDP was about 46% in 2001.[27] In 1990, it was estimated that the monetary value of other benefits realized from conservation was put at well over $6 billion. With the increase in bio-prospecting and bio-discovery activities in Nigeria and the growth in biotechnology-related industries that utilize indigenous genetic materials as feedstock, the 2002 estimate for the benefits of biodiversity to Nigeria was over $8 billion per annum.[28] The strategic plan therefore provides for a significant increase in the national expenditure on biodiversity conservation in order to ensure the continuous availability of these resources.

B. Carbon sequestration

In addition, wetlands act as carbon 'sinks' and are therefore a fundamental asset in our efforts to reduce levels of greenhouse gases in the atmosphere.[29] It has been submitted in some quarters that wetlands may account for as much as 40% of the global reserve of terrestrial carbon and can make an important contribution towards climate change mitigation.[30]

C. Flood buffers

Wetlands are also an effective buffer against flooding and storms, as they can store water in their soil or retain it as surface water, thereby slowing down the rate of flooding.[31] They are also nature's chemical filter 'factories' and encourage the breakdown of nutrients and sediments carried in water. They play a role in sewage treatment, and often act as a source of hydroelectricity. For untold generations, we have also used them to support our transportation and communication needs, and as a very important space for recreation and tourism activities.

D. Source of water

Wetlands also provide a wide range of other services such as water provisioning, management and purification. They allow farmers to irrigate farmland and harvest many species of fish and shellfish – more than two-thirds of the fish we consume are dependent on coastal and inland wetland areas and offer recreational and tourism opportunities.

E. Support to agriculture

Wetlands also support agriculture through the maintenance of water tables and nutrient retention in floodplains. Rice, a common wetland plant, is an important agricultural product in southern European countries.[32] A field visit to the Lagos Lagoon wetland areas revealed that the wetlands are used for the cultivation of different staples like plantain (*Musa sapientum var paradisiacal*); banana (*Musa sapientum*); sugar cane (*Saccharum officinarum*); bitter leaf (*Vernonia amygdalina*); red spinach/plumed cock's comb/silver cock's comb, locally called soko (*Celosia argentea*); West African mallow leaves, locally called Ewedu (*Corchorus olitorius*); cocoyam (*Colocasia esculenta*); and fruits and vegetables.

F. Reservoirs of fish

Wetlands and other freshwater habitats in Lagos, as elsewhere, are important reservoirs of fish and other aquatic food items for people. They provide habitat for a myriad of other diverse species. The marine and coastal environment of Nigeria (including Lagos) is rich in resources and species diversity. The mangroves in these areas are the largest remaining tracts in Africa and also the third largest in the world, covering about 9,723km².[33] The mangrove ecosystem provides a nursery and breeding ground for many of the commercial fishery species taken in the Gulf of Guinea. Nigeria's coast is said to have about 199 species of finfish and shellfish, a number of which are used commercially.[34] The Nigerian shrimp fishery is especially strong, and shrimp are now being exported to other countries, including the United States. Artisanal fisherfolk harvest a large variety of fish, crustaceans and molluscs from the estuaries and channels and utilize mangrove and swamp forest products for a variety of domestic uses.[35] A variety of birds, mammals and reptiles inhabit the mangroves and swamp forests of the coast, including a few endemic species like the Sclater's guenon and the Nile Delta red colobus monkey.[36] Although a few species of sea turtles lay eggs on Nigerian beaches, they are rare and under threat from human predation.[37]

The benefits discussed above are quite enormous and a great boost to the economy of such countries with wetlands. However, in the case of Nigeria, the benefits of these wetlands are not being fully maximized. On the contrary, the Nigerian wetlands are being destroyed and converted to other uses, resulting in loss of the wetlands as exemplified in the Lagos Lagoon wetlands.

III. Effects of loss of wetlands

There is evidence that the loss and degradation of wetlands increases hazards from coastal storms and tidal surges. The result is the ecosystem services that wetlands provide to people are compromised, leading to unintended but foreseeable consequences for both the natural systems and humankind. In Lagos State, more than half of the store of mangroves and swamps in the study area has been decimated, consumed ostensibly by urban development. At the local council level, the surviving mangroves in order of magnitude are found in Ikorodu, Etiosa, Kosofe and Epe LGAs, all on the Lagos Lagoon. Likewise, the remaining swamp wetlands are in order of magnitude also largely in four local councils. These are Epe, Ibeju–Lekki, Ikorodu and Eti–osa LGAs. The wetlands of the Lekki Conservation Center run by the Nigerian Conservation Foundation (NCF), which is now totally rimmed by urban development, is part of the remaining wetlands in Etiosa LGA.[38]

In Nigeria, on a fairly consistent basis, the news is replete with reports about devastating effects of flooding on properties. For example, Punch Newspaper[39] reported that whenever Oyan Dam is opened, the storm usually affects the floodplains and wetlands covering 2,800ha of River Ogun catchment within Lagos, comprising Ikosi–Ketu, Mile 12, Agiliti, Thomas Laniyan Estate, Owode–Onirin, Agboyi, Owode–Elede, Maidan and Isheri North Scheme. On the 10th of July 2011, the entire Lagos State was flooded. In some areas of Lagos State, the floodwater mark on the houses was about 3.5 metres. Conservatively, about 25 persons were said to have lost their lives in the Lagos flood, which caused the state to close down schools. In 2012, about 21 states in Nigeria, including Lagos State, were declared disaster zones by the Federal Government of Nigeria when they were over-run by flooding.[40] All these are the outcome of the unconcerned attitude of both the people and the Government of Lagos State to protect the wetlands, which are incessantly converted to uses with economic gains. These effects must be addressed for sustainable use of Lagos Lagoon using appropriate legal instruments. The next section therefore examines the regulation of the wetlands in Nigeria.

IV. The regulation of wetlands in Nigeria: issues and challenges for Lagos Lagoon

The Federal Government's mission is that Nigeria's rich biological endowment together with the diverse ecosystems will be secured, and its conservation and management assured through appreciation and sustainable utilization. This section attempts to review the major past and present efforts of the Government in biodiversity conservation, especially as it concerns the Lagos Lagoon wetland.

A. Policy

The 1999 national policy on the environment, among other objectives, focuses on securing a quality environment that is adequate for good health and well-being; conserving and using the environment and natural resources for the benefit of present and future generations; and restoring, maintaining and enhancing the ecosystems and ecological processes for the conservation of biological diversity. With respect to the CBD, the policy is complemented by the policy on biodiversity.[41] Some other policies that have bearing on the protection of the wetlands include the National Forest Policy,[42] the purpose of which is to ensure sustainable forest management and promote a participatory process of development while facilitating private sector forestry development and adopting an integrated approach to forestry development. There is also the National Policy on Erosion, Flood Control and Coastal Zone Management.[43] This policy is positioned to initiate and execute measures that curtail or minimize the effects of factors that accelerate or aggravate the processes of soil erosion, flooding and coastal degradation; execute projects that ameliorate the impacts of erosion, flooding and coastal degradation; initiate and promote the capture and efficient utilization of rain water; create awareness on practices that minimizes erosion, flooding and coastal degradation; and collate and disseminate data in its areas of mandate.

B. Legal framework

The Constitution[44] provides the fundamental legal principles for environmental protection in Nigeria. It states that the Government shall protect and improve the environment and safeguard the water, air and land, forest and wildlife. In an effort to achieve environmental sustainability in Nigeria, the Nigerian Government established the Federal Environmental Protection Agency (FEPA)[45] which was later scrapped in 1999. This Act was replaced in 2007 by a more comprehensive Act.[46] Whereas the new Act focuses on the enforcement of environmental guidelines and policies,[47] its Agency[48] is charged with responsibility for the protection and development of the environment, biodiversity conservation and sustainable development of Nigeria's natural resources as well as environmental technology. In addition, the Federal Government through the Agency has developed 24 Environmental Regulations which have been published in the Official Gazette and are now in force. Amongst the Environmental Regulations, six are directly or indirectly linked to management in Nigeria at various capacities.[49]

Furthermore, there are some existing laws and legislations on natural resources conservation in Nigeria which complement the federal laws on the conservation of natural resources. Some of these laws are the Natural Resources Conservation Act,[50] the Federal Environmental Protection Agency Act,[51] the Environmental Impact Assessment Act,[52] the Endangered Species (Control of International Trade and Traffic) Act[53] and the National Parks Decree.[54]

Nigeria is a signatory to several international treaties, conventions and multilateral agreements for the conservation and sustainable use of biodiversity, which demonstrates her commitment to the conservation of natural resources.[55] Consequently, the country took an active part in all the negotiation processes leading to the adoption of the Convention on Biological Diversity in Rio de Janeiro, 1992. Subsequently, the country ratified the convention in 1994 and thereafter started the process of preparing her Biodiversity Strategy and Action Plan. In 1993, "A Country Study Report" prepared by the Federal Environmental Protection Agency (FEPA) documented the status of Nigeria's biological diversity, policies, laws and conservation programmes.[56]

In practice, this legal framework and efforts to implement them by stakeholders have, over time, precipitated issues and challenges[57] which must be resolved for enhanced biodiversity management.

C. Institutional framework

Because wetlands protection is an environmental matter, the principal institution charged with the protection of wetlands is the Federal Ministry of Environment. The Ministry is to achieve this through the implementation and enforcement of the NESREA Act, its supporting regulations and other relevant laws. Apart from the above structure, Nigeria has several linkage centres in Nigerian institutes and universities created in 2001 as part of the activities undertaken in support of the Convention on Biological Diversity.[58]

In addition to domestic efforts, there are international collaborative efforts like the Canadian International Development Agency (CIDA). CIDA is the single donor in Nigeria with a stated environmental mandate.[59] Most of its support goes to capacity-building and policy efforts of Federal and State Governments in the natural resources sector. The United Kingdom's Department for International Development (DFID) has also supported community work around protected areas in Cross River State. Furthermore, the World Bank, through its Global Environment Facility 'Local Empowerment and Environmental Management Project,' is working with the Department of National Parks and Bauchi State in Nigeria to develop capacity for protected area management components (among other activities).[60]

There are also some regional activities supported by donor agencies which have components in Nigeria. These include the United Nations Environment Programme (UNEP)[61] and the United Nations Food and Agriculture Organization (UNFAO).[62] Some Non-Governmental Organizations (NGOs)[63] in this area also include the Nigerian Field Society, the Nigerian Conservation Foundation International, Wetlands International, the Wildlife Conservation Society and several others.

Ordinarily, these institutions are expected to assist in the management of the Lagos Lagoon wetlands for the benefit of the immediate community and Nigeria at large. However, this has not been the case due to some challenges discussed below.

V. Challenges to the protection of Lagos Lagoon wetlands

A. Absence of law

Nigeria has ratified the Convention on Biological Diversity but is yet to transmit it into its body of laws. By virtue of section 12 of the

Nigerian Constitution, such conventions cannot have the full force of law in Nigeria except the Convention on Biological Diversity is first domiciled.[64] Although there are some related laws on the subject matter, the absence of a specific law on wetlands is a big challenge to the institution charged with enforcement. In addition, there is an absence of unique vulnerable-wetlands Geographical Information System (GIS) maps for each community, and technical training and outreach to MassDEP staff on how to use the maps for stormwater-management planning. These maps show where each community's vulnerable wetlands are located relative to developed sites so that they can better determine and plan the appropriate placement for and type of stormwater-control options to better protect these resources and reduce phosphorus loading to the river.[65]

B. Weak enforcement

No matter what laws and policies are in the books about biodiversity and natural resource conservation, Nigeria will still have a major difficulty in enforcing them. The institutional framework for such enforcement is weak. The Ministry of Environment, as the coordinating Ministry, has mainly the NESREA Act and some other regulations listed above to rely on for the management of the wetlands. The regulations are neither comprehensive[66] nor backed by adequate deterrent measures. Most of the regulations are also not replicated in the various States' laws in Nigeria. Furthermore, some of the wetlands are not yet placed on the gazette, and are neither well documented nor adequately protected by regulation, so the result is weak enforcement of the available regulations.[67] In Nigeria, information about the component of biodiversity is incomplete; at best, only a provisional identification can be made of the components of biological diversity requiring special conservation measures. Data derived from the identification and monitoring of biological diversity, and of activities having or likely to have adverse impacts on biological diversity, are scattered among a large number of organizations. In practice, proper enforcement will require the collaboration of other ministries. Eliciting this cooperation is usually not easy because of inter-ministerial squabbles and function overlap, and generally do not have a strong capacity to organize, analyze, evaluate and disseminate data.

C. Capacity-building and lack of awareness

The laws are not properly understood by many of the enforcement agencies and even the communities. There are few trained and qualified people in the field to enforce compliance. In addition, an oversight function at all levels is sorely lacking, making way for 'entrepreneurial opportunities' to make money through unsustainable use of the wetlands, e.g. over-harvesting of seafoods and conversion of wetlands for construction purposes.[68]

Effective management of natural resources requires accurate data on the distribution and abundance of the resources involved, the ecological parameters of sustainability for each, the amount harvested from year to year, the benefits of the Lagos wetlands and other relevant factors. There is no such information on the Lagos wetlands; no data on the actual content of the wetlands, their peculiarities and the challenges of the habitat. Without proper understanding of existing regulations and information about the present status of most habitats and species, effective management becomes extremely difficult and challenging.

D. The top-down approach

The top-down approach[69] of the Government to biodiversity management and in particular wetlands conservation is a big challenge to the system. Despite the current limited recognition of wetlands benefits, many potentially conflicting interests still exist, such as that between the interests of landowners (the Government), the local community (the original owners of the land before the advent of the Land Use Act which divested them of such ownership rights) and the general public, and between developers and conservationists.[70] Part of the challenge is also how to integrate biodiversity concerns in sectoral policies and programmes and modify existing government policies and regulations to achieve consistency, e.g. the Land Use Act, the national policy on forestry and the policy on biodiversity conservation. The Government must find ways of carrying the local community along as the custodians of the various wetlands and compensate them for playing such roles.

E. Invasive species

In Nigeria, there are some exotic and non-indigenous species that are invading habitats from where they were previously unknown.[71] In the Lagos Lagoon coastal mangrove swamps, the Nipa palm (*Nypa fruticans*) is displacing native species. The water hyacinth (*Eichornia crassipes*) is another exotic plant species rapidly clogging up waterways in Lagos State, as is the cattail (*Typha* spp). Increased ponding and fluctuating water levels have been shown to aid conditions that contribute to the spread of invasive species and dominance of invasive species due to their level of tolerance of hydrologic change, loss of sensitive species and loss of species richness.[72]

F. *Widespread poverty*

Despite Nigeria's rich renewable and non-renewable resources, poverty in Nigeria is widespread and rated among the world's worst. A 1996 survey showed that about 67% of Nigerians (mostly women) live below the poverty level.[73] This is an indication that the country's natural resources are being poorly harnessed and demonstrates the need for environmental policies that are designed to reconcile conservation and development in a sustainable manner. In Nigeria, poverty is directly linked to biodiversity loss. This is because rural livelihoods depend almost entirely on biodiversity.[74] This condition is exacerbated by exponential increases in population accompanied by intensified industrial activities for economic development. In order to address biodiversity concerns and wetland conservation in particular, the problem of poverty must be addressed by providing alternative livelihood options to rural communities.

G. *Urbanization*

This is a major cause of loss of coastal wetlands. Urbanization impacts wetlands in numerous direct and indirect ways. There could be direct habitat loss, suspended solids additions, hydrologic changes, alteration of water quality, increased runoff volumes, diminished infiltration, reduced stream base flows and groundwater supplies, or prolonged dry periods, to mention just a few.[75] The ecosystem services provided by these wetlands are, however, being threatened by an increased rate of changes in land use and land cover, as well as global climate change, as many cities grow very rapidly, spreading into critical habitats and sites of conservation interest. For example, large areas of the Lekki Peninsula, which is an important mangrove and lowland swamp forest habitat for many threatened hydrophilic species, have been cleared and 'reclaimed' to create more space for the rapidly expanding city of Lagos.[76] Degradation of coastal wetlands through land development and water management thus reduces the capacity of wetlands to provide significant ecosystem services that reduce the risks of living and working in coastal landscapes.

According to the Millennium Ecosystem Assessment, climate change will not only further worsen the loss and degradation of many wetlands and cause the extinction of or decline in their species, but the human populations that are dependent on their services will also be negatively impacted. Based on the Reactive Global Orchestration and Order from Strength scenarios of the Millennium Ecosystem Assessment, the degradation of these wetlands is expected to increase while the extent of the wetlands is expected to decrease due to increases in human population.[77] There is therefore a need to devise ways of protecting wetlands from the impact of urbanization and climate change.

H. Flooding and erosion

The loss of wetlands will result in the impairment of ecological services like flood retention, leading to increased flooding and erosion.[78] The degradation of wetlands, with their filtering and purifying mechanisms, poses grave threats to both the surface and substrate water quality and volume in the Lagos Lagoon area.[79] This is believed to be exacerbated by the borrowing of sea sand for swamp reclamation from the nation's continental shelf and the overall impact of climate change. The challenge therefore is for the Government (federal and state) to develop the necessary political will to implement its existing relevant policies, laws and regulations on coastland management until the establishment of a specific law on the protection of wetlands.

I. Impact of population

Up to the year 1965, the extent of the Built-Up Area (BUA) in Lagos increased at an average annual rate of 321.06ha (7.57%), while the extent of BUA in inland wetlands and coastal wetlands was 1,840.41ha and 2,074.22ha, respectively.[80] However, between 2008 and 2014, the extent of the BUA in Lagos increased to 3,346.20ha, while the extent of BUA in inland and coastal wetlands was reduced to 222.37ha and 667.12ha, respectively.[81] The increase in the extent of BUA is indicative of the conversion of wetlands in the Lagos Lagoon area to other uses.[82] These land use–land cover changes for 1965, 2008 and 2014 show that about 80% of the area has been built-up. Some of the flora expected to be lost from the wetlands and creeks based on land use changes include *Thuja* sp and *Ficus* sp. This would affect some of the ecological services offered by these plants. The challenge therefore is for the Government to devise ways of managing the population to ensure that it does not render the wetlands vulnerable.

J. *Recognition of market or economic value*

Although the commercial value of biological diversity in Nigeria exceeds the cost of conservation measures by more than $3 billion at 1993 values ($3.75 billion versus $0.37 billion), biodiversity conservation has not been recognized as a feasible investment in Nigeria's economic development, and consequently natural resources valuation has not been fully incorporated into the national economic planning.[83] The challenge, therefore, is that until biodiversity conversion is accorded such recognition, it may be difficult to incorporate it into national economic planning.

The above challenges call for a review of government positions to be able to address the issues. The next section therefore examines the strategies for conserving the Lagos wetlands.

VI. Strategies for conserving Lagos Lagoon wetlands

A. *Legal framework*

Ratification of the Convention is not enough. There is an urgent need to transmit the Convention into the Nigerian body of laws. Furthermore, it is impossible to guarantee adequate conservation or the sustainability of the Lagos wetlands without a law on wetlands that will galvanize enforcement of such a policy initiative. In addition, the law should provide for the mapping of wetlands to be able to determine and plan the appropriate placement for and type of stormwater-control options to better protect the resources.

B. *Enforcement*

A strong institutional framework is essential for proper enforcement of law. This will entail having the correct mix of infrastructure and facilities necessary for such operations, proper funding, capacity development, data availability, collaboration with sister ministries or agencies, etc. It is equally necessary to ensure that other relevant/complementary laws and regulations,[84] like the Environmental Impact Assessment Act, are co-opted as part of the enforcement mechanism.

C. *Building capacity and awareness*

Critical to the conservation of biodiversity and sustainable wetlands in particular is the need for a systematic and long-term approach to building the knowledge base of Lagos' biological diversity and the integration of this knowledge into the decision-making process.

There is a need to improve the knowledge of change in wetland areas in Lagos State, and to improve the consistency of data on change in wetland areas in published papers and reports. Perhaps the best way to protect wetlands is to educate the public of their benefits. According to the NBSAP[85] by 2020, 30% of Nigeria's population should be aware of the importance of biodiversity to the ecology and economy of the country. This government strategy appears not to be sufficiently proactive. In a country of over 170 million people, and considering the benefits that Nigeria stands to lose from such lack of awareness exacerbated by the impact of climate change, the 30% awareness by 2020 appears too low. If the public does not recognize the benefits of wetlands preservation, wetlands will not be preserved. A more proactive measure will be to incorporate this information into the schools' curriculum from primary to tertiary education.

D. Economic value

The Government of Nigeria should recognize biodiversity conservation as a feasible investment in Nigeria's economic development and consequently fully incorporate natural resources valuation into the national economic planning. This calls for a review of government positions to be able to address the issues. The wetlands in Nigeria are not all placed on the gazette and well documented; the few that are well documented and recognized should be adequately protected through proper implementation and enforcement of the relevant law.

Thus, in line with article vi of the Convention on Biological Diversity,[86] the Federal Ministry of the Environment initiated the Strategy and Action Planning process in order to guarantee the conservation of Nigeria's biological diversity. This strategy should be adopted by other ministries and the communities.

E. Population, poverty and urbanization

The strategy for population explosion and urbanization is for the Government to develop the necessary political will to implement its biodiversity policy and related regulations while at the same time embark on the establishment of a law to protect the Lagos wetlands. A situation where the Government will permit reclamation of wetlands for construction purposes in outright disregard of its policies and other laws/regulations, like the Environmental Impact Assessment Act, is embarrassing to the system. The implementation of government policy on rural development is one sure way of controlling urban migration and poverty.

F. Public participatory approach

Many conservationists are of the view that the best hope for protecting and conserving natural resources is through a public participatory approach which will entail carrying the local communities along and also by compensating them for preserving the wetlands.[87] Communities near protected areas and any other remaining wild areas in Nigeria rely on these resources for their existence, and it is to their advantage to conserve them for future uses.[88] Carrying the local communities along will require the establishment and formalization of the 'Development Triumvirate' that will comprise the Government, the private sector, and other stakeholders like the indigenous community, the civil society, etc. For example, in Costa Rica and Peru, the governments adopted the Payment for Environmental Service system which entails paying the community members compensation for forest conservation, reforestation and agro-forestry.[89] The system was such a huge success that it was recommended for other countries. Protection can be accomplished only through the cooperative efforts of citizens.

G. Flooding and erosion

Various governments (federal and state) must develop the necessary political will to implement its existing policies, regulations and laws as a way of encouraging citizens to comply with rules. Government should show a good example by complying with existing policies and laws while also making efforts to establish a specific law on wetland management.[90] Indeed, many services delivered by wetlands (such as flood mitigation, climate regulation, groundwater recharge and prevention of erosion) are not marketed and accrue to society at large at the local and national scale. The Government is aware of these benefits and has put some policies and complementary laws in place to ensure that the ecosystem is protected, but on the other hand does not passionately ensure that these policies are implemented. For example, the Gas Re-Injection Act, which allows oil companies to flare gas on payment of a paltry sum, and continual shifting of the anti-flaring date[91] have been adjudged glaring evidence of the Government's unwillingness to confront the impacts of climate change and wetlands protection.[92]

H. *Invasive species*

The Government's strategy going by the new NBSAP 2016[93] is that by 2020, invasive alien species and pathways are identified and prioritized, priority species controlled or eradicated, and measures put in place to manage pathways in the six ecological zones of the country. As laudable as this strategy appears, Nigeria is yet to take concrete steps that will suggest that the target will be met. More specifically, therefore, the Government should adopt mechanisms that will use invasive species – particularly the water hyacinth, Nypa palm and cattail (*Typha* sp) – as raw materials for production, integrated management approach for the control and water utilization.[94]

VII. Conclusion

This chapter has attempted to examine the role of wetlands as a resource of great economic, cultural, scientific and recreational value to human life. It finds that although wetlands and people are ultimately interdependent, the unbridled rate of increase in the anthropogenic exploitation of biodiversity globally has not been commensurate with the rate of replacement. In the case of the Lagos lagoon wetlands, the scenario is worse. In spite of the huge benefits that the wetlands provide to Lagos State, there is no regulation on the protection of the wetlands nor is there any strong enforcement mechanism for their management. The implication is that the State stands to lose the wetlands with all their attendant benefits if no proactive steps are taken to check the current trend. As such, the progressive encroachment on and loss of wetlands needs to be stopped, and measures must be taken to conserve and make wise use of wetlands resources. To achieve this at a national level, Nigeria must transmit the Ramsar convention into its body of laws, put in place a strong enforcement mechanism and address the factors that militate against the conservation of biodiversity and protection of sustainable Lagos wetlands.

In the light of the above discussions, the Nigerian Government should enact a comprehensive wetlands protection legislation, and in addition place Lagos Lagoon wetlands on the gazette as a conserved urban ecosystem that should not be developed. The proposed law should work in synergy with other complementary laws and regulations on ecosystem protection while insisting on mandatory environmental impact assessment reports for every project with ecological implications. It should also strengthen the existing institutions and organizations that are capable of translating the complex knowledge of coastal wetlands and general biodiversity conservation to action, to engage in bold priority setting and transforming the Nigerian economy into a sustainable knowledge-based economy. Furthermore, to ensure ease in the implementation of the proposed law, the Government should not only integrate community management of wetlands and compensation for environmental services into national development policies and the proposed wetlands law, but equally recognize and maintain local community rights to wetlands resources. Specifically, this will require the establishment and formalization of the 'Development Triumvirate' that will comprise the Government, the private sector and other stakeholders like the indigenous community, civil society, etc. Finally, as a further step in improving the sustainability of the Lagos Lagoon wetlands, there is a need to improve both the knowledge of change in wetlands areas, and the consistency of data on changes in wetlands areas in published papers and reports.

Notes

1 'LIFE and Europe's Wetlands–Restoring Vital Europa', http://ec.europa.eu/environment/life/publications/lifepublications/lifefocus/dhttp://ec.europa.eu/environment/life/publications/lifepublications/lifefocus/ . . . (accessed on 2 January 2016).

2 'How Much Wetland Has the World Lost? Long-Term and Recent Trends in Global Wetland Area', www.publish.csiro.au/paper/MF14173.htm (accessed on 2 January 2016).

3 O. H. Chidi and O. E. Ominigbo, 'Climate Change and Coastal Wetlands: Nigeria in Perspective', *International Journal of Environmental Issues*, 2010, 7(2): 216–223.

4 UN–Habitat, 'Urban Development, Biodiversity and Wetland Management: Expert Workshop Report', Kenya Wildlife Training Institute, Naivasha, Kenya, 2010.

5 'Wetlands International', www.wetlands.org (accessed on 1 January 2016).

6 *Ibid.*

7 'The Status of the Nigerian Coastal Zones', 2007, United Nations Environmental Programme (UNEP), www.unep.org/abdjanconvention/docs (accessed 2 August 2011).

8 Spain, for example, has lost more than 60% of all inland freshwater wetlands since 1970. More northerly regions have also suffered, however. In France, 67% of wetlands have disappeared within the last century. Similarly, since the 1950s, 84% of peat soils have been lost in the United Kingdom, and 57% in Germany due to drainage for agriculture activities, forestry and landfilling for urban development. Lithuania has lost 70% of its wetlands in the last 30 years and the open plains of southwestern Sweden have lost 67% of their wetlands and ponds to drainage in the last 50 years.

9 This loss and degradation is caused by causes like flooding and erosion, desertification and drought, and deforestation. However, the principal

cause of loss of biodiversity is anthropogenic causes like drainage for agriculture, infrastructure developments, deforestation and malaria control, blocking and extraction of the water inflow, over-exploitation of groundwater resources, or the building of dams. Pollution from agricultural and industrial sources can increase the level of nutrients, pesticides or heavy metals, seriously impairing ecological processes, introduction of invasive species, urbanization and solid waste wildfires, and weak implementation of environmental policies.

10 H. O. Nwankwoala, 'Case Studies on Coastal Wetlands and Water Resources in Nigeria', *European Journal of Sustainable Development*, 2012, 1(2): 113–126.

11 Dependency of the rural population on biodiversity resources is about 70%. See O. Godswill and P. G. Tamnnobiekiri, 'Wetland Inventory and Mapping for Ikorodu Local Government Area, Lagos', www.fig.net/resources/proceedings/fig_proceedings/fig2015/papers/t (accessed on 17 January 2016).

12 In this paper, the reference to 'Lagos Lagoon' here shall include a reference to Lekki Lagoon since the two wetlands are connected at some point.

13 J. N. Obiefuna et al., 'Spatial Changes in the Wetlands of Lagos/Lekki Lagoons of Lagos, Nigeria', *Journal of Sustainable Development*, 2013, 6(7): 123–133 at 127.

14 L. Oyebande et al., 'An Inventory of Wetlands in Nigeria: World Conservation Union–IUCN', 2003, www.ccsenet.org/journal/index.php/jsd/article/viewFile/22379/17081 (accessed on 10 January 2015).

15 Chidi and Ominigbo, 'Nigeria in Perspective'.

16 This is the Biodiversity Convention's official recognition of a wetland.

17 G. Asibor, 'Wetlands: Values, Uses and Challenges', paper presented to the Nigerian Environmental Society at the Petroleum Training Institute, Effurun, 21 November 2009.

18 Dependency of the rural population on biodiversity resources is about 70%. See Godswill and Tamnnobiekiri, 'Wetland Inventory'.

19 M. O. Ajibola, B. A. Adewale and K. C. Ijasan, 'Effects of Urbanisation on Lagos Wetlands', *International Journal of Business and Social Science*, September 2012, 3(17): 310–318 at 314.

20 I. Aigbedion and S. E. Iyayi, 'Environmental Effects of Mineral Exploitation in Nigeria', *International Journal of Physical Sciences*, 2007, 2(2): 33–38.

21 'The Ramsar Convention on Wetlands: Its History and Development', www.ramsar.org/sites/default/files/documents/pdf/lib/Matthews–history (accessed on 1 January 2016).

22 J. P. Silva et al., LIFE 111 ('The Financial Instrument for the Environment') is a programme launched by the European Commission and coordinated by the Environment Directorate-General (LIFE Unit – E.4) ISBN 978-92-79-07617-6. The EU has been a major provider of funds for wetland conservation projects, both within and outside the Union. The Commission's financial instrument for the environment, LIFE, has been contributing to a large number of projects supporting the conservation of wetland ecosystems within the Natura 2000 network; see Article 1.1 of the Convention.

23 The principal substantive obligations are contained in Articles 3 and 4.

24 'CBD Strategy and Action Plan – Nigeria (English version)', www.cbd.int/doc/world/ng/ng-nbsap-01-en.doc (accessed on 20 January 2016).
25 L. M. Talbot, 'The World Conservation Strategy', *Environmental Conservation*, 1980, 7: 259–268.
26 'Discourse Analysis of Nature: Conservation Policies in Africa: a Beninese Case Study', https://echogeo.revues.org/13964 (accessed on 2 January 2016).
27 'Nigeria – Convention on Biological Diversity', www.cbd.int/doc/world/ng/ng-nr-04-en.doc (accessed on 3 January 2016).
28 *Ibid.*
29 *Ibid.*
30 'Wetlands: Unexpected Treasures', *Copernicus*, Issue 55, www.copernicus.eu/sites/default/files/documents/Copernicus_Briefs (accessed on 3 January 2016).
31 'Mapping and Protecting Vulnerable Wetlands and Stormwater Management Planning Project', www.mass.gov/eea/agencies/massdep/water/watersheds/mapping-vulnerable-wetlands.html (accessed on 6 October 2016).
32 'LIFE and Europe's Wetlands–Restoring Vital Europa.
33 'Nigeria Biodiversity and Tropical Forestry Assessment', 2008– pdf.usaid.gov/pdf_docs/Pnadn536.pdf (accessed on 3 January 2016).
34 *Ibid.*
35 *Ibid.*
36 *Ibid.*
37 *Ibid.*
38 Obiefuna et al., 'Spatial Changes in the Wetlands of Lagos'.
39 *The Punch Newspaper* p. 2, 2 August 2010; *The Punch Newspaper* p. 4, 23 August 2010.
40 E. G. Orie, 'The Clean Development Mechanism as a Tool for Sustainable Development: A Case for Regulatory Action', unpublished Ph.D. dissertation, Nigerian Institute of Advanced Legal Studies, University of Lagos Nigeria, 2013.
41 'CBD Strategy and Action Plan – Nigeria'.
42 National Forest Policy, 2006.
43 Erosion, Flood Control and Coastal Zone Management (National Policy on Erosion, Flood Control and Coastal Zone Management, 2005).
44 The Constitution of the Federal Republic of Nigeria 1999, Section 20.
45 Federal Environmental Protection Act, 1989.
46 This is the National Environmental Standards and Regulations Enforcement Agency (Establishment) Act, 2007.
47 For example, the Nigerian National Policy on Environment, 1991.
48 The Act established the National Environmental Standards and Regulations Enforcement Agency (NESREA). Section 8(k) of the statute mandates the agency to present for the Minister's approval proposals for guidelines, regulations and standards on environment matters (excluding the oil and gas sector), such as: atmospheric protection; air quality; ozone-depleting substances; noise control; effluent limitations; water quality; waste management and environmental sanitation; erosion and flood control; coastal zone management; dams and reservoirs; watersheds; deforestation and bush burning.

49 The Regulations include: National Environmental (Wetlands, River Banks and Lake Shores) Regulations, 2009. S. I. No. 26 which provides for the conservation of wetlands and their resources in Nigeria; National Environmental (Watershed, Mountainous, Hilly and Catchments Areas) Regulations, 2009. S. I. No. 27, which provides for the protection of water catchment areas, identification of major watersheds, restriction on the use of watersheds, mountainous and hilly areas; delineation of roles, prevention of fires in watersheds, afforestation and reforestation as well as grazing of livestock and the National Environmental (Access to Generic Resources and Benefit Sharing) Regulations, 2009. S. I. No. 30, which regulates the access to and use of generic resources. Others are: the National Environmental (Soil Erosion and Flood Control) Regulations, 2010. S. I. No. 12, which checks all earth-disturbing activities, practices or developments for non-agricultural, commercial, industrial and residential purposes to protect human life and the environment; the National Environmental (Desertification Control and Drought Mitigation) Regulations, 2010. S. I. No. 13, established to provide an effective and pragmatic regulatory framework for the sustainable use of all areas already affected by desertification and the protection of vulnerable lands and National Environmental (Protection of Endangered Species in International Trade) Regulations, 2010. S. I. No. 16, for the protection of endangered wildlife species of fauna and flora.

50 The Natural Resources Conservation Act, 1989 is the most recent legislation on natural resources conservation. The Act establishes the Natural Resources Conservation Council, which is empowered to address soil, water, forestry, fisheries and wildlife conservation by formulating and implementing policies, programmes and projects on conservation of the country's natural resources.

51 The Federal Environmental Protection Agency Act (Chapter 131, Laws of the Federation, 1990). This was promulgated to protect the country's environment from degradation.

52 The Environmental Impact Assessment Act (No. 86 of 1992): This Act requires that environmental impact assessment must first be carried out before any project likely to impact the natural environment could be undertaken.

53 The Endangered Species (Control of International Trade and Traffic) Act 11 of 1985: This Act makes provision for conservation and management of the country's wildlife and protection of some of the country's rare and endangered species.

54 The National Parks Decree (Decree No. 36 of 1991): As one of the principal laws on biodiversity conservation, the Act was promulgated to provide a protective sanctuary for wildlife species as well as to promote and preserve the beauty and conservation of the country's natural vegetation. This Act led to the establishment of six national parks in Nigeria. This has been subsumed in the National Parks Act of 1999, which created two additional national parks. On the other hand, the fairly comprehensive Draft National Forestry Act, put together in 2001, is yet to go through the processes of becoming a law. This Act is designed to provide for the conservation and sustainable management of the nation's forests, wildlife

resources and rich biodiversity, and to conform to international processes and initiatives on global forests and the environment.

55 Some of the international agreements include the United Nations Convention on the Law of the Sea, Convention on the Control of Trans-boundary Movements of Hazardous Wastes and their Disposal (Basel Convention), Desertification Convention, Protocol on Substances that deplete the Ozone Layer (Montreal Protocol), Convention on the Conservation of Migratory Species of Wild Animals, African Convention on the Conservation of Nature and Natural Resources, Convention on International Trade in Endangered Species of Wild Fauna and Flora (CITES), and Convention on Fishing and Conservation of the Living Resources of the High Seas.

56 'CBD Fourth National Report – Nigeria – Convention on Biological Diversity', www.cbd.int/doc/world/ng/ng-nr-04-en.pdf– Convention on Biological Diversity, (accessed on 20 January 2016).

57 Some of the challenges as discussed below include absence of specific law on conservation, few trained and qualified people in the field to enforce compliance, poverty and urbanization.

58 These include Linkage Centers for Arid Environments (in Maiduguri), for Freshwater Environments (in Minna), for Highlands/Montane Environments (in Jos), for Delta Environments (in Port Harcourt), and for Marine and Coastal Environments, in conjunction with the Nigerian Institute for Oceanography and Marine Biology (in Lagos) and the Linkage Center for Forests, Conservation, and Biodiversity at the University of Agriculture in Abeokuta, which is designed to focus on coordinating data and research relevant to biodiversity conservation. Regrettably there is very little information on how well these centres have done in fulfilling their mandates.

59 'Nigerian Biodiversity', www.scribd.com/document/175097158/Nigerian–Biodiversity, (accessed on 3 January 2016).

60 'Nigeria Biodiversity and Tropical Forestry Assessment', 2008.

61 UNEP through the Global Environment Facility (GEF) is funding the Wings over Wetlands program, a joint program with Wetlands International, Birdlife International, the Ramsar Convention, and other groups to conserve African-Eurasian migratory birds.

62 The UNFAO program contains elements aimed at poverty alleviation that may divert communities from unsustainable resource extraction in environmentally sensitive areas.

63 For further reading on NGOs in this area, see 'Nigerian Biodiversity and Tropical Assessment', 2008, pdf.usaid.gov/pdf_docs/Pnadn536.pdf (accessed on 3 January 2016).

64 Dr (Mrs) E. G. Orie, 'Climate Change and Sustainable Development: the Nigerian Legal Experience', *NOUN Current Issues in Nigerian Law Volume*, 2015, 5: 68–114 at 98.

65 'Mapping and Protecting Vulnerable Wetlands and Stormwater Management Planning Project'.

66 For instance, there is no provision for the mapping of the waters, which is a major way of protecting wetlands by preventing flooding. For further reading, see T. H. Votteler, 'Wetland Protection Legislation Water Resources of the United States,' https://water.usgs.gov/nwsum/WSP2425/legislation.html (accessed on 13 January 2016).

67 Nwankwoala, 'Case Studies on Coastal Wetlands'; O. Adekola, S. Whanda and F. Ogwu, 'Assessment of Policies and Legislation That Affect Management of Wetlands in Nigeria', *Wetlands*, 2012, 32: 665–677.

68 'CBD Fifth National Report – Nigeria (English version)', www.cbd.int/doc/world/ng/ng–nr–05–en.pdf (accessed on 20 January 2016).

69 Decisions concerning the fate of wetlands, however, are often made through processes that are unsympathetic to local needs or that lack transparency and accountability. Precisely, often times the Government would establish the policies and laws but only bring it to the notice of the local community at the point of implementation. This method disconnects the communities from such regulation, makes partnership with the Government difficult and implementation almost impossible. The scenario would be different where the communities as stakeholders are allowed to participate in the making of those policies and laws. With such participation, they will feel a sense of obligation towards the successful implementation of the policies and laws.

70 Nwankwoala, 'Case Studies on Coastal Wetlands'; Adekola, Whanda and Ogwu, 'Assessment of Policies and Legislation'.

71 Some of these species come through the ballast waters of the foreign ships while others are carried by the tide along the West African coast.

72 S. Odunuga, *Land Use Dynamics: Climate Change and Urban Flooding: A Watershed Synthesis*, Saarbrücken: Lambert Academic Publishing, 2010.

73 'CBD Strategy and Action Plan – Nigeria'.

74 Dependency of the rural population on biodiversity resources is about 70%. See Godswill and Tamnnobiekiri, 'Wetland Inventory and Mapping'.

75 Ajibola, Adewale and Ijasan, 'Effects of Urbanisation on Lagos Wetlands'.

76 *Ibid.*; Obiefuna et al., 'Spatial Changes in the Wetlands of Lagos'.

77 *Millennium Ecosystem Assessment: Ecosystems and Human Well-Being: Wetlands and Water Synthesis*, Washington, DC: World Water Resources Institute, 2005, p. 7.

78 'UN-Habitat, Urban Development, Biodiversity and Wetland Management: Expert Workshop Report, 16–17 November' 2009. Kenya Wildlife Training Institute, Nairobi, Kenya.

79 N. O. Uluocha and I. C. Okeke, 'Implications of Wetlands Degradation for Water Resources Management: Lessons from Nigeria', *GeoJournal*, 61: 151–154, http://dx.doi.org/10.1007/s10708-004-2868-3 (accessed on 17 January 2016).

80 O. Adegun, S. Odunuga and Y. Appia, 'Dynamics in the Landscape and Ecological Services in System I Drainage Area of Lagos', *Ghana Journal of Geography*, 2014, 7(1): 83–84.

81 *Ibid.*

82 J. N. Obiefuna et al., 'Land Cover Dynamics Associated With the Spatial Changes in the Wetlands of Lagos/Lekki Lagoon System of Lagos, Nigeria', *Journal of Coastal Research*, 2012, http://dx.doi.org/10.2112/JCOASTRES-D-12-00038.1 (accessed on 2 January 2016).

83 'CBD Fourth National Report – Nigeria', p. 12.

84 See notes 49–55 for the list of laws and regulations.

85 See target 1 of the NBSAP 2016 submitted by Nigeria, www.cbd.int/nbsap (accessed on 15 January 2016).

86 The Federal Ministry of Environment is merely complying with the provisions of article vi, which encourages each Contracting Party, bearing in mind its particular circumstances, to develop national strategies, plans or programmes for the conservation and sustainable use of biological diversity or adapt for this purpose existing strategies, plans or programmes which shall reflect, among other things, the measures set out in the Convention relevant to the Contracting Party concerned; and also to integrate, as far as possible and as appropriate, the conservation and sustainable use of biological diversity into relevant domestic sectoral or cross-sectoral plans, programmes and policies. In other words, such national strategies, plans or programmes should not be restricted to a particular Ministry.

87 E. G. Orie, 'Climate Change and Forest Management in Nigeria: The Case for Regulatory Action', paper delivered at the IUCN International Academy for Environmental Law Jakarta Indonesia, 7–12 September 2015.

88 *Ibid.*

89 O. A. Fagbohun and E. G. Orie, 'Nigeria: Law and Policy Issues in Climate Change', *ELRI*, 2015, ELRI Monograph Series, p. 20.

90 The word *management* is used here in the sense of reactive and proactive ecosystem administration.

91 Associated Gas Re–Injection Amendment Act of 2010 amended the Associated Gas Re–Injection Act (AGRA) 1979. The amendment set December 31, 2012 as the deadline for the abatement of gas flaring, but went ahead to provide a new section that permitted companies to continue the flaring of gas on the payment of a temporary gas flaring penalty of $5.00 per 1000scf of gas flared. The deadline for the abatement of gas flaring has finally been shifted to 2030. Dennis Otiotio, Gas flaring regulation in the Oil and Gas Industry: A Comparative Analysis of Nigeria and Texas Regulations, www.academia.edu/. pp. 25–30.

92 Orie, 'The Clean Development Mechanism as a Tool for Sustainable Development', p. 259.

93 Target 9 of the new NBSAP submitted February 2016.

94 Presently there is a move to commence the use of water hyacinth as raw material for fertilizer production. This should be done speedily and in such a commercial quantity that will make the hyacinth both scarce and of high economic value.

2

MARINE SPATIAL PLANNING AS A KEY INSTRUMENT FOR THE SUSTAINABLE USE OF MARINE RESOURCES

The case of Mauritius

Marie Valerie Uppiah

Since time immemorial, the sea has been used as the backbone of the economic, social, cultural and environmental life of many countries. Throughout centuries, people have always used the sea to satisfy their curiosity. The sea has been used to discover new land, to fish, to conduct research, and to find new resources and exploit them.[1]

However, today it is unanimously recognised by the international community that much damage has been done to the sea and its resources. There are various ways in which the sea and its biodiversity have been degraded. Firstly, with climate change, there has been an increase in the sea temperature and the acidification of sea water. This has caused the migration of many marine species, which has led to a disturbance in the marine ecosystem of many regions of the world.[2]

Secondly, damage to the sea and its flora and fauna is caused by maritime pollution, particularly through oil spills. According to Hermien Roelvert, it is estimated that the amount of oil spilt into the ocean for the period between 1970 and 2009 is about 5.65 million tons.[3] His analysis does not include oil spills which are less than 7,000 tons; therefore, it is stated that the real amount of oil spilt is more than 5.65 million tons during this period. The consequences of these oil spills are not only disastrous for the marine animals living in the sea but represent also a danger for marine animals living onshore. For instance, many coastal birds have lost their lives as they got stuck in the oil and were not able to get out. Other marine animals, like penguins, seals and fish, may drown or consume some of the oil present in the water and later die.

Thirdly, another man-made adverse impact on the marine ecosystem of many coastal States is the phenomenon of overfishing. Overfishing is causing great disturbances in the marine ecosystems of many regions in the world. One good example to illustrate how overfishing can destroy an important feature of the marine ecosystem of a country is through the analysis of the collapse of the Caribbean coral reef.[4] Due to the overfishing of herbivorous fish, only sea urchins grazed the micro algae living in the area. A disease affected the sea urchins which transferred the virus to the algae. When the algae deposited themselves on the reef, they started to replace and contaminate the coral, causing, after some time, the coral reef to collapse.[5]

In order to regulate the damage caused to the marine ecosystem, a panoply of international as well as regional conventions have been adopted by States. Concerning the regulation of climate change, in 2015, the Conference of Parties (COP 21) was held in Paris where the issue of climate change was discussed and countries took the decision to work together in order not to increase the world temperature by 2 degrees.[6] If this is maintained, it can have a positive effect on the marine ecosystem as there will be the stabilisation or even reduction of sea acidification, thus reducing the amount of species migration. With less species migration, marine ecosystems may rejuvenate and after a period of time regain their initial flora and fauna.

Concerning the regulation of marine pollution, in 1972, the United Nations held a conference in Stockholm where the Declaration on Human Environment and an Action Plan were adopted to integrate approaches to solve global environmental issues. Principle 7 of the Declaration requests States to take the necessary measures to prevent marine pollution by hazardous substances which represent a threat to human, animal and plant life.[7]

For the regulation of overfishing, various international instruments have been developed to deal with this issue. The most prominent one is the United Nations Convention on the Law of the Sea (UNCLOS) 1982. With regards to the phenomenon of overfishing, Part V of the UNCLOS 1982 provides for the rights and duties of States in their Exclusive Economic Zone (EEZ). According to Article 56 of the Convention, a coastal State has the sovereign right to exploit, explore, conserve and manage the resources found in its EEZ. Coastal States are therefore required to devise measures to control fishing activities in their waters and hence control and combat overfishing.

One way of regulating overfishing in a country's EEZ and enhance a sustainable use of marine resources is through the application of Marine Spatial Planning (MSP). Marine Spatial Planning refers to the sustainable use of marine resources according to the place they are located in the sea. For instance, a particular area of the sea is divided into several sub-areas, where each sub-area is exploited according to the resources present therein.[8]

The aim of this chapter is to analyse how Marine Spatial Planning (MSP) can help States to sustainably explore and exploit the marine resources present in their EEZ and other parts of their marine territory. It studies the case of Mauritius, which has recently devised a roadmap for the development of its ocean economy. This roadmap involves elements of MSP which are being implemented for an effective and sustainable use of the marine resources. The chapter also advocates for the implementation of a policy to promote and regulate MSP at a local, regional or even international level.

I. The United Nations Convention on the Law of the Sea 1982[9]

According to Douvere, Marine Spatial Planning refers to the creation and establishment of

> a more rational organization of the use of marine space and the interactions between its uses, to balance demands for development with the need to protect the environment and to achieve social and economic objectives in an open and planned way.[10]

In broader terms, MSP can be considered as the division of a marine area into different sectors whereby each sector can be sustainably developed or exploited according to the ecosystem present there.

Before indulging into an in-depth analysis of the features and discussions around MSP, an overview of the United Nations Convention on the Law of the Sea (UNCLOS) 1982 will be done in order to examine how the Convention regulates the use of the sea.

The UNCLOS was first signed in December 1982 and then came into force 12 years later, that is, in November 1994. The reason why the UNCLOS 1982 was delayed in its application was because many States refused to sign it. For the UNCLOS 1982 to become effective, it needed the signature of 60 States and their ratification. However, many States refused to sign the Convention as it did not allow them to take out reservations to any article of the Convention.[11] A reservation means that a State has the possibility, when signing a treaty, to remove or modify certain legal provisions of the treaty. However, for the UNCLOS, taking out reservations is prohibited by its Article 309, which states that:

> *Reservations and exceptions: No reservations or exceptions may be made to this Convention unless expressly permitted by other articles of this Convention.*

This article has caused a delay in the application and implementation of the Convention in many countries.[12] The reason for this delay is linked to the fact that many countries wanted to modify some aspects of the Convention before signing or even ratifying it. However, in 1994, 60 States signed and ratified the Convention and immediately after it became effective.

The striking feature of the UNCLOS 1982 is the fact that it divides the sea into several layers, each having its own distinct features and the degree of sovereignty. The UNCLOS 1982 divides the sea into these seven distinct areas: the baseline, the internal waters, the territorial sea, the contiguous zone, the exclusive economic zone, the continental shelf and the high seas.

Article 5 of the UNCLOS 1982 defines the baseline as the low water mark alongside the coast where a State can start the measurement to establish which portion of the sea belongs to it and over which it can exercise sovereignty. For the internal waters, Article 8 of the Convention describes them as all the waters that are on the landwards side of the coastline and over which the coastal State has full sovereignty. For example, if a criminal offense occurs on board a foreign vessel within the internal waters of a particular State, the latter will have full criminal jurisdiction to try the matter.[13]

The territorial sea is the third layer of the division of the sea. Article 3 of the Convention provides that the coastal State has the right to establish a territorial sea which shall extend to up to 12 nautical miles (nm) from the baseline. Article 17 requests States to grant the right of innocent passage to ships of all States through their waters. Passage is considered innocent as long as the foreign vessel passing through the territorial sea of the coastal State does not hinder the peace, security and good order of the State.[14] If the foreign vessel represents a threat to the coastal State, the latter can stop it and does not allow it to pass through its waters.[15] The contiguous zone is the area of the ocean where the coastal State can exercise its control 'to prevent infringement of its customs, fiscal, immigration or sanitary laws and regulations within its territory or territorial sea' and punish those infringement.[16]

The Exclusive Economic Zone (EEZ) is the largest section of the division of the sea under the UNCLOS 1982. Article 57 provides for the breadth of the EEZ, which extends to up to 200 nm from the baseline. Article 56 allows the coastal State to undertake a series of activities in its EEZ. The State can explore and exploit all the resources, living and non-living, found in the waters; and it can establish artificial islands, conduct marine scientific research and ensure the protection and preservation of the marine environment.

The main reason why the EEZ was given a breadth of 200 nm is due to the fact that most of the richest phytoplankton, which is also known as the food of the fish, lies 200 nm from the shore. Therefore, this region contains a high concentration of fish and thus many coastal States decided to extend their right to control and fish in this region.[17]

In a study conducted by Hollis and Rosen, it was stated that

> the creation of the EEZ gave coastal nations jurisdiction of approximately 38 million square nautical miles of ocean space. The world's EEZs are estimated to contain about 87% of all of the known and estimated hydrocarbon reserves as well as almost all offshore mineral resources. In addition, the EEZs contain almost 99% of the world's fisheries, which allows nations to work to conserve the oceans vital and limited living resources.[18]

It has to be noted that one major problem that many coastal States have to face with the large area of EEZ that they own and the rich resources it contains is linked to illegal, unregulated and unreported fishing.[19] Many States, especially developing countries, face capacity constraints in regulating and policing their fishing rights over those illegal activities. It is recommended that developed States who are signatories or not of the UNCLOS give a helping hand to developing member States in order to combat illegal activities taking place in the developing States' EEZ.

The continental shelf is the sixth layer in the division of the sea provided by the UNCLOS 1982. The shelf is a land slope that descends beneath the continental plate where the ocean starts. The main characteristic of the continental shelf is that it hosts the world's richest marine ecosystem. In addition to this, the shelf can be used for energy production, be it renewable or non-renewable.[20] The right to explore and exploit the resources found on the continental shelf by the coastal State is granted by Article 77 of the UNCLOS 1982.

Finally, the last layer of the division is the high seas. As per Article 87 of the Convention, the high seas are open to all States. Whether a State is landlocked or coastal, it has the right, *inter alia*, to navigate, fish or fly over the high seas.[21] It has to be noted that no State can claim sovereignty over the high seas.[22] Hence, it can be deduced that the high seas belong to no one and everyone at the same time, since no State can claim sovereignty over it; however, they all can enjoy the resources that it contains.

The United Nations Convention on the Law of the Sea 1982 can be considered as the backbone of maritime law. The Convention provides for the delimitation of the sea surrounding States and establishes the extent of sovereignty that each State may have over the portion of the sea allocated to them. The usefulness of having such a delimitation and allocation of sovereignty reduces the risks for States to enter into conflicts over sovereignty of the sea. Furthermore, the Convention puts constant focus on the preservation of the environment in the exploitation and exploration of marine resources. It can be stated that the Convention encourages the sustainable use of marine resources by landlocked as well as coastal States. Moreover, by not allowing States to take out reservations under Article 309 of the Convention, it allows for a degree of uniformity in the maritime law of all member States. Finally, all countries in the world, landlocked or coastal, have the right to enjoy the resources of the sea through the freedom of the high seas.

Despite all these advantages, the Convention has undergone many criticisms. The first criticism relates to the fact that the implementation of the Convention depends on its introduction in national legislation.[23] The consequence of this dependence is that governments may be selective on which aspect of the Convention they want to implement in their national legislation and when to implement it. Governments may implement aspects of the Convention which would benefit them and neglect other aspects. Another drawback of the Convention is the way in which delimitation of sovereignty has been done. The division of the sea and its resources has been done by man-made delimitation rather than through the natural delimitation of the marine environment.[24] This is said to lead to a mismanagement of the various marine ecosystems, as an area which contains much biodiversity may fall outside the jurisdiction and sovereignty of a State and thus is not protected.

The UNCLOS 1982 gives the green light for States to explore and exploit the resources found in their marine territory. However, it does not explain how such exploitation and exploration can be done in a sustainable way. This is where Marine Spatial Planning comes into play, whereby it provides for the technical aspects of the sustainable use and management of the ocean resources.

II. Marine Spatial Planning

The United Nations Convention on the Law of the Sea 1982 grants States the right to explore and exploit the resources found in their marine territory. The Convention provides for the right to exploit marine resources found in the Exclusive Economic Zone as well as on the continental shelf. It has to be noted that the Convention does not provide for how States should exploit and/or manage the resources found in these regions. This is where Marine Spatial Planning provides for a useful way through which coastal States can sustainably exploit and manage the marine resources surrounding them.

Spatial planning refers to the creation of a rational arrangement of economic, environmental and social activities in a given area in order to avoid tensions and conflicts among these sectors.[25] For example, a bank is built in an area where there is a rich natural ecosystem. This might create a tension between members of the business community and environmentalists. Hence, the aim of spatial planning is to allocate areas in a particular region that will be suitable for economic, environmental or social activities.

The concept of spatial planning has been introduced in the ocean resources management arena as there has been an overlapping of activities which were having adverse effects on the marine environment.[26] For instance, overfishing in various areas of the world has led, not only to a reduction in the fish population in these regions, but has also caused partial destruction of the ecosystem of the region. In the Caribbean Small Islands for instance, due to a rise in fishing activities between 1950 and 2000, some commercial fish species are in danger of extinction, not only because of overexploitation but because their natural habitats have been destroyed as well.[27] The need to have a new system of regulating and managing human activities within the marine environment was felt.

Marine Spatial Planning comes as the solution to the problem of human activities within the marine environment. MSP, as described by Douvere, 'is a process that can influence *where* and *when* human activities occur in marine spaces'.[28] The rationale behind MSP, just like for spatial planning, is to establish a rational arrangement or organization of human activities within the marine environment in order not to disturb it.

The way MSP is applied is by analysing the features of the marine ecosystem. For instance, regions of the sea where the marine ecosystem is dense will receive little or no human activities. On the other hand, a region which does not contain many living or non-living organisms will be used in order to allow human activities, such as fishing, drilling for oil, producing energy and other activities. Therefore, it can be said that the aim of Marine Spatial Planning is to provide for an ecosystem-based management of the different human activities taking place in and on the ocean.[29]

Through MSP, particularly through the ecosystem-based management, focus is being placed on a specific area of the sea and its ecosystem and then, according to the ecosystem, economic, environmental or social activities will be undertaken there. The main objective of MSP is that the management of the marine ecosystem comes first and then the organisation and allocation of human activities in the marine environment is done. MSP departs from the traditional approaches of using the sea. One traditional approach to sea use is the fact that focus is placed on the different human activities taking place in the sea with little or no consideration to the location where those activities are taking place. However, with MSP, there has been a shift in focus whereby the management of the area comes first, then activities are carried out in it.[30]

For MSP to be effective, Ehler and Douvere have identified three phases that are important during its implementation.[31] These phases are: planning and analysis, implementation, and monitoring and evaluation.[32]

For the first step, which consists of planning and analysis, the human and environmental factor should be placed at the heart of any action plan to be devised. At this stage, an integrated and comprehensive spatial plan should be worked out that shall strike the balance between protecting, enhancing and providing for a sustainable use and development of marine resources. During the implementation stage, the execution of the spatial plan is done. However, it has to be noted that changes and/or improvements can be brought during the implementation stage through regulations and/or incentives.

Finally, during the monitoring and evaluation phase, an assessment of the effectiveness of the spatial plan is done to analyse its outcome and suggest possible room for improvement for future plans.[33] Marine Spatial Planning provides for many positive outcomes if a country wants to apply it in regulating the activities and use of its EEZ and the resources found therein. Firstly, MSP strikes the right balance between protection of the marine environment and the conduct of economic activities offshore. For example, in order to reduce overfishing, MSP can restrict fishing activities to be conducted in a certain area of the EEZ in order to protect the fish and marine resources of this area.

Secondly, MSP allows for a long-term plan to be devised which shall contain the planning, processes and allocation of a specific area to be used for development. This allows for a level of transparency with regards to which area of the sea is being used and for which purpose.

Furthermore, with the ecosystem-based approach encompassed in Marine Spatial Planning, this allows for a sustainable use and development of the coastal and marine environment. The ecosystem-based approach of MSP enables the assessment of marine biodiversity of a specific region in the sea, particularly in the EEZ, and suggests the most suitable activity to be conducted in the different sectors of the area. It can therefore be stated that the use of the ecosystem-based approach of MSP allows for an integrated and environmentally friendly way of conducting business activities at sea. Moreover, the Convention on Biological Diversity (CBD) promotes the application of the ecosystem-based approach of the MSP by stating that

> The ecosystem approach is a strategy for the integrated management of land, water and living resources that promotes conservation and sustainable use in an equitable way. Thus, the application of the ecosystem approach will help to reach a balance of the three objectives of the Convention: conservation; sustainable use; and the fair and equitable sharing of the benefits arising out of the utilization of genetic resources.[34]

Finally, it can be stated that MSP and its ecosystem-based approach recognises the existence of some important ecological areas in the sea and that human activities cannot be carried out there. For example, the Great Barrier Reef Marine Park in Australia illustrates how MSP is applied whereby the sea is divided into various areas, and whereby some human activities can take place and others are protected because of their rich ecosystem and human activities are prohibited.[35]

MSP allows for the division of the sea, especially in the EEZ and continental shelf where most of the resources are found, and allows for sustainable development and use of the sea to be carried out in the various divisions or sectors assigned as 'business friendly' without affecting the whole marine environment.

Despite those advantages, MSP and the ecosystem-based approach are said to contain some flaws. It has to be highlighted that there is a lack of operational tools for the implementation of the concept.[36] Many governments, especially in developing countries and Small Islands Developing States (SIDS), do not have the tools and expertise required to devise and implement the ecosystem-based approach of

MSP. Although the CBD praises the concepts of Marine Spatial Planning and the ecosystem-based management approach of the sea and its resources, it also points out the difficulties that the application of the concept presents. The CBD recognises the complexity of the application of the ecosystem-based approach, and the lack of tools or expertise required to apply the concept is voiced by many.[37]

Despite some of MSP's implementation difficulties, some countries have been able to successfully implement MSP and ecosystem-based management into their local sea use exploitation and management strategy. For instance, in Australia, the Great Barrier Reef Marine Park was created. With the use of Marine Spatial Planning and zoning, in some areas of the park, commercial sea-based activities such as tourism and fishing take place, while on the other hand, some areas of the ecosystem are granted a high level of protection.[38]

Another illustration of the application of MSP and the ecosystem-based approach is in China, where in 2002 the government adopted the Law on the Management of Sea Use. One feature of this law is that it divides the sea into various sectors under criteria related to ecological functions and then regulates the use of those sectors.[39]

In Europe, three countries, namely, the Netherlands, Germany and Denmark, have developed a trans-boundary management of the sea surrounding them. These countries have adopted a similar zoning system which consists of a zero-use zone, high-level protection zones and general-access zones in order to regulate the use and management of the sea.[40]

Marine Spatial Planning in an effective way for coastal States to manage their EEZ and their continental shelf. Articles 56 and 77 of the UNCLOS 1982 allows the coastal State to exploit the resources, living and non-living, that are found in the EEZ and on the continental shelf for their economic sustainability. By implementing the concept of Marine Spatial Planning and ecosystem-based management of marine resources, coastal States divide their EEZ or the continental shelf into various sectors whereby in some, commercial and other forms of activities can be done, while others are protected for marine purposes. By doing so, the States strike the right balance between economic development and the protection of their marine flora and fauna.

III. The case study of Mauritius

Located in the Indian Ocean, Mauritius is a small island State with a population of approximately 1.2 million inhabitants. Since the island was discovered, it has throughout time been used as a strategic location by the various colonizers who settled in the country.

In 1598, the Dutch came to the island and started to grow sugar cane. When the Dutch left, the French colonized the island and brought slaves from Africa to work in the sugar cane fields. In 1814, Mauritius and its outer islands became British colonies and in 1968, Mauritius gained its independence.

When Mauritius gained its independence from the United Kingdom, it also got its first written Constitution. Article 1 of the Constitution states that Mauritius shall be a sovereign democratic State.[41] Article 2 points out that the Constitution shall be considered as the Supreme law of the country and any other law inconsistent with it shall, to the extent of its inconsistency, be void.

Furthermore, the Constitution at its Chapter XI provides for the islands that are comprised in the State of Mauritius. Section 111 of the Constitution of Mauritius reads as follow:

'Mauritius' includes–

(a) the Island of Mauritius, Rodrigues, Agalega, Tromelin, Cargados Carajos and the Chago Archipelago, including Diego Garcia and any other island comprised in the State of Mauritius;

(b) the territorial sea and the air space above the territorial sea and the island specified in the paragraph (a);

(c) the continental shelf; and

(d) such places or areas as may be designated by regulations made by the Prime Minister, rights over which are or may become exercisable by Mauritius;

This section of the Constitution allows the government of Mauritius to have sovereignty over its outer islands as well as their territorial sea and air space.

Before an analysis of the importance of Marine Spatial Planning is done for Mauritius, an examination of the evolution of the Mauritian economy will be done in order to examine the actual status of the Mauritian economy and why MSP is of relevance to the country.

In 1968, when Mauritius got its independence, the main source of income and economic activity of the country was the exploitation of sugar cane. During the 1960s and 1970s, sugar was the backbone of the Mauritian economy and amounted to 90% of the country's export.[42]

Furthermore, in order to strengthen its sugar sector, the Mauritian government signed the Sugar Protocol in 1975. Through this Protocol, the European Economic Community agreed to buy sugar from countries of the African Caribbean and Pacific region at a price which was several times higher than the normal price of sugar on the international market.[43] It has to be noted that Mauritius got the largest quota share to sell sugar to the EU.[44] Hence, from 1975 to 2009, Mauritius had to supply this amount of sugar to the EU until the abolition of the Sugar Protocol in 2009. With the abolition of the Protocol, Mauritius lost a significant market and source of revenue for its sugar.

Along with the sugar sector, during the late 1970s and early 1980s, the textile sector started to find its place as one of the economic pillars of the country. The textile sector increased the rate of employment in the country, generated income and boosted the level of local as well as foreign investment in the country.[45] In 1982, Mauritius signed the Multi-Fiber Agreement which allowed the country to sell its textile products to the European Economic Community at a preferential rate. In addition to this, in 2000, the island signed the African Growth and Opportunities Act 2000 (AGOA 2000) which allowed it to sell some of its textile products as well as agricultural products at a preferential rate to the United States. Even though Mauritius was able to get preferential market access in many countries for the expansion of its textile sector, the abolition of the Multi Fiber Agreement in 2004 led Mauritius to lose another source of revenue for its products. Recently the island has been facing another threat to its textile sector. The new threat is that investors are opting to invest in Madagascar rather than in Mauritius as the former country offers more comparative advantages, such as cheap labour and availability of raw materials, compared to Mauritius. Hence, in recent years, Mauritius has been facing some challenges with its textile sector.

Another pillar of the Mauritian economy is the tourism sector. For years, the tourism sector has been a secure source of income for Mauritius. K. Jaufeerally estimates this income comes from two main sources, namely through tourists spending in the country and investment from international hotel chains.[46] Another study done by the Ministry of Tourism and External Communication estimated that the gross tourism receipts for Mauritius in the year 2000 were $508.3 million and contributed to about 11% of the Mauritian gross domestic product (GDP).[47] Just like the two other sectors mentioned above, the tourism sector of the island was affected by external circumstances, namely the 2008 financial crisis, and this led to a decline in the number of tourists coming to the country.

With a fall in the sugar, textile and tourism sectors, beginning in 2000, the various governments at the head of the country took the initiative to further diversify the economy. As of 2002, there was a boost in the development of the financial services sector in Mauritius. Mauritius signed Double Taxation Agreement Treaties with various countries in order to enhance investment in the country. This sector has also been prone to major criticisms from the international community. Mauritius has been taxed as a tax haven by the European Union and the Organisation for Economic Cooperation and Development.[48] This has been a blow to the financial services sector of Mauritius as the range of investment in the financial sector of the country has decreased.

Hence, in order to help Mauritius face its economic challenges as well as develop a new source of income, the government of the country, in 2012, launched the ocean economy concept. This concept allows Mauritius to explore and exploit the marine resources found in its Exclusive Economic Zone as per Article 56 of the UNCLOS 1982. It has to be pointed out that the overall size of the EEZ of Mauritius is 1.9 million km². The EEZ encompasses the coasts of the main island of Mauritius, Rodrigues, St Brandon, Agalega, Tromelin and Chagos Archipelago.

In addition to this, in 2009 Mauritius and the Seychelles presented a joint submission to the United Nations Commission on the Limits of the Continental Shelf (CLCS) whereby the two countries claimed for an extended continental shelf beyond their normal 200 nm EEZ.[49] The size of the continental shelf that Mauritius and the Seychelles was claiming is equivalent to the size of Germany. After much negotiation, the Commission recommended in favour of the two countries, which now have the sovereign rights to explore and exploit the resources present on the seabed and subsoil of the continental shelf.[50] It has to be underlined that if a third-party country wishes to explore or exploit the resources present on the continent shelf that is jointly managed by Mauritius and the Seychelles, that country will have to seek the express consent of the two States before undertaking any activity therein.[51] Furthermore, this extended continental shelf has been added to the EEZ of Mauritius. An additional 396,000 km² has been added to the existing 1.9 million km² of EEZ that Mauritius already has, and hence the overall size of the EEZ over which Mauritius has sovereign rights is now 2.3 million km².

The implementation of the ocean economy concept will create seven new investment opportunities in the EEZ of Mauritius. These opportunities are, first, seabed exploration for hydrocarbons and minerals.[52]

The continental shelf between Mauritius and the Seychelles, also known as the Mascarene Plateau, has some intrinsic geological features which suggest the possibility that oil is present in this region. Therefore, through the seabed exploration for hydrocarbons and minerals aspect of the ocean economy, facilities and regulations will be set up by the government to enhance investment in this sector.

The second business opportunity provided by the Mauritian ocean economy is fishing, seafood processing and aquaculture.[53] With its large span of water, Mauritius has, throughout the years, undergone various forms of fishing activities, ranging from artisanal fishing to commercial fishing in the Mascarene Plateau. Recently, aquaculture has been developed and 20 sites, in-lagoon, have been granted for this activity.[54]

Third, there is the process of Deep Ocean Water Application (DOWA). The commercial activities under the DOWA can be categorized into two streams, namely, the upstream and the downstream. Activities taking place in the upstream involve mainly the production and preservation of energy. The downstream commercial activities will involve, *inter alia*, the production of cosmetics and pharmaceutical products with the marine resources available of high-end aquaculture and seaweed culture.

The fourth element is marine services. Here, most of the activities will be done on land where facilities such as marine finance, ship registration, marine ICT or marine biotechnology will be available. Seaport-related activities are the fifth aspect of the Mauritian ocean economy concept. The Mauritian government is undertaking steps to enlarge the port so that it can undertake further activities. These activities include ship repairs, the accommodation of larger vessels and bunkering, among others. If Mauritius extends its port and provides for these facilities, especially for bunkering, the island can become a bunkering hub in the Indian Ocean.[55] Marine renewable energies are the sixth focus of the ocean economy whereby offshore wind and ocean waves can be used to produce energy for Mauritius and its outer islands.[56]

Finally, ocean knowledge is the last cluster of the Mauritian ocean economy concept whereby academics, practitioners and all stakeholders involved in the maritime activities of Mauritius are invited to share their knowledge and experience with the population.

If all of the clusters of the ocean economy concept are done in an effective and efficient way, the outcome will be an increase in the Mauritian GDP up to 20% by 2025, job creation and the creation of further investment opportunities for the country.[57] What can be learnt from this Mauritian experience is that the island, although not naming it as such, is using Marine Spatial Planning in order to implement its ocean economy concept. By dividing its Exclusive Economic Zone and continental shelf for various economic purposes, such as seabed exploration for hydrocarbons and minerals, deep ocean water applications or marine renewable energies, the island is segmenting its EEZ so as to allow for the development of these sectors. However, what is lacking in this analysis of the segmentation of the Mauritian EEZ for business purposes is whether the marine ecosystem has been taken into consideration during the segmentation phase.

The three phases of implementation suggested by Ehler and Douvere[58] can be used to examine the implementation of the seven clusters forming part of the Mauritian ocean economy concept.

Concerning the first phase, which is planning and analysis, the government and its various stakeholders have to adopt an integrated and comprehensive marine spatial plan whereby the development of a particular sector, for instance the introduction of wind turbines in the sea to produce electricity, will not destroy the marine ecosystem of this particular area. The government and the stakeholders should be able to have the information required about the biodiversity of a particular area and analyse whether the implementation of wind turbines or drills, or any other man-made material, would not cause any danger to the ecosystem of this area.

With regards to the implementation phase, the actors have to implement the plan that they have devised in the planning and analysis phase. For instance, concerning aquaculture, a study has been done to locate 20 sites that would be suitable for this activity. After an analysis of the water and an evaluation of the location of the sites have been done, those 20 sites were selected and 15 of them are already operational.[59] If changes or improvements need to be done, it is during the implementation phase that they have to be carried out.

Finally, monitoring and evaluation is done to assess the effective implementation of the plan. During this phase, when the clusters have been implemented, the various actors have to evaluate what has been done and improvements should be provided for future developments. If a plan of action has gone wrong or was not effective, it is advisable to change the activity and propose a new one which will be 'environmental friendly'.

The potential benefits for Mauritius to use Marine Spatial Planning can be analysed from various perspectives. Firstly, it will lead to a reduction in conflicts among the various users of the sea. If Mauritius divides the sea around it into various segments, this will allow for a better allocation and indication about where and what kind of activity can be done at sea. For example, if Mauritius follows the example of Australia or China and divides the sea into various sections – for instance, one area for commercial activities on a constant basis, another one for *ad hoc* commercial activities and preservation of the ecosystem and a final one for preservation and conservation of the ecosystem solely – it will give a clear indication to the business community as well as environmentalists about where and when some business activities can and cannot be done. Therefore, this segmentation of the sea will avoid an overlapping of activities in the same area and also reduce the probability of conflicts among the various users of the sea.

The second advantage of MSP for Mauritius is that it will allow for a better coordination of activities among the various stakeholders involved in regulating and managing the activities at sea. Currently, there are various organisations involved in the management and the regulation of marine activities in Mauritius. These institutions are, *inter alia*, the Ministry of Ocean Economy, Marine Resources, Fisheries and Shipping; the World Bank; the Mauritius Oceanography Institute; and the University of Mauritius. It is recommended that a unique platform be created so that these stakeholders and others can collaborate together in order to come up with common objectives for a smooth and proper administration and implementation of MSP in Mauritius.

The third benefit of the implementation of MSP for the island is that it can act as a catalyst for investment in the maritime sector. Under MSP, the sea will be divided into several sectors under which different forms of activities will take place. The advantage of this division, from an investment perspective, is that it will give investors an indication as to the fields open for investment. For instance, development and investment will be needed in the different ocean economy clusters that Mauritius is currently undertaking. Hence with MSP, investors will know when, where and how to invest in the various clusters. Along with the development of commercial and business activities, laws will also be developed in order to provide for an effective management and regulation of the activities. Therefore, with a proper action plan for MSP as well as transparent and clear rules for its implementation, these can act as a magnet for investment in the maritime sector of Mauritius.

Finally, at the regional level, islands of the Indian Ocean as well as the group of SIDS can collaborate in order to devise a plan of action with regards to Marine Spatial Planning that will meet their interests and benefit them. Mauritius is in the early stage of developing its ocean economy concept. It is recommended that the government and the stakeholders apply the notion of MSP and ecosystem-based approach of management in their planning and analysis phase as well as in their implementation phase. This will allow them to strike the right balance between marine environment protection and economic development through marine economic activities.

IV. Conclusion

The United Nations Convention on the Law of the Sea 1982, through its Parts II, V, VI and VII, divides the sea into seven layers, each having their distinct features. This division allows States to have greater access to the world's marine resources. The Convention allows States, whether landlocked or coastal, to enjoy the benefit of these marine resources by exploring and exploiting them. However, it is of utmost importance that the States, while doing so, take due consideration for the preservation of the marine environment. It has to be noted that although the UNCLOS 1982 does give permission to States to exploit marine resources and preserve the environment, it does not provide for how this should be done. This is where Marine Spatial Planning shows its importance and relevance for the sustainable use of marine resources.

Marine Spatial Planning, through an ecosystem-based approach, provides for the division of an area of the sea and then provides for how the area should be managed and exploited. MSP puts at its heart the marine ecosystem and then the development of commercial or business activities around this ecosystem. In this way, there is the preservation of the environment as well as a sustainable use of the available marine resources for the economic development of the coastal States.

This research paper has shown that many countries have applied MSP, along with an ecosystem-based approach, in their national agenda for the exploitation and exploration of their marine resources. Australia, China and some European countries have implemented this framework in their national legislations and have been able to strike the right balance between marine ecosystem preservation and economic development in the maritime sector.

Mauritius, although not describing its ocean economy concept as Marine Spatial Planning, has used MSP in the development of its ocean economy concept. The Mauritius Exclusive Economic Zone and continental shelf are divided into various segments so as to allow the implementation of the various clusters forming part of the ocean economy concept. It is being recommended that the island devise a plan of action that will take into consideration the marine ecosystem of its waters and developments to be made therein. By having consideration for both, the island will be able to implement the seven sectors of the ocean economy concept in a marine environment-friendly way.

For centuries, the sea has been used as a cornerstone for the economic development of many countries. Thanks to its resources, the sea has been able to provide food for all nations and created employment for many people in coastal States. In order to allow the sea to continue to provide us with its considerable benefits, coastal States should take the appropriate steps when developing economic activities in their waters. The most appropriate way to protect the marine environment and at the same time enhance economic development at sea is to use Marine Spatial Planning, with its ecosystem-based approach.

Notes

1 Secretariat of the Convention on Biological Diversity, 'One Ocean, Many Worlds of Life', 2012, 1–77, www.cbd.int/idb/doc/2012/booklet/idb-2012-booklet-en.pdf (accessed on 17 September 2016).
2 Scott C. Doney et al., 'Climate Change Impacts on Marine Ecosystems', *Annual Review of Marine Science*, 2012, 4: 11–37.
3 H. Roelvert, 'Oil Spills and their Effects on the Ocean', 2016, www.divestyle.co.za/index.php?option=com_content&view=article&id=516:oil-spills-and-their-effect-on-the-ocean&catid=911:environment&Itemid=126#page (accessed on 1 October 2016).
4 T. P. Hughes, 'Catastrophes, Phase Shifts, and Large-Scale Degradation of a Caribbean Coral Reef', *Science*, 1994, 256(5178): 1547–1551.
5 Marten Scheffer et al., 'Cascading Effects of Overfishing Marine Systems', *Trends in Ecology and Evolution*, 2005, 20(11): 579–581.
6 Conference of Parties, Paris 2015, www.cop21paris.org/about/cop21 (accessed on 28 December 2015).
7 United Nations Conference on Human Environment, 'Declaration of the United Nations Conference on the Human Environment', 1972, www.unep.org/documents.multilingual/default.asp?documentid=97&articleid=1503 (accessed on 1 October 2016).
8 F. Douvere, 'The Importance of Marine Spatial Planning in Advancing Ecosystem-Based Sea Use Management', *Marine Policy*, 2008, 32(5): 762–771.
9 The United Nations Convention on the Law of the Sea of 10 December 1982 (hereinafter as the UNCLOS).
10 Douvere, 'The Importance of Marine Spatial Planning', p. 766.
11 The UNCLOS, Article 309.
12 D. Hollis and T. Rosen, 'The United Nations Convention on the Law of the Sea (UNCLOS) 1982', *The Encyclopedia of Earth*, 2010: 22, www.eoearth.org/view/article/156775/ (accessed on 29 December 2015).
13 Vide the *Wildenhus (1887) 120 US 1* in which the US Supreme Court stated that an American Court had jurisdiction to try the case of a murder committed by a member of the crew of a Belgian ship in a US port.

14 The UNCLOS, Article 19.
15 The UNCLOS, Article 25.
16 The UNCLOS, Article 33.
17 Hollis and Rosen, 'The UNCLOS 1982'.
18 *Ibid.*
19 R. Beckman and T. Davenport, 'The EEZ Regime: Reflections After 30 Years', *Proceedings From the 2012 LOSI-KIOST Conference on Securing the Ocean for the Next Generation*, 2012, pp. 1–41.
20 Hollis and Rosen, 'The UNCLOS 1982'.
21 The UNCLOS, Article 87(1).
22 The UNCLOS, Article 89.
23 H.-S. Bang, 'Port State Jurisdiction and Article 218 of the UN Convention on the Law of the Sea', *Journal of Maritime Law and Commerce*, 2009, 40(2): 291–313.
24 J. C. Kunich, 'Losing Nemo: The Mass Extinction Now Threatening the World's Ocean Hotspots', *Columbia Journal of Environmental Law*, 2005, 30(1): 1–133.
25 United Nations Economic Commission for Europe, 'Spatial Planning Key Instrument for Development and Effective Governance', 2008, 1–43, at 1, www.unece.org/fileadmin/DAM/hlm/documents/Publications/spatial_planning.e.pdf (accessed on 4 January 2016).
26 Douvere, 'The Importance of Marine Spatial Planning', p. 763.
27 United Nations Environment Programme, 'Overfishing and other Threats to Aquatic Living Resource', 2006, 55–69, www.unep.org/dewa/giwa/publications/finalreport/overfishing.pdf (accessed on 4 January 2016).
28 Douvere, 'The Importance of Marine Spatial Planning', p. 765.
29 M. M. Foley et al., 'Guiding Ecological Principles for Marine Spatial Planning', *Marine Policy*, 2010, 34(5): 955–966.
30 *Ibid.*
31 C. Ehler and F. Douvere, 'Vision for a Sea Change', *Report of the First International Workshop on Marine Spatial Planning: Intergovernmental Oceanographic Commission and Man and the Biosphere Programme*, 2007, pp. 1–71, www.belspo.be/belspo/northsea/publ/sea%20change%20vision%20.pdf (accessed on 4 January 2016).
32 *Ibid.*
33 *Ibid.*
34 The Convention on Biological Diversity, The Ecosystem Approach, www.cbd.int/doc/publications/ea-text-en.pdf (accessed on 8 January 2016).
35 J. C. Day, 'Zoning Lessons From the Great Barrier Reef Marine Park', *Ocean and Coastal Management*, 2002, 45: 139–156.
36 A. K. Abramson and S. Dewsbury, 'Marine Ecosystem–Based Management: From Characterization to Implementation', *Frontiers in Ecology and the Environment*, 2006, 4(10): 525–532.
37 Convention on Biological Diversity (CBD), 'In-depth Review of the Application of the Ecosystem Approach: Barriers to the Application of the Ecosystem Approach', in *Proceedings of the 12th Meeting of the Subsidiary Body on Scientific, Technical and Technological Advice*, Paris: UNESCO, 2007.
38 Day, 'Zoning Lessons From the Great Barrier Reef Marine Park', p. 141.

39 H. Li, 'The Impacts and Implications of the Legal Framework for Sea Use Planning and Management in China', *Ocean and Coastal Management*, 2006, 40(9–10): 717–726.

40 J. Enemark, Secretary of the Common Wadden Sea Secretariat, Personal Communication. February 2007.

41 Article 1 of the Mauritian Constitution: 'Mauritius shall be a sovereign democratic State, which shall be known as the Republic of Mauritius'.

42 Rojid Sawkut et al., 'Trade and Poverty in Mauritius: Impact of EU Sugar Reforms on the Livelihood of Sugar Cane Workers', Trade and Pro-Poor Growth Thematic Working Group, 2009, 1–57, www.tips.org.za/ research–archive/trade–and–industry/southern–african–development–research–network–sadrn (accessed on 12 January 2016).

43 'The Sugar Protocol', http://eeas.europa.cu/delegations/barbados/eu_bar bados/development_coop/sugar_protocol/index_en.htm (accessed on 12 January 2016).

44 Sawkut et al., 'Impact of EU Sugar Reforms', p. 2.

45 G. Joomun, 'The Textile and Clothing Industry in Mauritius', in H. Jauch and R. Traub–Merz (eds.), *The Future of the Textile and Clothing Industry in Sub-Saharan Africa*, Boon: Friedrich–Ebert–Stiftung, 2006, pp. 193–211.

46 K. Jaufeerally, 'An Analysis of Tourist Arrivals, Spending and Other Trends in the Tourism Industry of Mauritius', *Institute for Environmental and Legal Studies*, 2000, http://iels.intnet.mu/tap1_99.htm (accessed on 18 January 2016).

47 Website of the Ministry of Tourism and External Communications Mauritius, http://tourism.govmu.org/English/Pages/Overview-of-Tourism-Sec tor-in-Mauritius.aspx (accessed on 18 January 2016).

48 'EU Tax Haven Blacklist Disputed by OECD and Listed Countries', http:// mnetax.com/9490-9490 (accessed on 18 January 2016).

49 'Commission on the Limits of the Continental Shelf', www.un.org/depts/ los/clcs_new/submissions_files/submission_musc.htm (accessed on 17 September 2016).

50 'Recommendations of the Commission on the Limits of the Continental Shelf', 2011, 1–37, www.un.org/depts/los/clcs_new/submissions_files/musc 08/sms08_summary_recommendations.pdf (accessed on 17 September 2016).

51 'Mauritius and Seychelles to Joint Manage Extended Continental Shelf', The International Union for Conservation of Nature, www.iucn.org/con tent/mauritius–and–seychelles–jointly–manage–extended–continental–shelf (accessed on 17 September 2016).

52 Board of Investment Mauritius, 'Ocean Economy', www.investmauri tius.com/investment–opportunities/ocean-economy.aspx (accessed on 20 January 2016).

53 *Ibid.*

54 *Ibid.*

55 Hellenic Shipping News Worldwide, 'Mauritius Takes Next Step to Become Global Bunker Hub', 2015, http://business.mega.mu/2015/10/21/ mauritius-takes-next-step-become-global-bunker-hub/ (accessed on 20 January 2016).

56 Board of Investment Mauritius, 'Ocean Economy'.
57 'Roadmap of the Mauritian Ocean Economy', 2013, 1–46, www.ocean economy.mu/PDF/Part1.pdf (accessed on 20 January 2016).
58 Ehler and Douvere, 'Vision for a Sea Change'.
59 Board of Investment Mauritius, 'Ocean Economy'.

Part II

BIODIVERSITY GOVERNANCE AND DIPLOMACY

3

BIODIVERSITY CONSERVATION AND MANAGEMENT IN INDIA, BRAZIL AND SOUTH AFRICA

Law, policy and diplomacy in the contemporary age

Amrendra Kumar

Biodiversity and its conservation from various perceptions and purposes has been on the international agenda for many years, but it has only been in the last twenty or more years that focus has moved towards its maintenance and management as well. Biodiversity conservation and management has now been altogether recognized as a 'process which involves promotion, protection, management, conservation and sustainable use of various biological resources to ensure continued provisions of the ecosystem'.[1] This is largely due to the fact that biodiversity contributes diversity through provisioning and regulating ecosystem services and indirectly supports ecosystem services for many aspects of human well-being. Agenda 21 states that

> Our planet's essential goods and services depend on the variety and variability of genes, species, populations and ecosystems. Biological resources feed and clothe us and provide housing, medicines and spiritual nourishment. The natural ecosystem of forest, savannah, postures, and rangelands, deserts, tundra, rivers, lakes and seas contains most of the Earth Biodiversity.[2]

In other words, it provides an ecosystem where resources and other intangible services are provided to mankind.

Biological resources have been considered very significant for the overall development of any country, especially to biodiversity-rich countries like India, Brazil and South Africa. The physical features and climatic situations of these countries across the continents have helped to host, harbour and sustain immense biodiversity under diversified ecological habitats. Consequently, they are accountable and responsible for its conservation and management through principles, practices and procedures emanating from national, regional and international law, policy and diplomacy. These countries have shown their international commitment and cooperation in this regard since the Rio Earth Summit held under the United Nations Conference on Environment and Development.[3] They have signed accordingly several multilateral environmental agreements under the auspices of the United Nations Environment Programme (UNEP) along with the Convention on Biological Diversity (CBD, 1992) and its protocols, like the Cartagena Protocol on Biosafety (2000) and the Nagoya Protocol on Access and Benefit Sharing (2010), in pursuance of biodiversity conservation and management.[4] They have consequently implemented these international principles and procedures in their domestic jurisdictions by enacting the legislations and adopting the policies in this regard. Not only this, these countries are politically and strategically important partners as well as parties of BASIC, BRICS, IBSA and other groups of countries for strategic partnership, mutual cooperation and regional diplomacy for not only political and economic but also ecological issues, including biodiversity conservation and management.[5]

In this context, the role of these countries, being biodiversity-rich countries, has been continuously looking forward in the global governance of biodiversity. Hence, this chapter aims to explore their laws, policies and diplomacy addressing biodiversity conservation and management in the contemporary age. It tries to examine the constitutional provisions, legislations, regulations, plans and policies of these countries in this regard. This chapter proposes that these countries have greater responsibility to conserve and manage their biodiversity for the overall development in the contemporary period. They should amend the existing legislations and improve the plans and policies for better biodiversity conservation and management; actively participate in the events and negotiations formalizing the practice and procedure for conservation, access, transfer and equitable sharing of the benefits of the biological resource; and take initiative and leadership through diplomacy in attaining the objectives of the CBD and other multilateral environmental agreements (MEAs) at the regional and national levels.

I. Biodiversity conservation and management

Biodiversity refers to the variability among living organisms from all sources, including terrestrial, marine and other aquatic ecosystems.[6]

It is estimated that there are about 'fourteen million plant and animal species of which only 1,750,000 are known to science'.[7] These are considered immensely valuable, especially in maintaining ecological balance and fulfilling human needs. It is said that biodiversity on the planet is a precious gift from God to humans. Human and animal are directly dependent on biodiversity, which could be sustained for the future only when it is well conserved and adequately managed. However, the conservation as well as management of biodiversity has been transformed recently in the international agenda, as evident from the practices of community-based conservation and management in developing nations.[8]

Biodiversity conservation and management practices are basically a 'social process in which women and men in the community across the classes, castes, ages, occupations and power groups are important actors to conserve, manage and use the biodiversity in a sustainable way maintaining the ecological system'.[9] Not only that, the people of the community build up strong cultural and spiritual connotations with the biodiversity by worshipping leaves, trees, rivers, ponds and mountains, and associating the animals and birds with gods and goddesses. There are thousands of indigenous communities with distinct language, tradition, custom and culture across the nations who are largely dependent on biodiversity for basic resources such as food, fodder, fuel, fibers, medicines and intangible sources in diversified ecological habitats.[10] There is a wide range of industries which try to access the biological resources for different purposes using biotechnology for 'pharmaceutical, seed, crop protection, horticulture, cosmetic and personal care, food and beverage products'.[11] Through bioprospecting, the abovesaid industries get these resources, which are 'either accessed directly from diversified ecological habitats or acquired indirectly through manmade habitats i.e. gene bank, seed bank, botanical garden and research institution.'[12] Hence, biodiversity conservation and management has been very important for all the stakeholders for its continued utilization and consumption.

Biodiversity conservation and management has been defined as 'the process involved in the promotion, conservation and sustainable use of various biological resources to ensure continued provisions of ecosystem services'.[13] The process, access, transfer, utilization and management of biological resources by humans, either directly or indirectly, are also included. It also has been called 'community-based conservation' and 'collaborative management' to refer to the same broad range of situations.[14] Community-based conservation includes 'a whole range of situations from one extreme in which officials/private agencies predominantly retain control but consult with local communities in planning or implementation to the other extreme in which communities are completely in control'.[15] Collaborative management has been defined as a 'partnership in which government agencies, local communities and resource users, non-governmental organizations and other stakeholders negotiate, as appropriate to reach on context, authority and responsibility for the management of a specific area or set of resources'.[16] Certain terms like 'co-management', 'joint management' and 'participating management' have been used synonymously with 'collaborative management'.[17] Thus, there are a variety of options among which a choice can be made to suit the context of conservation and management. However, legal discourse and endorsement of such collaborative management in many countries is sparse, but growing with the support of international non-governmental organizations (NGOs), United Nations agencies and multilateral development banks. Their research and studies have tried to promote the conservation and sustainable development by endorsing collective and integrated management of land, water and biological resources for domestic laws and regulations.[18] They further helped by designing and implementing a range of norms, practices and directives to address the needs for conservation and management of biodiversity. However, we will have to first look into the international legal instruments in this regard advocating for the conservation and management of biodiversity.

II. International law and agreements

The initial efforts for international conservation law were either regional, dealing with specific geographic areas, or sectoral, for particular subjects of species. The momentum developed in the early 1980s, concluding with several multilateral treaties or agreements on the basis of biological resources, specific species and geographic areas. There are now around 500 or more multilateral environmental agreements (MEAs) dealing with biodiversity conservation and management directly or indirectly.[19] However, the Convention on Biological Diversity (1992) has been specifically adopted as the prominent international legal instrument or treaty dealing directly with the conservation and management of biodiversity under national jurisdiction.[20] The basic objectives of this treaty are the 'conservation of biological diversity, sustainable use of its components and the fair and equitable sharing of benefits derived from the use of biological resources'.[21] It reaffirms the principle of state sovereignty to exploit the resources pursuant to their own environmental policies within their own jurisdiction without causing damage to the environment of other states.[22] It includes obligations to develop national laws, plans and policies for the conservation and management of biological diversity. Further, it provides a general legal framework regulating access to biological resources and sharing of benefits arising from their use.[23] Therefore, it requires the member states to provide access on 'mutually agreed terms (MATs)' and is subject to the 'prior informed consent (PIC)' of the country of origin and holders of those resources and associated traditional knowledge.[24] For the better execution of benefit-sharing arrangements, there has been accepted the 'Bonn Guidelines on Access to Genetic Resources and Fair and Equitable Sharing of the Benefits Arising out of Their Utilization (2002)'[25] which assists the member states and other stakeholders in developing an overall access and benefit-sharing (ABS) strategy through legislative, administrative and policy measures or in negotiating contractual arrangements. Being non-binding and voluntary in nature, it became the matter of criticisms among the user and provider countries in the due course of time. Consequently, a binding and effective international legal instrument within the CBD regime was felt among the member states to be an effective implementation and development of the ABS strategy.

Then the 'Nagoya Protocol on Access to Genetic Resources and the Fair and Equitable Sharing of Benefits Arising From Their Utilization to the Convention on Biological Diversity (2010)'[26] was adopted along with the 'Strategy for Resource Mobilization', '27 Achi Targets' and 'the Strategy Plan for Biodiversity 2011–2020'. The Nagoya Protocol is comprised of the objective of fair and equitable sharing of the benefits arising from the utilization of genetic resources, including by appropriate access to genetic resources and by appropriate transfer of relevant technologies, and by appropriate finding.[27] This objective is also directed towards contributing to the conservation of biological diversity and sustainable use of its components.[28] The basic principles under this protocol have been provided on the access, benefit sharing and compliance measures, in this regard dealing with genetic resources and/or traditional knowledge and associated with genetic resources.[29] However, this protocol would require implementing those principles and procedure on ABS through legislative, administrative and policy measures subject to domestic ABS legislation and regulation requirements.[30]

Apart from this, there is also a special protocol under this convention, namely the 'Cartagena Protocol on Biosafety (2003)',[31] which is closely linked to ensure the safe transfer, handling and use of living modified organisms (LMOs) – thus serving the purpose of the safety and management of living biological resources. The CBD contains many provisions directly relating to LMOs which provide a mandate to include the new rights and obligations of parties relating to the trans-boundary movements and the handling and use of LMOs. The objective of the protocol is to contribute to ensuring an adequate level of protection for the safe transfer, handling and use of LMOs resulting from modern biotechnology that may have adverse effects on the conservation and sustainable use of biological diversity, and also taking into account the risk to human health and specifically focusing on trans-boundary movements.[32] The parties are required under these obligations to take necessary and appropriate legal, administrative and other measures to implement this protocol in their domestic sphere.[33]

Besides the CBD regime, there are also several other international legal instruments addressing broadly the concept of biodiversity conservation and management. There has been the 'International Undertaking on Plant Genetic Resources (1983)'[34] adopted by the United Nations Food and Agriculture Organisation, which called upon its member states to share their genetic resources with the rest of the world. But after the adoption of the CBD, it was renegotiated with a new 'International Treaty on Plant Genetic Resources for Food and Agriculture (PGRFA, 1999)'[35] for agro-biodiversity as a component for overall biodiversity management. It is the first treaty to provide a legal framework which not only recognizes the need for conservation and sustainable use of plant genetic resources, but also delineates a regime for access and benefit sharing for agro-biodiversity management. Apart from the PGRFA Treaty, 'the International Convention for the Protection of New Varieties of Plants (1991)'[36] establishes a legal protection regime for the plant varieties developed by commercial breeders that seeks to influence the management of agricultural biodiversity.

Apart from this, there are also certain conventions outside or before the CBD addressing concerns of biodiversity conservation and management indirectly. First, the 'Convention on International Trade in Endangered Species of Wild Fauna and Flora (1973)'[37] supports conservation to more than 30,000 plant and animal species with the aim to ensure that international trade in specimens of plants and wild animals does not threaten their survival. Second, the 'Convention on Migratory Species of Wild Animals (1979)'[38] basically provides for the conservation of terrestrial, marine and avian migratory species and their habitats by giving strict protection to them across the nations. Third, the 'Convention on Wetlands of International Importance (1971)' hosted by the International Union for Conservation of Nature

(IUCN) provides that the conservation of wetlands and their flora and fauna can be ensured by combining far-sighted national policies with coordinated international action.[39] Fourth, the 'Convention for the Protection of the World Cultural and Natural Heritage (1972)' led by the United Nations Educational, Scientific, and Cultural Organization (UNESCO) acknowledges that specific parts of the cultural or natural heritage are of outstanding universal value and therefore need to be preserved as part of the world heritage of mankind.[40] This provides duties to the member states to take appropriate legal measures necessary for the identification, conservation and rehabilitation of these heritages.[41] There has also been evaluation of norms from statements in the Stockholm Declaration 1972,[42] the World Charter for Nature 1982[43] and the Rio Declaration 1992[44] for the conservation and management of biodiversity. Each of these legal instruments, conventions, protocols and declarations provides the legal concepts of protection, preservation, maintenance, and restoration of the biological resources, sites and heritages with the view of biodiversity conservation and management.

III. Regional law and diplomacy

At the regional level, the international community has tried to evolve certain regional laws to protect and manage the biodiversity and associated traditional knowledge. Regional laws actually promote common standards for the protection and management of biodiversity shared between countries or trans-boundary existences.[45] It may also be a model instrument that can guide the countries in developing their national laws. There are currently four model legal regimes at the regional level for the purpose of conservation and management of genetic resources. Firstly, there is the Andean Pact Decision 391 on the Common Regime on Access to Genetic Resources (1996).[46] Secondly, there is the Central American Protocol on Access to Genetic and Biochemical Resources and Associated Traditional Knowledge (2000).[47] Thirdly, there is the ASEAN draft Framework Agreement on Access to Biological and Genetic Resources (2000).[48] Finally, there is the African Model Law for the Protection of the Rights of Local Communities, Farmers and Breeders for the Regulation of Access to Biological Resources (2000).[49] These regional legal measures have been useful in the sense that they allow neighbouring provider countries with similar types of standards to set the same access and benefit-sharing conditions to user countries in trans-boundary existences of genetic resources and traditional knowledge.[50] But all such regional agreements have been initiated with neighbouring countries within the continents, i.e. Asia, America and Africa. As far as inter-continent initiatives are concerned, there is no remarkable affordance to make regional agreements on biodiversity conservation and management. However, environmental diplomacy has taken place on different environmental issues like climate change, sustainable development, and conservation and management of biodiversity.[51] India, Brazil and South Africa, being biodiversity-rich countries across the continents, have played a greater role together at the global forum with strong and efficient diplomacy and cooperation in this regard.

Nowadays, regional diplomacy has been a special mechanism basically used by regional organizations to advocate or engage for political, economic, environmental and cultural benefits[52] as globalization and liberalization have promoted interdependence among states, bringing the consciousness for mutual assistance and cooperation for mutual benefits. In line with this, India, Brazil and South Africa have been geographically, politically and strategically important partners as well as parties of different regional organizations such as BASIC, BRICS, IBSA and other groups of countries for strategic partnership, mutual cooperation and global engagement for not only political and economic but also ecological issues, including biodiversity conservation and management.[53] These regional groups that include India, Brazil and South Africa have shown common interests and the ability to engage with each other through cooperation, coordination and consultation to set their claims and priorities in a global forum.

For biodiversity conservation and management, these countries are members of the Like-Minded Megadiverse Countries (LMMC)[54] that harbour the majority of the Earth's biodiversity and associated traditional knowledge. Being biodiversity-rich countries, these countries, with 15 other like-minded countries, set up this LMMC in 2002 for consultation and cooperation so that their interests and priorities related to the preservation and management of biological diversity could be promoted.[55] They effectively joined efforts in negotiating the development of the Nagoya Protocol under the Convention on Biological Diversity.[56] At the same time, they have also called upon to do the same in the future on other multilateral agreements, such as Biosafety, Climate Change and Trade. There have been numerous meetings held since its inception, and they have adopted several declarations as well in this regard.

First of all, these like-minded mega-diverse countries, including India, Brazil and South Africa, adopted the 'Cancun Declaration' on

February 18, 2002, in Mexico.[57] It reaffirmed their sovereign rights over their natural resources as per the CBD and acknowledged that their natural heritage, which represents nearly 70% of the planet's biological diversity associated with our cultural wealth and diversity, must be preserved and utilized in sustainable manners.[58] It also recognized the urgent need to develop human resources, institutional capabilities as well as appropriate legal frameworks and public policies to enable member countries to take an active part in the new economy associated with the use of biological diversity and biotechnology.[59] They also decided to take an active part in the discussion of issues related to biological diversity within other regional and international forums for a common benefit.[60] In the 'Cusco Declaration on Access to Genetic Resources, Traditional Knowledge and Intellectual Property Rights', like-minded mega-diverse countries shared similar feelings and concerns in Peru the same year.[61] It reaffirmed the 'Cancun Declaration' as well as a commitment to promote negotiations within the framework of the CBD for an international regime to promote and safeguard the fair and equitable sharing of benefits arising out of the utilization of genetic resources.[62] It called upon the member states to strengthen national and regional processes in order to incorporate elements contained in this declaration and in national policies and regulations, especially regarding genetic resources, traditional knowledge and intellectual property rights (IPRs).[63]

Then the 'New Delhi Ministerial Declaration on Access and Benefit Sharing' was adopted by the like-minded mega-diverse countries in Delhi, 2005.[64] It recalled the Cancun and Cusco Declarations, and agreed to renew the commitment as a group to consolidate consultation and cooperation and to develop various strategies on various issues relating to the conservation of biological diversity and sustainable use of its components for the benefits of their countries and peoples.[65] They agreed to ensure that an

> international regime to be developed on access and benefit sharing, includes inter alia the following elements: prior informed consent of the country of origin; mutually agreed terms between the country of origin and user country, mandatory disclosure of the country of origin of biological material and associated TK in IPR application, along with undertaking that prevalent laws and practices of the country of origin have been respected, mandatory specific consequences in the event of failure to disclose the country of origin in the IPR application.[66]

At last, they agreed to explore the possibility of developing coherent national legislations for regulating an ABS system on genetic resources and associated traditional knowledge. Even after the conclusion and adoption of the 'International Regime on ABS' as the Nagoya Protocol, their purpose has been half achieved and led to the adoption of a national implementation by law and policy. Consequently, there is a need to explore the national law and policy of the member states among the group, especially India, Brazil and South Africa.

IV. National laws and policies

The responsibility for achieving the goals and objectives of such international legal instruments, declarations and strategy rests largely with the countries themselves. Consequently, the implementation and development of international and regional principles and procedures largely depends upon the domestic legislations and regulations. Under CBD and other international instruments, member states undertake to develop national biodiversity laws, policies, strategies and action plans on biodiversity conservation and management for a better environment and sustainable development. Due to their huge biological diversity and bioprospecting potentiality, India, Brazil and South Africa have enacted their biodiversity conservation laws and built up strong institutional setups to implement in their provinces or municipalities.[67] In a recent development for the effective implementation of an ABS regime under the CBD regime, they have also ratified the Nagoya Protocol and assured further to implement it by incorporating or amending their domestic legislations and regulations as well.[68] Their overall biodiversity conservation laws, policies and plans may be dealt with in view of biodiversity conservation and management as described below.

A. India

The Republic of India is host to rich natural and biological resources due to its bio-geographic location, diversified climatic condition and ecological diversity. India is one of the twelve major biodiversity countries of the world embracing three major biological realms: the Indio-Malaya, Euro-Asian and Afro-tropical regions.[69] India accounts for around 8% of the recorded species of the world, which includes millions of races, sub-species and local variants of species.[70] It is estimated that

India has approximately 45,000 species of plants representing as much as 11% the world's flora, which includes about

17500 species of flowering parts, 48 species of gymnosperms, 1200 species of pteridophytes, 1980 species of mosses, 845 species of liverworts, 6500 species algae, 2050 species of lichens, 14500 species of fungi and 850 species of bacteria.[71]

The total estimates of animal species in India has been outlined 'about 80,450 which include mammals (372 species), bird (1230 species), reptiles (428 species) with over 300 species of amphibians and 500 species of molluscs'.[72] Indian cultural and ethnic diversity includes over '550 tribal communities of 227 ethnic groups spread over 5,000 villages'[73] that have traditionally protected and regularly contributed in the management of the biodiversity.

In view of this, India has been a pioneer in the formulation and implementation of international and national laws, policies and plans relating to environmental protection and biodiversity conservation. Under the Indian Constitution, Article 253 provides legislative power to parliament to pass laws for implementation of any treaty, agreement, or conventions and decisions made at any international conference or forum.[74] Further, Article 21 of the Constitution provides fundamental rights to life and livelihood, which has been enlarged to the right to get pollution-free water and air along with protection of the ecosystem and biodiversity along with the directive principles of state policy under Article 48A for the protection and improvement of the environment and safeguarding of forests and wildlife.[75] To value and preserve the rich heritage of our composite culture, to protect and improve the natural environment including forests, lakes, rivers and wildlife, and to have compassion for living creatures are fundamentals duties of the citizens of India.[76]

Taking into account the constitutional provisions and international commitments, India has enacted several domestic laws and developed many policies from time to time for the protection of the environment and biodiversity.[77] Among them, the Forest Conservation Act, 1980 has been enacted for the conservation of forests in the country.[78] Further, the National Forest Policy, 1988 was adopted for afforestation and maintenance of ecological diversity.[79] There is also the Environment Protection Act, 1985 backed by the National Environment Policy (2006) for the protection and improvement of the environment and biodiversity.[80] The Wildlife Protection Act, 1972 was also enacted with the objective of effectively protecting wildlife and controlling illegal trade in wildlife.[81] However, the enactment of the Biological Diversity Act, 2002 came later with the aim of conservation, sustainable use, and access and benefit sharing of biological resources.[82] For that, the National Biodiversity Action Plan (2008) also has been prepared to identify the threats to biodiversity and constraints in biodiversity conservation, taking account of the existing legislations, strategies, plans and action points to achieve the purpose.[83] In addition to this, the Protection of Plant Varieties and Farmer's Rights Act, 2000,[84] the Geographical Indication of Goods Act, 1999,[85] the Scheduled Tribes and other Traditional Forest Dwellers Act, 2006[86] as well as the Panchayat Extension of Scheduled Areas Act, 1996[87] complement the efforts of conservation, use and management of biological resources and associated traditional knowledge.

Biodiversity conservation and management in the territorial jurisdiction of India has been specifically addressed in the Biological Diversity Act, 2002.[88] The objectives of this Act have been the conservation of biological diversity, sustainable use of its components and fair and equitable sharing of the benefits arising out of the use of biological resources and associated traditional knowledge.[89] The Act insists upon incorporating appropriate benefit-sharing provisions in the access agreements and mutually agreed terms related to access and transfer of biological resources or knowledge occurring in or obtained from India for commercial use, bio-survey, bio-utilization or any other monetary purposes.[90] This Act provides an institutional framework through a three-level hierarchical system: National Biodiversity Authority (NBA), State Biodiversity Boards (SBBs) and Biodiversity Management Committees (BMCs) for biodiversity conservation and management.[91] Above all, this Act provides a progressive and facilitative legal framework as well as a regulatory and advisory role for authorities at each level to ensure conservation of all the biological resources and biodiversity heritage sites. For proper implementation and operation of the ABS regime, there has been notified 'The Biological Diversity Rules, 2004' to address the access and benefit-sharing practices and procedures for the management of the biological resources.[92] Not only this but the new 'Guidelines for Selection and Management of Biodiversity Heritage Sites' (2010) was issued for the identification and conservation of biodiversity heritage sites.[93] Recently, there has also been issued new 'Guidelines on Access to Biological Resources and Associated Knowledge and Benefit Sharing Regulations, 2014'[94] in pursuance of the Nagoya Protocol for access requirements and benefit-sharing mechanisms in India.

B. Brazil

The Republic of Brazil is also considered very rich for natural and biological diversity in the world. It is one of the mega-diverse countries embracing seven distinct biomass reserves: Amazonia, Coatinga, Carrado, Partanal, Mata Atlantic, Campos Sulinos and Costerios.[95] Amazonia is the biggest reserve for biological diversity, Coatinga is the home of the biggest reserve of prehistoric biocultural sites, and Partanal is one of the largest wetlands areas in the world.[96] Brazil is endowed with around 20% of the world's flora and fauna in its biodiversity regions in land, water, wetlands and coastal zones within three distinct climatic regions.[97] There has been estimated approximately '517 species of amphibians, 1,677 species of birds, 518 species of mammals, 550 species of reptiles and more than 5,000 species of freshwater fish and 5 million insect species in different habitats of Brazil'.[98] Brazil is also rich in cultural and ethnic diversity where 'around 345,000 indigenous people in 250 indigenous communities live in the country'.[99]

Like India, Brazil also has been at the forefront in the formulation and implementation of national laws, policies and plans relating to biodiversity conservation and environmental protection. In this regard, the Federal Constitution of the Republic of Brazil (1988) denotes an entire chapter on the environment under Article 225.[100] Further, there are several legislations which make provisions for a National Environmental Policy, a National Council for the Environment and a National Policy for Water Resources. There has also been specifically enacted a Land Statute, a Forest Code, an Act for the Protection of the Fauna, a Decree-Law for the Protection and Promotion of Fisheries, an Act for Protection of Cultivators, an Act of Industrial Property, an Act of Biosafety and an Act of Environment, Crime, and Protection of Protected Areas under the Act of National System of Conservation Units.[101]

The oldest legislation for the protection of national resources is the Forest Code (1966) which regulated these resources under 'permanent preservation areas' and 'legal reserve areas' of forest.[102] However, the first initiative for the systematic protection of biodiversity was established under the National Policy of Environment (1981).[103] It plays the central role in the management and control of the activities causing environmental impacts through environmental zoning, environmental impact assessment, environmental licensing and defining of protected areas.[104] For implementation of the CBD, Brazil has specifically enacted the National Policy of Biodiversity (2002) which establishes principles and guidelines for biodiversity conservation and management.[105] It basically provides several principles and directives which also include the principles of precaution and prevention; the participation of society in decisions about the management of natural resources; the integration of sector and cross-sector programs and plans in the management of resources; the approach of decentralization in list management; and the emphasis on the need to consider the causes of sensitive reduction or loss of biological diversity.[106] Further, the Biosafety Act (2005) provides provisions for safety, inspection and control of the activities that involve genetically modified organisms such as manipulation, consumption, production, transport, transfer, import, export, research and release in the environment.[107]

A provisional law attributes to the Genetic Heritage Management Council (CGEN) the implementation of policies for the management of the genetic heritage.[108] This provisional measure further establishes the ABS legal framework in Brazil for access to genetic resources, protection of and access to associated traditional knowledge, sharing of benefits, and access to technology and also its transfer.[109] It also recognizes the right of indigenous peoples on the access and use of their traditional knowledge and also provides for the alternative of sharing benefits through contracts or MATs.[110] It requires previous authorization of CGEN in order to access the genetic resources and associated traditional knowledge for research, bioprospecting technical development, and prior informed consent (PIC) for indigenous peoples and communities as a necessary condition for accessing their genetic resources or associated traditional knowledge, along with benefit sharing with the providers when any product or process that results from the access to genetic resources or associated traditional knowledge arrives at market.[111] The national authority for regulation of the ABS regime in Brazil is CGEN, established in April 2002, which approved a number of norms to promote biodiversity conservation and management including 40 resolutions and eight technical orientations.[112] Recently, Brazil has also amended their biodiversity law to incorporate certain clarity and certainty on the ABS regime for better conservation and management of biodiversity.[113]

C. South Africa

The Republic of South Africa is also the custodian of rich natural and biological resources due to its distinct location and climatic variations. It is also a mega-diverse country and considered to be the third largest biodiversity-rich country in the world. This country has been divided into certain biomes such as: Deserts, Fynbos, Nama Karoo, Grassland, Savana, Albany Thicket, Forest and Wetlands.[114] South Africa is home for approximately 10% of the world's flora and 7% of the world's fauna. The genetic diversity has been estimated at nearly 24,000 species of plants, 300 species of mammals, 800 species of birds and

50,000 species of insects in the diversified habitats in South Africa.[115] Marine biological diversity has been estimated at over 11,000 species found in South African waters, which is about 15% of the global species in the coastal ecosystem.[116] The cultural diversity is also richest in the world, with millions of indigenous peoples and communities with a distinct identity and diversity in South Africa.

Consequently, South Africa has also formulated and enacted several legislations, policies and plans governing biodiversity conservation and management. South Africa's Constitution provides the overall framework for environmental governance by establishing the right of a clean environment, which is to be protected for the benefit of present and future generations through reasonable legislative and other regulatory measures.[117] Accordingly, South Africa has enacted certain legislations and developed some policies governing biodiversity, such as the White Paper on the Conservation and Sustainable Use of Biological Diversity 1997, National Environment Management Act 1998, the Protected Areas Act 2003, the Biodiversity Act 2004, National Biodiversity and Action Plan 2005, National Spatial Biodiversity Assessment 2006, National Biodiversity Framework 2008 and National Protected Area Expansion Strategy 2008.[118] Apart from these, there are several other legislations relating to water, forest and marine resources that are relevant to biodiversity conservation.

But it was the White Paper that was the first national policy explicitly providing the need for a legislative and administrative mechanism for biodiversity conservation and management.[119] Later on, South Africa specifically promulgated the Biodiversity Act 2004 with the rationale to resolve the fragmented nature by biodiversity-related legislation and to facilitate cooperation between the different levels of government in this regard.[120] The basic objective of this Act is to provide fair and equitable sharing among the stakeholders of benefits arising from bioprospecting.[121] Under the Chapter 'Bioprospecting Access and Benefit Sharing', it sets out a legislative framework and regulations for the purpose of regulating bioprospecting and to provide for a fair and equitable sharing by stakeholders in benefits from bioprospecting involving indigenous biological resources.[122] For that, the principles have been set out as a) a permit is required before anyone can carry out bioprospecting involving indigenous biological resources or before anyone can export these resources for the purpose of bioprospecting or other research; b) a permit will only be issued if there has been material disclosure to stakeholders and if their consent to the bioprospecting has been obtained; and c) cause must be reflected in a benefit-sharing agreement that allows for sharing between stakeholders in any future benefits that may result from the bioprospecting or research.[123] Apart from this, there has been passed the 'Biodiversity Conservation Regulation (2008)' for the institutional and procedural aspects on biodiversity conservation and management.[124] In short, this Act along with the regulation regulates biodiversity outside the protected areas of South Africa, providing a national biodiversity framework to ensure an integrated, coordinated and consistent approach to biodiversity conservation and management. Besides this, the Protected Areas Act 2003 establishes a consistent set of legal requirements for the biodiversity conservation and management of national, provincial and local protected areas of South Africa.[125] It aims to balance the relationship between biodiversity conservation and human settlement and promotes the sustainable utilization of protected areas for human benefits. This Act provides for the declaration of protected areas on not only state land, but also private or community land, and requires all protected areas to be managed according to ministerial approved management plans.

Along with these legislations, there are also four important national policies that guide the efforts for biodiversity conservation and management in South Africa. Firstly, the National Biodiversity Strategy and Action Plan (2005) sets out a comprehensive framework and long-term plans of action for conservation and sustainable use of biodiversity and the equitable sharing of benefits derived from such use.[126] Secondly, the National Spatial Biodiversity Assessment (2004) assesses the threat status and protection level of South Africa's ecosystem and identifies national areas of biodiversity importance.[127] Thirdly, the National Biodiversity Framework (2008) provides the policy framework to coordinate with NGOs and individuals involved in conserving and managing biodiversity.[128] Lastly, the National Protected Area Expansion Strategy (2008) was developed as a tool to achieve cost-effective protected areas expansion that enhances ecological sustainability and resilience to climate change.[129]

However, the actual result depends upon the proper implementation of such principles and procedures by these countries through their domestic policies and regulatory mechanisms. They have strong domestic laws on biodiversity and wildlife backed by the multitier authorities to facilitate and regulate biodiversity conservation and management. The objective of biodiversity conservation and management could be achieved by strengthening the national system through enacting the domestic laws, institutional effectiveness, cultural integration and cooperative governance.

V. Conclusion

As biodiversity provides a number of benefits to plants, animals and peoples, it needs to be conserved and managed to ensure that these benefits continue for present and future generations. The conservation and management of biodiversity sometimes comes into conflict with the crucial requirements of ecological survival, development needs and population pressures, which lead to the sacrifice of priceless biological resources and natural heritage.[130] Besides, there also have been certain ill-effects that cause most of the loss of biodiversity such as land degradation, deforestation, impact of climate change, loss of ecosystem resilience, loss of freshwater resources, increased infestation by invasive alien species and introduction of large living modified organisms through biotechnology.[131] This has necessitated the serious efforts across the globe towards biodiversity conservation and management for the continued provisions of the ecosystem services. It is our generation of people who are undoubtedly the last to have the opportunity to make affordances in conservation and management of biodiversity and to take steps before we cross the threshold of no return, beyond which our future generations could be in danger without fault.

In this context, India, Brazil and South Africa hold greater responsibility and accountability as biodiversity-rich countries in the world. In view of this, they have actively participated in the formulation and implementation of international conventions, protocols and declarations relating to biodiversity consideration and environmental protection. Accordingly, biodiversity laws and policies have been made along with rules, regulations and institutions for effective protection and management in their provinces and states. However, there are certain challenges in the execution and operation of such plans and policies in the contemporary age due to economic aspirations and technological development under the persistent restraints of insufficiency of funds and technology as well as lack of awareness and political will towards biodiversity conservation and management. Besides, there are some inherent problems prevalent in their legal systems which contain issues like ownership of biological resources, definition of certain terms, conflict of norms with existing laws, exceptions to academic and scientific research on ABS, bioprospecting and bio-trade, sharing of the equitable benefits, identification of the rightful holder of the resources, rights of the indigenous peoples and local communities, intellectual property rights, institutional literacy, and lack of capacity development methods and information management mechanisms for biodiversity conservation and management.[132]

Hence, there is a need to review the domestic laws and policies by these countries for coherent, clear and certain rules, procedures and institutions in the time of contemporary development. Additionally, they must strengthen their purpose through cultural integration, institutional effectiveness, public participation and cooperative governance in their national and regional spheres. For that, there is also a need for a strong commitment and cooperation through regional diplomacy and strategic partnerships to promote the agenda of mutual concerns, such as biodiversity conservation, environmental protection and climate change. Therefore, they should actively participate in the events, talks and negotiations formalizing the norms, practices and institutions for conservation, access, transfer and equitable sharing of the benefits of the biological resources, and further, take initiative and leadership through diplomacy in attaining the objectives of the CBD and other MEAs as well as protect the interest of mutual concerns in the global forum.

Notes

1 M. Khedka and R. Verma, *Gender and Biodiversity Management in the Greater Himalayas: Towards Equitable Mountains Development*, Kathmandu: ICIMOD, 2012, p. 25, www.icimod.org/mountaindevelopment/eng/pdf (accessed on 30 October 2015).

2 Agenda 21 was adopted for environmental protection and sustainable development at the Earth Summit held in Rio de Janeiro (1992). The Conservation of Biological Diversity was described under chapter 15 para 15.2. http://sustainabledevelopment.org/contents/documents/agenda2f.pdf (accessed on 30 October 2015).

3 'United Nations Conference on Environment and Development', 1982, www.unep.org/document.multilingual/default/eng/asp (accessed on 30 October 2015).

4 'United Nations Environment Programme (UNEP)', www.unep.org/biodiversity/cbd/members/html (accessed on 30 October 2015).

5 S. Biswal, 'Regional Diplomacy Through the Prism of BRICS', *World Focus*, September 2015: 123–129.

6 E. Barucha, *Textbook for Environmental Studies*, New Delhi: UGC, 2nd ed, 2013, p. 85.

7 *Ibid.*

8 A. Kothari, *Communities and Conservation: National Resources Management in South and Central Asia*, New Delhi: Sage pub, 1998, p. 257.

9 O. Lynch and K. Talbott, *Balancing Acts: Community Based Forest Management and National Law in Asia and the Pacific*, Washington, DC: World Resource Institute, 1995, p. 25.

10 K. ten Kate and S. Laird, *The Commercial Use of Biodiversity: Access to Genetic Resources and Benefit Sharing*, London: Earthscan, 1999, p. 3.

11 *Ibid.*, p. 9.

12 K. ten Kate and S. Laird, 'Biodiversity and Business: Coming to Terms With the Grand Bargain', *International Affairs*, 2000, 76(2): 214–264.
13 Khedka and Verma, *Biodiversity Management in the Greater Himalayas*, p. 26.
14 M. I. Jeffery, *Biodiversity Conservation Law + Livelihoods: Bridging the North– South Divide*, New York: Cambridge University Press, 2008, p. 393.
15 *Ibid.*, p. 397.
16 *Ibid.*
17 G. Burini–Feyerabend, *Collaborative Management of Protected Areas: Tailoring the Approach to the Context*, Gland: IUCN, 1996, p. 12.
18 See FAOLEX, which contains a substantial and updated electronic legislative and policy database on agriculture, food and natural resources for conservation and management, www.faolex.fao.org/faolex/eng/country profiles/asp (accessed on 15 March 2015).
19 B. Desai, *Multilateral Environmental Agreement: Legal Status of Secretariat*, New York: Cambridge University, 2010, p. 48.
20 'Convention on Biological Diversity (1992) (hereinafter CBD)', http://cbd. int/doc/publication/cbd.en.pdf (accessed on 20 December 2015).
21 CBD, Article 1.
22 CBD, Article 3.
23 CBD, Article 15.
24 CBD, Article 15(2).
25 'Bonn Guidelines on Access to Genetic Resources and Fair and Equitable Sharing of the Benefits Arising Out of Their Utilization', 2002, http://cbd. int/doc/publication/cbd.bonn.gds.en.pdf (accessed on 20 December 2015).
26 'Nagoya Protocol on Access to Genetic Resources and the Fair and Equitable Sharing of Benefits Arising from their Utilization to the Convention on Biological Diversity (hereinafter as Nagoya Protocol)', 2010, http://cbd. int/doc/protocol/Nagoya.pdf (accessed on 20 December 2015).
27 Nagoya Protocol, Article 1.
28 Nagoya Protocol, Article 9.
29 Nagoya Protocol, Articles 5, 6 and 15.
30 Nagoya Protocol, Article 18.
31 'Cartagena Protocol on Biosafety', 2003 (hereinafter Cartagena Protocol), http://cbd.int/doc/legal cartegna_protocol.en.pdf (accessed on 20 December 2015).
32 Cartagena Protocol, Article 1.
33 Cartagena Protocol, Article 62.
34 'International Undertaking on Plant Genetic Resources', 1983, www.fao. org/Ag/Cgfra/iu/eng/pdf (accessed on 22 December 2015).
35 'International Treaty on Plant Genetic Resource', 1999, www.planttreaty. org/contents/text.treaty.official/eng/pdf (accessed on 22 December 2015).
36 'International Convention for the Protection of New Varieties of Plants', 1991, www.upov.int/texts/legal/eng/pdf (accessed on 22 December 2015).
37 'Convention on International Trade in Endangered Species of Wild Fauna and Flora', 1973, www.cites.org/eng/disc/txt.php (accessed on 20 December 2015).
38 'Convention on Migratory Species of Wild Animals', 1979, www.cms.int/ en/convention_text.php (accessed on 20 December 2015).

39 'Convention on Wetlands of International Importance', 1971, known as 'Ramsar Convention', www.ramsar.org/sites/default/files/document/library/convention_text.pdf (accessed on 23 December 2015).

40 'Convention for the Protection of the World Cultural and National Heritage', 1972, www.whc.unesco.org/uploads/activites/documents/562.pdf (accessed on 23 December 2015).

41 Convention for the Protection of the World Cultural and National Heritage (1972), Article 4.

42 'Stockholm Declaration', 1972, www.unep.org/documents_multilingual/defaults.asp (accessed on 25 December 2015).

43 'World Charter for Nature', 1982, www.un.org/documents/ga/res/3a37r007.html (accessed on 25 December 2015).

44 'Rio Declaration on Environment and Development 1992', www.un.org/documents_multilingual/default/asp?documented78.html (accessed on 25 December 2015).

45 J. C. Medaglia, *Overview of National and Regional Measures on Access to Genetic Resources and Benefits Sharing: Challenges and Opportunities in Implementing the Nagoya Protocol*, Geneva: CISDL, 2nd ed., 2012, p. 75.

46 'Andean Pact Decision 391 on Common Regime on Access to Genetic Resources', 1996, www.cbd.int/doc/measures/abc/msr.abs_ach_en.pdf (accessed on 25 December 2015).

47 'Central America Protocol on Access to Genetic Resource and Biochemical Resources and Associated Traditional Knowledge', 2000 (draft), www.cbd/int/due/meetings/hgj./09/en.doc (accessed on 25 December 2015).

48 'ASEAN Draft Framework Agreement on Access to Biological and Genetic Resources', 2000, www.aseanbiodiversity.org/index.php (accessed on 25 December 2015).

49 'OAU Model Law for the Protection of the Rights of Local Communities, Farmers and Breeders for the Regulation of Action to Biological Resource', 2000, www.cbd/int/doc/measures/abs/msr_abs-_oau.en.pdf (accessed on 25 December 2015).

50 Medaglia, *Overview of National and Regional Measures*, p. 74.

51 G. Poyyamoli, 'Environmental Diplomacy and Sustainable Development', *World Focus: Environment Diplomacy and Sustainable Development*, 1 February 2015: 14–32.

52 Biswal, 'Through the Prism of BRICS'.

53 *Ibid.*, p. 123.

54 Like-Minded Megadiverse Countries (LMMC) Groups were set up in February 2002 for consultation and cooperation in a global forum for mutual interests on biodiversity, biosafety and climate change, www.lmmc.org/about/history.aspx (accessed on 26 December 2015).

55 'Like-Minded Mega-diverse Countries (LMMC) Groups, Preamble', www.lmmc.org/about/history.aspx (accessed on 26 December 2015).

56 'LMMC Proposals on the Negotiations of Inter–Governmental Groups Meetings Held During 2006–2010', www.cbd.ent/negotiateions/ad.doc/meeting/en/proposals/pdf (accessed on 28 December 2015).

57 'Cancun Declaration (February 18, 2002, Mexico) (hereinafter Cancun Declaration, Mexico)', www.environment.gov.za/sites/~/Cancun–lmmc/declaration.pdf (accessed on 28 December 2015).

58 Cancun Declaration, Mexico, Preamble, para 3.

59 Cancun Declaration, Mexico, Preamble, para 5.
60 Cancun Declaration, Mexico, Decision, para 1.
61 'Cusco Declaration on Access to Genetic Resource Traditional Knowledge and Intellectual Property Rights (hereinafter Cusco Declaration)', October 8, 2002, www.environment.gov.za/sites/casco-lmmc-declaration-traditiona knowledge.pdf (accessed on 28 December 2015).
62 Cusco Declaration Preamble, para 6.
63 Cusco Declaration, Declaration for Actions, para 4.
64 'New Delhi Ministerial Declaration on Access and Benefits Sharing (hereinafter as New Delhi Ministerial Declaration)', November 5, 2015, www.environment.gov.za/~/newdelhi_declaration_access-benefits.html (accessed on 28 December 2015).
65 New Delhi Ministerial Declaration, Preamble, para 4 and para 8.
66 New Delhi Ministerial Declaration, Declaration, para 3.
67 Medaglia, *Overview of National and Regional Measures*, p. 74.
68 See 'Parties to Nagoya Protocol', 2010, www.cbd.int/abs/nagoya-proto col/signatories /default.html (accessed on 30 July 2016).
69 K. Venkataraman, 'Access and Benefit Sharing and Biological Diversity Act of India: A Progress Report', *Asian Biotechnology and Development Review*, 2010, 10(3): 69–80.
70 *Ibid.*, p. 70.
71 K. Venkataraman, 'India's Biodiversity Act 2002 and its Role in Conservation', *Tropical Ecology*, 2009, 50(1): 23–30, 24.
72 *Ibid.*, p. 24.
73 *Ibid.*, p. 25.
74 'The Constitution of India', 1950, Article 253, www.lawmin.nic.in/consti tution/eng.pdf (accessed on 28 November 2015).
75 The Constitution of India, Article, 21 and 48A.
76 The Constitution of India, Article, 51A.
77 A. B. Majumdar, D. Nandy and S. Mukherjee, *Environment and Wildlife Laws in India*, Gurgaon: LexisNexis, 2013, p. 23.
78 'The Forest (Conservation) Act', 1980, www.mha.nic.in/acts /English/for est.conservation.pdf (accessed on 28 November 2015).
79 'The National Forest Policy', 1988, www.mha.nic.in/rules_regulations/ forest/NFP.pdf (accessed on 28 November 2015).
80 'The Environment (Protection) Act', 1986, www.mha.nic.in/acts/english /environment_protection.pdf (accessed on 28 November 2015). See also 'National Environment Policy', 2006, www.moef.gov.in/sites/default/files/ en/nep-2006.pdf (accessed on 28 November 2015).
81 'The Wildlife (Protection) Act', 1972, www.mha.nic.in/acts/english/wild life_protection.pdf (accessed on 28 November 2015).
82 'The Biological Diversity Act', 2002, (hereinafter The BD Act), www.mha. nic.in/acts/english/biological_diversity.pdf (accessed on 28 November 2015).
83 'National Biodiversity Action Plan', 2008, www.nba.int.in/plans/english/ biodiversity/pdf (accessed on 28 November 2015).
84 'The Protection of Plant Varieties and Farmer's Right Act', 2001, http://mha. nic.in/acts/English/plant_varieties.pdf (accessed on 28 November 2015).
85 'The Geographical Indications of Goods Act', 1999, www.mha.nic.in / acts/English/geographis-indiaction.pdf (accessed on 28 November 2015).

86 'The Scheduled Tribes and Other Traditional Forest Dweller's Act', 2005, http://mha.nic.in/acts/en/forest_rights.pdf (accessed on 28 November 2015).
87 'The Panchayat Extension of Scheduled Areas Act', 1996, http://mha.nic.in/acts/english/ panchayat_ schuleled.pdf (accessed on 28 November 2015).
88 The BD Act.
89 The BD Act, Preamble.
90 The BD Act, Section 3.
91 The BD Act, Sections 8, 22 and 41.
92 'The Biological Diversity Rules', 2004, www.nba.india.org/rules/eng.pdf. (accessed on 30 November 2015).
93 'Guidelines for Selection and Management of Biodiversity Heritage Sites', 2010, www.nba.india.eng/guidelines/bans/eng:pdf (accessed on 30 November 2015).
94 'Guidelines on Access to Biological Resources and Associated Knowledge and Benefits Sharing Regulations', 2014, www.nba.india.eng/guidelines/arbs/eng.pdf (accessed on 30 November 2015).
95 'Brazil Institute of Environment and Renewable Natural Resources (IBAMA)', Brazilian Biomass, http://siscon.ibama.gov.br (accessed on 5 November 2015).
96 IBAMA.
97 'Brazil Has Three Distinct Climatic Regions: Tropical, Equatorial and Sub-Tropical', http://mma.gov.br/home.html (accessed on 5 December 2015).
98 J. R. Morato-Leite, 'Experience, Mistakes and Challenges: The Implementation of the Convention on Biological Diversity in Brazil', in Jeffery, *Conservation Law + Livelihoods*, New York: Cambridge University Press, 2008, pp. 155–180, 156.
99 *Ibid.*, p. 155.
100 'Federal Constitution of the Republic of Brazil', 1988, Article 225, www.brazil.gov.br/constitution.en/html (accessed on 28 December 2015).
101 J. R. Morato-Leite, 'Experience, Mistakes and Challenges', p. 161.
102 'The Forest Code (Act of 4771/65)', www.brazil.gov.br/forest_code.en/html (accessed on 28 December 2015).
103 'The National Policy of Environment (Act of 6938/81)', www.brazil.gov.br/enviornment_policy.en/html (accessed on 28 December 2015).
104 The National Policy of Environment, Article 9.
105 'The National Policy of Biodiversity (Decree No. 4399/02)', www.brazil.gov.br/biodiversty_policy.en/html (accessed on 28 December 2015).
106 The National Policy of Biodiversity, Objectives.
107 'The Biosafety Act (Act of 11105/05)', www.brazil.gov.br/biosafety_act.en/html (accessed on 28 December 2015).
108 'The Access to Genetic Heritage and Associated Traditional Knowledge Under Provisional Measures (Act of 2186-16/01)', www.brazil.gov.br/provisional_measures.en/html (accessed on 28 December 2015).
109 The Access to Genetic Heritage and Associated Traditional Knowledge Under Provisional Measures, Article 1.
110 The Access to Genetic Heritage and Associated Traditional Knowledge Under Provisional Measures, Article 8.
111 The Access to Genetic Heritage and Associated Traditional Knowledge Under Provisional Measures, Article 9.

112 'Genetic Heritage Management Council (Decree No. 3945/01)', www.brazil.gov.br/genetic_hertiage.en/html (accessed on 28 December 2015).
113 'New Biodiversity Law (Decree No. 13.123/15)', www.brazil.gov.br/bio divesity_law.en/html (accessed on 28 December 2015).
114 South Africa National Biodiversity Institute (SANBI), 'Biodiversity for Development: South Africa Landscape Approach to Conserving Biodiversity and Promoting Ecosystem Resilience', 2010, www.sanbi.org/about/information/new.book.highlights.conervation.pdf (accessed on 10 November 2015).
115 *Ibid.*
116 'South Africa National Biodiversity Strategy and Action Plan', 2005, www.gov.za/documents/plans/downloads.pdf (accessed on 10 November 2015).
117 'The Constitution of Republic of South Africa (1996) adopted on 11 October 1996', www.gov.za/documents/constitution/republic–South Africa_1996.html (accessed on 10 November 2015).
118 SANBI, www.sanbi.org/about/information /documents.conervation.pdf (accessed on 10 November 2015).
119 'South Africa White Paper on the Conservation and Sustainable Use of Biological Diversity', 1997, www.gov.za/documents/white_paper/down loads.pdf (accessed on 10 November 2015).
120 'The South Africa's Biodiversity Act', 2004, www.environment.gov.za/../nampa/act no.51of 2004.pdf (accessed on 10 November 2015).
121 *Ibid.*, Objectives.
122 *Ibid.*, Chapter III.
123 *Ibid.*, Principles.
124 'South Africa's Biodiversity Conservation Regulations', 2008, www.environment.gov.za/hempaa–regulations/documents/pdf (accessed on 10 November 2015).
125 'South Africa's National Environment, Management: Protocol Areas Act', 2003, www.environment.gov.za/../nampa.actno.57of2003/pdf (accessed on 10 November 2015).
126 SANBI, www.sanbi.org/about/information/documents.national_plan.pdf (accessed on 10 November 2015).
127 'South Africa's National Spatial Biodiversity Assessment', 2004, www.bgis.snabi.org/nsba/project.asp. (accessed on 10 November 2015).
128 'South Africa's National Biodiversity Framework', 2008, www.environ ment.gov.za/ . . . /nampa_ biodiversity framework.pdf (accessed on 10 November 2015).
129 'South Africa's National Protected Areas Expansion Strategy', 2008, www.environment.gov.za/~/nampa-protected area expansion/strategy.pdf (accessed on 10 November 2015).
130 B. H. Desai, 'Environment and Development: Making Sense of Predicament of Developing Countries', *World Focus: Environment and Sustainable Development*, May 2013: 3–8.
131 A. Kiss and D. Shalton, *Guide to International Environmental Law*, Leiden: Martinus Nijhoff Publishers, 2007, p. 178.
132 E. Kamau and G. Winter, *Common Pools of Genetic Resources: Equity and Innovation in International Biodiversity Law*, London: Routledge, 2013, p. 16.

4

BIODIVERSITY CONSERVATION AND MANAGEMENT

The Ugandan experience

Moses S-N Watulo

Uganda, divinely referred to as the Pearl of Africa by Prime Minister Sir Winston Churchill, is one of the countries making up the East African bloc, comprising Uganda, Kenya, Tanzania, Rwanda, Burundi and South Sudan. Geographically, Uganda is a land-linked country, that is, it has no access to the sea, being bordered by five countries. This geographical location has repercussions as far as the conservation and management of biodiversity is concerned because of the interlinked nature of the environmental masses that traverse the national boundaries. For instance, the great River Nile that has its source at Jinja in Uganda on Lake Victoria flows all the way to the Red Sea through South Sudan, Sudan and Egypt on its way, and this forms the Lake Victoria Basin composed of a network of rivers and the freshwater inland lake shared between Uganda, Kenya and Tanzania. Within the diverse geographical environment marked by breathtaking physical features like mountains, lakes, rivers, tropical forests and vast wetlands which form an intricate biodiversity universe, vast species of flora and fauna life thrive, notwithstanding the glaring negative effects of human activity that has already rendered some species extinct.

Conservation and management of biodiversity in East Africa and Uganda in particular becomes critical in light of the ongoing oil and gas exploration, and impending production in the Albertine Graben (Lake Albert and Western Rift Valley region), in addition to the existing environmental footprint of the heavy industries. Biodiversity conservation and management is a very topical issue in East Africa. East Africa is traversed by the Equator and two rift valleys, namely the Eastern Rift Valley (runs through Kenya) and the Western Rift Valley (runs through Uganda). These rift valleys are marked by impressive water bodies like Lake Albert and Lake Edward in Uganda, the whole area forming what is called the Albertine Graben. In the western rift valley, we have Lake Naivasha and Lake Baringo. The Albertine Graben in Uganda also features the famous Queen Elizabeth and Muchson Falls National Parks, which are home to diverse species of animal, plant, bird and other species. Besides these, there are other land masses in the form of dense tropical forests, like the impressive Bwindi Impenetrable forest which is home to the world-famous mountain gorillas.

As a natural consequence, most of the region's wildlife habitats in the Albertine Graben had been protected in the form of national parks and forests. The inventory of biodiversity species in the Albertine Graben and other protected habitats like Mount Elgon and Kidepo National Parks is tremendous. But the Albertine Graben now faces serious biodiversity risks occasioned by the discovery of oil and the resultant oil production activities. In this chapter, the status of biodiversity conservation in Uganda, biodiversity risks in Uganda, the role of various agencies, the legal framework, and finally, recommendations for achieving an effective biodiversity conservation and management regime are discussed.

I. Biodiversity at the species level and based on taxa

Uganda is exceptionally rich in biodiversity with surveys reporting the occurrence of over 18,783 species of flora and fauna. The distribution of fauna species by taxa shows that liverworts have to make up the highest percentage of the species at 46%, followed by fungi (poly pore) at 16%, birds at 10.2% and mammals at 7.5%.[1] Due to various threats, several Ugandan species have qualified to be included on the International Union for the Conservation of Nature (IUCN) Red Data list. The list shows that as of 2016, one species was already extinct, 45 species critically endangered, 35 species endangered and 100 species vulnerable.[2]

The key fauna and flora biodiversity resources in Uganda may be described under the following categories: mammals, birds, fishes, reptiles, amphibians, plants and insects.

Uganda has approximately 380 mammal species and is ranked 13 in the world in terms of mammal species richness.[3] Uganda has approximately 1,016 species of birds (10% of the world total)[4] with the largest species density in the Albertine Graben.

The Albertine Graben forms part of the western arm of the Great Rift Valley system in East Africa. It is classified as a global biodiversity hotspot. The rich biodiversity can be attributed to a number of factors, including great contrasts in relief (from 620–5,100 masl), variable climatic conditions and diverse habitat types and ecosystems. The Albertine Graben is naturally endowed with biodiversity in terms of species richness and abundance, species of high conservation value, and a rich and varied landscape with many ecosystems. The ecosystems range from savannahs and wetlands to forests. The species biodiversity includes 14% of all African reptiles (175 species), 19% of Africa's amphibians (119 species), 35% of Africa's butterflies (1,300 species), 52% of all African birds (1,061 species), 39% of all African mammals (402 species of mammals) and about 128 species of fish. It has been described as being the most vertebrate species-rich country on the African continent.[5]

The fish biodiversity in Uganda is dominated by the cichlid family, consisting of 324 species of which 292 are endemic to Lake Victoria. Of the over 600 fish species found in Uganda, the only commercial fish species is the Nile perch (*Lates niloticus*) found in all the major lakes except Edward and George. Other commercially exploited species include the Nile tilapia (*Oreochromis niloticus*) found in all major water bodies, mukene (*Rastrineobola argentea*) from Lakes Victoria and Kyoga, muziri/mukene (*Neobola bredoi*) of Lake Albert, catfish (*Clarias gariepinus*) and the silver catfish (*Bagrus documak*) from all major water bodies. *Alestes Baremose, Brycinus nurse* and *N. bredoi* currently constitute about 80% of the fish biomass in Lake Albert. The most common fish species to almost all the water bodies is the lungfish (*Protopterusaethiopicus*).[6]

The Albertine Rift is home to approximately 40% of the mammals, 50% of the birds, 14% of the reptiles, and 19% of the amphibians of Africa. The species that are endemic to the Nile Basin part of the rift are mainly small mammals such as shrews, rats and bats, as well as frogs and toads, chameleons, butterflies and dragonflies. Some of the larger endemic animals include the mountain gorilla, Rwenzori duiker, owl-faced monkey, L'Hoest's monkey, and the Rwenzori turaco.[7]

Fungi are generally poorly known or documented in Uganda. However, available records show that there are 420 species of fungi[8] in Uganda. Fungi hosts include the reptiles, lichens, insects and plants which also exist in vast varieties throughout the country.

II. Biodiversity in protected areas

Protected areas (PAs) in Uganda mainly fall under two resources, namely forestry and wildlife. Out of a total surface area of 241,551 sq. km (both land and water), 25,981.57 sq. km (10%) is gazetted as wildlife conservation areas, 24% is gazetted as forest reserves and 13% is wetlands.[9]

A. Biodiversity in wildlife conservation areas

Uganda has 10 national parks, 12 wildlife reserves, 10 wildlife sanctuaries, 5 community wildlife areas, 506 central forest reserves and 191 local forest reserves. However, it is estimated that over 50% of Uganda's wildlife resources still remain outside designated protected areas, mostly on privately owned land, which is of the most urgent concern for protection and development.[10]

Uganda's wildlife conservation areas are very rich in biodiversity. There are 405 species of mammals, 177 species of reptiles, 119 species of amphibians and approximately 1,000 bird species. In Uganda's wildlife conservation, some mammal species are restricted in their distribution. For example, zebras are restricted to Lake Mburo and Kidepo National Parks, giraffes to Murchison Falls and Kidepo National Parks, and mountain gorillas to Bwindi Impenetrable and Mgahinga National Parks. There are three local extinctions among the large mammals, namely oryx, black rhino and Derby's eland.[11]

B. Biodiversity in forest reserves

Uganda's tropical forests are also very rich in biodiversity. Central forest reserves (CFRs) are known to house some 1,259 species of trees and shrubs; 1,011 species of birds, 75 species of rodents; 12 species of diurnal primates; and 71 butterfly species. Four species of mammals (chimpanzee, L'Hoest's monkey, elephant and leopard), one species of bird (Grauer's rush warbler) and one species of butterfly (cream-banded swallowtail butterfly) are also listed as "vulnerable". Four species of forest birds (Nahan's francolin, African green broadbill, flycatcher and forest ground thrush) are classified as "rare". The Uganda red colobus monkey and Kibale ground thrush are categorized as "intermediate" species since not enough information is available about them.[12]

III. Biodiversity outside protected areas

Uganda's present policies and legislation for the management of terrestrial biodiversity outside PAs is inadequate. The existing land tenure systems of land holdings, leasehold and customary holdings offer little incentive for protection and management of biodiversity outside PAs.

Maintenance of habitats and species are at the mercy of individual landowners, while wildlife is under considerable pressure and requires more attention for conservation. A few areas outside the PA system with considerable populations of mammals have been identified in several rangelands in Uganda, e.g. the former Ankole Ranching Scheme which has viable numbers of impala, zebra, waterbuck, bush pigs, buffaloes, warthogs, oribi, topi and hippos. Other areas in districts such as Kiboga and Luwero also have reasonable animal populations outside PAs.[13]

The bulk of the forests (64%) in Uganda are found on private land (National Forestry Authority or NFA, 2011) which is outside protected areas. These forests harbour the same extent of biological diversity as those inside the forest reserves. This situation shows that private landowners and communities could play a significant positive role in managing forest biodiversity in Uganda, given the right incentives to do so.[14] As with wildlife, the status of plants outside PAs is not known. However, there are some restricted range species that are critical; for example, *Rytgyinia sp.* is confined to Iganga District in eastern Uganda whereas *Aloe tororoana* is only known on Tororo rock, an area of only a few hectares. *Phoenix reclinata* is highly vulnerable outside PAs, as it is heavily harvested as poles for fencing, especially in urban areas.

A. Biodiversity in wetlands

Uganda's wetlands are known to support some 43 species of dragonflies (of which 20% are known to occur in Uganda only), 9 species of molluscs, 52 species of fish (which represent 18% of all fish species in Uganda), 48 species of amphibians, 243 species of birds, 14 species of mammals, 19 species of reptiles and 271 species of saprophytes. Papyrus and other wetland plants have commercial value, and many other plants are used for medicinal purposes.

B. Biodiversity in savannah ecosystems

Grasslands/savannas cover more than 50% of the land area of Uganda and are dominated in different locations by species of grasses, palms or acacias. A diversity of other plant and animal species are also closely associated with various natural savanna types. Much of this habitat has been converted to human use for agriculture and grazing. The remaining pockets of natural savannas and grasslands are primarily found in various protected areas in Uganda.

C. Biodiversity in aquatic ecosystems

About 20% of the surface area of Uganda is under water, comprising lakes (46,900 sq. km), swamps (7,300 sq. km) and rivers (2,000 sq. km). Uganda's fisheries landscape therefore includes the diverse resources ranging from the five large lakes – Victoria, Kyoga, Albert, Edward and George – and Kazinga Channel; over 160 small lakes; and a network of rivers, swamps and floodplains; all of which are critical habitats, breeding and nursery grounds for fish and potential sites for aquaculture development. The 160 small water bodies occur in eastern and western Uganda, but their potential for fish production is largely unknown.

Aquatic biodiversity is, to a large extent, outside the PA system. It therefore suffers direct human impacts as communities exploit it for their sustenance. For example, fish biodiversity has been adversely affected due to unregulated exploitation without adequate provisions for sustained renewal of the fish. There has also been a considerable change in fish species composition in lakes such as Victoria and Kyoga following the introduction of the Nile perch in the 1950s. Shoreline vegetation, such as papyrus, Vossia and Typha, which are under increasing threat, form an important habitat for fish biodiversity. Uganda has about 600 fish species in terms of biodiversity and all are edible, but those commonly encountered in trade are dominated by the Nile perch, Nile tilapia and small fishes (mukene, ragoogi and nkejje).

D. Below ground biodiversity

Little is known about the status of soil biodiversity because it has received less attention from researchers and planners. As far as biodiversity conservation is concerned, the most important of these is the soil bacteria.

IV. Habitat trends and implications for biodiversity

A. Forests

According to the annual report of the National Forestry Authority (NFA, 2011), forest land in Uganda is presently estimated at 3.3 million hectares or 16% of the total country area, declining from 4.9 million hectares or 20% in 2001. Of the total area of forests, 30% is in protected areas (forest reserves, national parks and wildlife reserves) while 70% is found on private and customary land. Uganda is estimated to be losing its forest cover at a rate of 80,000 hectares per year, implying a loss in forestry biodiversity as well. The size of forests and woodlands has significantly declined from 45% to 20% of the total land surface between 1890 and 1990. The majority of the forest loss has occurred outside of protected areas, largely due to conversion of forest lands into agriculture and over-harvesting of wood for energy supplies in the form of firewood and charcoal.[15]

B. *Wildlife protected areas*

Uganda's wildlife protected areas include 10 national parks, 12 wildlife reserves, 7 wildlife sanctuaries and 5 community wildlife areas. The biodiversity in the wildlife conservation areas has in some cases declined and in other cases increased over the years. The major threats to PAs are related to the seemingly high population growth rate of Uganda, which results in high demand for resources including land, fuel and income – but also to the failure by local communities to recognize the value of PAs and associated biodiversity. Population growth has increased the demand for agricultural land and fuel wood for domestic use. Although opportunities to ameliorate PA degradation exist through sound exploitation, rural poverty restricts the ability of local communities to invest in sustainable land use practices.

C. *Wetlands*

There is a fair level of complexity in categorizing Uganda's wetlands and inconsistency in their size. According to the Wetlands Management Department, wetland cover is presently estimated at 13% of the country's area, or about 30,000 km², of which one-third is permanently flooded.[16] In Uganda, most wetlands occur outside protected areas and their range and quality are rapidly being eroded for agricultural land, urban settlement and industrial development. In eastern Uganda alone, 20% of wetlands have been destroyed; in the central region, 2.8%; in the northern region, 2.4%; and in the western region, 3.6% of wetlands have been destroyed. This has implications on wetlands biodiversity, especially for wetland-dependent species such as Sitatunga.

V. Key biodiversity risks in Uganda

Uganda has made some progress in conservation and management of biodiversity, especially following the signing and ratification of the

Convention on Biological Diversity (CBD). However, the legislation on the subject is in bits and pieces, each talking about a specific area, and this does not support the biodiversity conservation and management regime effectively.

Uganda faces the following key biodiversity risks:

A. Ecological disasters and species loss

Uganda, just like other East African countries, is facing a glaring risk of extinction of various species of flora and fauna. The region is now experiencing unprecedented oil and mineral exploration activities that cause grave ramifications for biodiversity in terms of loss of ecological habitats. The oil companies pose the gravest biodiversity risks, but their pathetic mitigation plans are only exposed in the advent of an ecological disaster such as the one witnessed by oil spillage in the Niger Delta.[17] In this case, a vast ecological habitat was destroyed by the activities of Shell Corporation. Human communities were directly affected and a way of life destroyed completely. Regrettably, what was lost will never be replenished. On paper, these corporations appear to be at the top of their game, but in practical terms, the level of disaster preparedness related to their activities is very low or perhaps absent.

For Uganda, an ecological disaster in the Albertine Graben could wipe out completely the thriving populations of fauna and flora life. This area consists of two major national parks, several freshwater lakes, forests and sections of the River Nile. Yet even the oil production agreements are kept secretly away from the eyes of the public, which could have perhaps sought for inbuilt safeguards as a foresight; in hindsight of past disasters, such as the Niger Delta Disaster already adverted to and others like BP Oil Spill that threatened various species, especially birds, the Government should adopt a more transparent outlook in the management of the oil activities and put in place robust risk management frameworks to help mitigate environmental disasters of this kind.[18] A report of the Economic Policy Research Centre (EPRC, 2015) concluded that there is little trust in government's ability to effectively and efficiently manage oil resources.[19]

B. Proliferation of cancerous invasive species

Foreign species, alien to the East African habitats, are causing menacing dangers. The introduction of the Nile perch in East African waters in the 1950s has already wreaked havoc with mounting extinctions of many of the indigenous species.[20] But the 1990s witnessed the arrival of the water hyacinth. The water hyacinth covered much of Lake Victoria with dense vegetation, which spread rapidly over the entire surface, suffocating fish and marine life below its surface.

C. Massive industrialization and pollution

Countries were and are trying to out-compete each other in terms of attracting private investment, and in most cases they have ignored environmental concerns, perceiving them to be anti-development. East Africa has witnessed the destruction of wetlands on a massive scale due to indiscriminate industrialization.[21] In Uganda, especially in the Capital City of Kampala, where the industries have preference because of market considerations, the problem is manifest. The industries discharge the pollutants into the water channels and as the pollutants flow, they destroy downstream habitats as well. The three largest wetlands in Kampala that are under threat of total destruction are Kinawataka, Nakivubo and Lubigi. Kinawataka and Nakivubo are both part of a tributary system in which their waters flow south towards Lake Victoria. Lubigi is located on the northwest side and is part of a system in which its waters flow north towards Lake Kyoga. Kinawataka is also referred to as Kinawataka–Kawoya; it is 1.5 km^2 and is located in Nakawa sub-county, 6.5 km east of the city centre.[22]

VI. Legislative, strategic and judicial efforts in biodiversity conservation and management

A. Biodiversity legislation in Uganda

The role played by the legislature and the judiciary is of paramount importance in promoting conservation efforts, as it balances between the demands of industry and the fragilities of the environment. In East Africa and Uganda in particular, most of the current legislation has been framed with a view to environment protection and nature conservation. Biodiversity conservation and management has not been dealt with under one complete law, but with different bits of legislation addressing different components of biodiversity, such as the environment, plants, wildlife, wetlands, water masses and forests. Biodiversity includes diversity at the genetic level, for example, between individuals in a population or between plant varieties, the diversity of species, and the diversity of ecosystems and habitats.

The following are the existing legislations in the area of biodiversity conservation and management:

B. The Constitution of Uganda, 1995

Environmental protection and conservation is a matter of national importance. Like adverted to earlier, environmental protection and conservation is a corollary to biodiversity conservation and management, as there is no specific statutory law on biodiversity matters. This importance has been given effect in the Constitution of the Republic of Uganda, under the National Objectives and Directive Principles of State Policy.[23] It provides that the State shall promote sustainable development and public awareness of the need to manage land, air and water resources in a balanced and sustainable manner for the present and future generations. The Constitution further provides that the utilization of natural resources of Uganda is to be managed in such a way as to meet the development and environment needs of present and future generations of Ugandans. In particular, the State is required to take all possible measures to prevent or minimize damage and destruction to land, air and water resources due to pollution or other causes. It further provides that Parliament shall, by law, provide for measures intended to protect and preserve the environment from abuse, pollution and degradation; to manage the environment for sustainable development; and to promote environmental awareness.[24]

Additionally, an individual's right to a clean and healthy environment has been enshrined in the Constitution,[25] and individuals can enforce this right through judicial action. These provisions are important in broadening the *locus standi* of citizens to redress environmental wrongs. The State and local governments are required to create and develop parks, reserves and recreation areas; ensure conservation of natural resources; and promote the rational use of natural resources so as to safeguard and protect the biodiversity of Uganda. The public trusteeship of rivers, lakes, wetlands, national parks, game reserves and forest reserves is vested in the State.

C. The National Environment Act, 1995

This Act was enacted in 1995, and it establishes the National Environment Management Authority (NEMA) as the overall body and principal agency responsible for coordinating, supervising and monitoring all aspects of environmental management in Uganda.[26] In Uganda and East Africa in general, biodiversity conservation is understood in terms of environmental protection and management.

The Act also provides for the functions of the authority, the relationship with lead agencies and delegation.[27] Some of the functions include: to coordinate the implementation of Government policy and the decisions of the policy committee; to ensure the integration of environmental concerns in overall national planning through coordination with the relevant Ministries, departments and agencies of the Government; to liaise with the private sector, intergovernmental organizations, nongovernmental agencies and governmental agencies of other States on issues relating to the environment; to propose environmental policies and strategies to the policy committee; and to initiate legislative proposals, standards and guidelines on the environment in accordance with this Act.

The Act further establishes the Policy Committee on the Environment, composed of 10 ministers charged with various sectors of the environment, which is responsible for the formulation and implementation of policy guidelines, and coordinating environmental policies of various government agencies. It also establishes the Board of Directors, which is mandated to appoint technical advisory committees such as the one on environmental impact assessment and biodiversity.[28] The technical committee on biodiversity was charged with implementation of the National Biodiversity Strategy and Action Plan (NBSAP) in accordance with the requirements of the Convention on Biological Diversity (CBD) to which Uganda is a party.

D. The Agricultural Seeds and Plants Act, 2006

This Act came into force on 29 June 2007 to provide for the promotion, regulation and control of plant breeding and variety release, multiplication, conditioning marketing, importing and quality assurance of seeds and other planting materials, as its key objects.[29] The Act also establishes the National Seed Certification Service, which is responsible for the design, establishment and enforcement of certification standards, methods and procedures; registration and licensing of all seed producers, auctioneers and dealers; advising the National Seed Industry Authority on seed standards; and providing the Authority with technical information on any technical aspects affecting seed quality.[30] The Act imposes stringent requirements for variety testing. All imported and domestic varieties of seeds or breeding materials are required to be tested for a minimum of three generations before their releases; the Act also imposes licensing requirements for the importation and dealing in varieties of seeds and plants. The Act prescribes offences for the sale of prescribed seed under a different name, tampering with seed samples, altering official records, altering documents and marks, and breach of secrecy, and the concomitant penalties.[31]

E. *The Plant Protection and Health Act, 2015*

The rigour of the Agricultural Seeds and Plants Act, 2006 has been enhanced by the passing of the Plant Protection and Health Act, 2015. This Act seeks to consolidate and reform the law relating to the protection of plants against destructive diseases, pests and weeds; to prevent the introduction and spread of harmful organisms that may adversely affect Uganda's agriculture, the natural environment and livelihood of the people; to ensure sustainable plant and environmental protection; to regulate the export and import of plants and plant products and the introduction of plants in accordance with international commitments so as to protect and enhance the reputation of Uganda's agricultural imports and exports; and to entrust all plant protection regulatory functions to the Government, and for other related matters.[32]

The Act provides for powers to make rules for the prevention of pests, weeds and diseases; imposes duties on the occupiers of the land; and grants the right of entry for the destruction of infected articles.[33] Offences that may be committed by persons and criminal liability for corporate entities and their officers have been created in the Act, with penalties for committing offences being a fine of 250 currency points for corporate entities and 24 currency points or imprisonment not exceeding one year or both for individuals.[34]

F. *The Uganda Wildlife Act, 1996*

This Act was enacted in 1996 and came into force on 1 August 1996 to provide for the sustainable management of wildlife, to consolidate the law relating to wildlife management, and to establish a coordinating, monitoring and supervisory body for that purpose.[35] It fundamentally changes the way wildlife is managed. It repealed the National Parks Act and the Game (Preservation and Control) Act. The Act creates the Uganda Wildlife Authority (UWA)[36] as the principal body charged with the performance of, among others, these functions: ensuring the sustainable management of wildlife conservation areas; identifying and recommending areas for declaration as wildlife conservation areas and the revocation of such declaration; establishing the management plans for wildlife conservation areas and for wildlife populations outside wildlife conservation areas; proposing policies and procedures for the sustainable utilization of wildlife by and for the benefit of the communities living in proximity to wildlife; and promoting conservation of biological diversity ex-situ and contributing to the establishment of standards and regulations for that purpose.

Therefore, the relevant functions of UWA for the purposes of wildlife protected areas and wildlife management areas are, among others, to preserve selected examples of biotic communities in Uganda and their physical environment; to preserve populations of rare, endemic and endangered species of wild plants and animals; and to generate economic benefits from wildlife conservation for the benefit of the people of Uganda. The protection of wildlife under the Act is seen from two perspectives, conservation within conservation areas and conservation outside protected areas. Conservation areas are declared by the Minister[37] in consultation with the District Council in whose jurisdiction the proposed area is located. Parliament is empowered to approve such establishment by its positive resolution. Conservation areas are divided into two categories: wildlife protected areas and wildlife management areas. The wildlife management areas include wildlife sanctuaries, community wildlife areas, and such other areas as the Minister may declare. Wildlife protected areas include national parks, wildlife reserves and such other areas as the Minister may declare to be wildlife protected areas.[38] The Act preserves community rights.[39] Local communities and individuals who have property rights in land within the protected areas are permitted to carry on activities compatible with conservation principles and practices of wildlife resources.

Individuals and communities play a key role in conserving biodiversity. They are the principal actors along with corporate entities in creating biodiversity risks. In this regard, the individuals and communities who are enlightened tend to protect biodiversity masses over which they have control arising from their land rights. Those near forests do not actively engage in illegal deforestation activities. Those around wetlands do not destroy them by clearing vegetation cover and undertaking unguided land reclamation. These individuals and communities who are enlightened act as guardians of the biodiversity habitats. The Act stipulates that 'Any person who, except in accordance with this Act, enters into or resides in, or attempts to enter into or reside in, any national park, wildlife reserve or any other protected area declared under section 18(2) commits an offence'.[40]

The Act further creates general penalties[41] and stipulates that a person convicted of an offence under this Act for which no other penalty is provided is liable, in the case of a first offence, to a fine of not less than 30,000 shillings but not exceeding three million shillings or to imprisonment for a term of not less than three months or to both such fine and imprisonment; and in the case of a second or subsequent offence, to a fine of not less than 300,000 shillings but not exceeding six million shillings or to imprisonment for a term of not less than six months or to both such fine and imprisonment. This is one way of controlling access to species in protected areas.

A novel feature of the Act is the provision of wildlife use rights, which are tradable in the sense that they are transferable from one person to another, such as rights to hunt, farm, ranch, trade in or use wildlife for educational purposes. The Act provides for their management and transfer, and they are classified into classes A, B, C, D, E and F.[42] These wildlife use rights are transferable[43] and in some cases, a transfer permit is needed, especially for class A and class E. This kind of transfer is known as a permitted transfer. The person possessing the permit for wildlife use rights can transfer the same to another person.

All the foregoing is intended to conserve wildlife throughout Uganda so as to maintain the abundance of the diversity of species and to support sustainable utilization of wildlife for the benefit of the people of Uganda. The Act changes the philosophy of wildlife conservation in Uganda. It moves away from a state-centred management system to a system that encourages public participation and private sector involvement. It further updates and modernizes the law and goes a long way toward implementing the conservation philosophy of the Convention on Biological Diversity that was ratified by Uganda on 8 September 1993, Uganda becoming a party on 29 December 1993.

G. The Water Act, 1997

The Water Act is one piece of Uganda's environmental legislation with key provisions to enhance the sustainable development of water resources. Water is the habitat for vast species of flora and fauna life, and therefore its effective management is key in conserving and managing biodiversity. The Act provides for the establishment of the Water Policy Committee[44] and its functions,[45] which include: to assist the Minister in the coordination of hydrological and hydro-geological investigations; to coordinate the preparation, implementation and amendment of the water action plan and to recommend the water action plan to the Minister; and to advise the responsible Minister, as the case may require, on any dispute between agencies involved in water management that may be referred to it.

The Act provides for water easements[46] that enable a holder of a water abstraction permit to bring water to or drain water from his land over land owned or occupied by another person. In the same way, an easement may enable the holder of a waste discharge permit to drain waste from his land over the land owned or occupied by another person. All these aspects of the Water Act have the object of sustainable use of water resources, which runs through the entire Act. Waste, misuse and pollution, which may lead to unsustainable use of water, are prohibited.

VII. Strategic efforts

A. Convention on Biological Diversity

It is noteworthy that Uganda signed and ratified the Convention on Biological Diversity (CBD) on 12 June 1992 and 8 September 1993, respectively, and then became a party on 29 December 1993. The CBD has three objectives, namely the conservation of biological diversity, its sustainable use and the fair and equitable sharing of the benefits arising from the utilization of genetic resources. Article 6 (a) of the CBD requires Parties to the Convention to develop national strategies, plans or programmes for the conservation and sustainable use of biological diversity. However, only the National Biodiversity Strategy and Action Plan (NBSAP),[47] the main instrument for implementation of the Convention at the country level, exists in an explicit form. There has been no movement on the legislative front to constitute this into a self-contained law on biodiversity conservation and management, but separate bits of legislation exist.

B. National Biodiversity Strategy and Action Plan (2015–2025)

At its tenth meeting in Nagoya, Japan, the CBD Conference of the Parties (COP 10) adopted the new Strategic Plan for Biodiversity 2011–2020, with 20 Aichi Biodiversity Targets. The Parties then committed themselves to revising their NBSAPs and to adopt them as policy instruments by 2015. They also committed themselves to developing national targets that would support the achievement of the Strategic Plan and Aichi Targets, and to report thereon at COP 11 or 12 in 2012 or 2014. Uganda developed its first National Biodiversity Strategy and Action Plan (NBSAP1) in 2002. The process was coordinated by the National Environment Management Authority (NEMA), which is the institution coordinating the implementation of the CBD in Uganda. The NBSAP had an initial implementation period of 10 years with a major review after 5 years. The first review should have taken place in 2007, but this was not done due to lack of financial resources. The second review has been done simultaneously with the formulation of the second-generation NBSAP (NBSAP2).

The NBSAP is the main instrument for implementation of the Convention at the country level. The NBSAP provides Government with a framework for implementing its obligations under CBD as well as the setting of conservation priorities, channelling of investments and building of the necessary capacity for the conservation and sustainable use of biodiversity in the country.

VIII. Lessons learnt from implementing NBSAP1 for Uganda

A number of lessons were learnt from implementation of NBSAP1 (2002–2012).[48] The NBSAP was effective in addressing various biodiversity concerns in the country, such as: improving coordination among various agencies through the formation of a Technical Committee on Biodiversity Conservation (TCBDC); improving collaboration between the CBD and other international conventions at the national level; addressing a number of the Articles of the Convention, such as the CBD programme of Work on Protected Areas (PAs), formulation of Regulations on Access to Genetic Resources and Benefit Sharing, establishment of a biodiversity information sharing mechanism, preparation of a National Invasive Species Strategy and Action Plan, and promotion of public awareness on biodiversity as well as support to relevant areas of biotechnology and Biosafety; implementation of the Convention's Thematic Programmes of Work and Cross-Cutting Issues such as inland waters biodiversity, agro-biodiversity, identification, monitoring and assessment, development of biodiversity indicators and the expanded programme of work on forest biological diversity.

A. Key obstacles to NBSAP1 implementation

The key obstacles to NBSAP1 implementation included: inadequate financial resources for implementation of planned activities and programmes in the NBSAP; inadequate awareness of NBSAP1 among implementing partners and the general public; inadequate human and infrastructure capacity in relevant fields of biodiversity conservation, such as taxonomy, and capacity to carry out conservation and characterization of germplasm in the National Gene Bank; lack of a central node/Clearing House Mechanism (CHM) to facilitate information sharing among institutions involved in biodiversity conservation; limited information on indigenous farm plant and animal genetic resources; and inadequate managerial and technical capacity at the

District and lower local Government levels for implementation of the NBSAP.

B. NBSAP2 2015–2025

In line with the decisions of COP 10 on NBSAP review, Uganda has initiated the preparation of NBSAP2. As the focal point to the CBD, NEMA is coordinating the development and updating of NBSAP1. The process started when Uganda participated in the capacity building workshop for the review and updating of NBSAP for eastern Africa, which took place in Kigali, Rwanda, in June 2011. The workshop was organized by the CBD Secretariat and attended by representatives from Uganda. Uganda also benefited from the regional workshop for Africa for updating the NBSAP which took place in Addis Ababa, Ethiopia, in March 2012. Uganda was again well represented at the workshop.[49]

Using the knowledge and skills gained from the above workshops, Uganda began the process of reviewing and updating it NBSAP with a capacity building workshop. The purpose of the workshop was to create a clear understanding of NBSAPs, the NBSAP review process, the strategic plan for biodiversity 2011–2020 and its Aichi targets; to identify stakeholders to be involved/consulted during the review and updating of NBSAP1; and to develop a roadmap to guide the process as well as agree on the thematic areas for stocktaking/assessment of baseline information to feed into the NBSAP review and updating process.

The capacity building workshop achieved all the above objectives and in addition, initial national biodiversity targets and a provisional outline for the revised and updated NBSAP were developed.

Four Thematic Working Groups were identified (details on the CHM website) and became operational in December 2012. The thematic working groups carried out stocktaking to provide baseline information to feed into the NBSAP review and updating process. The reports of the Thematic Working Groups were also used to prepare the fifth national report to the CBD for Uganda. Terms of Reference for the Thematic Working Groups were developed taking into account the guidance by the CBB for the preparation of the fifth national report. The four Thematic Working Groups were: Aquatic and Terrestrial Biodiversity; Policy, Legislation and Institutional Framework for Biodiversity Management in Uganda; the Status of Biotechnology and Biosafety in Uganda; and Biodiversity for Poverty Eradication and National Development.

C. Overarching principles of NBSAP2

The CBD Strategic Plan (2011–2020) and the complementary Aichi Biodiversity Targets, Vision 2040 and the National Development Plan (NDP) have all closely guided the formulation of NBSAP2. NBSAP2 will be implemented in line with the following principles, which have been mainly derived from these instruments:[50] sustainable development and environmental sustainability; mainstreaming of biodiversity conservation, sustainable use of biological resources and equitable sharing of benefits from biological resources into existing policy, legislative, institutional and development frameworks as appropriate; stakeholder participation in the development and implementation of biodiversity strategy and action plans; awareness creation, education, training and capacity building at local, national and institutional levels to enhance effective participation and implementation of biodiversity measures; recognition, promotion and upholding of traditional and indigenous knowledge of biological resources and sustainable resource management, and where benefits arise from the use of this knowledge; engagement and collaboration with international partners to enhance conservation and sustainable use of Uganda's biological diversity; sustainable use; and benefit sharing arising from the use of biological resources.

IX. The judicial efforts

The Judiciary in Uganda has played an emphatic role by espousing the role of Public Interest Litigation and positively ruling in favour of relaxation of the rule of *locus standi* in the various court cases. Most of these cases have been brought under the caption of environmental protection in Uganda and East Africa in general because the biodiversity conservation concept is still in nascent stages. Nevertheless, biodiversity conservation is being addressed, albeit indirectly.

In Uganda, a landmark case on environmental protection as a necessary corollary on biodiversity conservation that laid a strong foundation for conservation efforts through individual or group efforts in the High Court is the *Advocates Coalition for Development and Environment (Acode) case*,[51] in which the Court had to deal with the question regarding the *locus standi* of the applicants (Acode) to bring the action. The applicants were challenging, among other grounds, whether the granting of the forest permit to Kakira Sugar Works Ltd by the first respondent amounted to de facto degazetting its statutory obligations when it permitted Kakira Sugar Works Ltd to occupy a forest reserve and change the land use without carrying out a full Environmental Impact Assessment Study; and whether de facto degazetting the Butamira Forest Reserve was in violation of the applicants' rights to a clean and healthy environment and protection of the country's natural resources.

In upholding the right of the applicants, the court ruled positively in favour of the applicants, observing as follows:

A. Locus standi

One of the most spirited arguments by the respondent in this case was that the applicants did not have locus standi to take up this action. It was contended that the applicants were mere impostors since they were not living near Butamira Forest Reserve. It was contended that people who live near Butamira, who would be directly affected if the environment were to be upset by the Government's dealing with the Reserve, were not complaining about the decision the Government had taken. It was concluded that the proprietors of Kakira Sugar Works Ltd to whom the responsibility of managing the Reserve was vested were living within its environs, and as such as reasonable and rational human beings, were not likely to endanger their own lives by polluting the environment in which they live. The applicant brought this action under Article 50 of the Constitution, claiming that their rights to a clean and healthy environment had been affected by the respondents' acts and omissions. That Article provides that

> any person who claims that a fundamental or other right or freedom guaranteed under this constitution has been infringed or threatened, is entitled to apply to a competent court for redress which may include compensation. (2) Any person or organisation may bring an action against the violation of another person's group's human rights.
>
> (Constitution of the Republic of Uganda)

The importance of the above law is that it allows any individual or organization to protect the rights of another, even though that individual is not suffering the injury complained of and does not know that he or she is suffering from the alleged injury. To put it in the biblical sense, the Article makes all of us our 'brother's keeper'. In that sense, it gives all the power to speak for those who cannot speak for their rights due to their ignorance, poverty or apathy.

In the *Greenwatch case*,[52] an action was taken against the Attorney General and NEMA under Article 50 of the Constitution for, among other things, failing or neglecting their duties towards the promotion or preservation of the environment. It was held that the State owes that duty to all Ugandans and any concerned Ugandan has the right of action against the Governance of the Republic of Uganda and against NEMA for failing in its statutory duty. In *the Environmental Action Network Ltd* v. *Attorney General and NEMA*,[53] Article 50 of the Constitution was again interpreted where it was observed inter alia that the Article does not require the applicant to have the same interest as the parties he or she seeks to represent or for whose benefit the action is brought.

Lastly, in the recent case of *British American Tobacco Ltd.* v. *The Environmental Action Network*,[54] Ntabgoba PJ (as he then was known) had a lengthy discussion of Article 50 of the Constitution of Uganda wherein he held that the said Article does recognize the existence of marginalized groups like children, illiterates, the poor and the deprived on whose behalf any person or a group of persons could take an action to enforce their rights.

It is very clear from the above authorities that the applicants in this case were clothed with legal standing to take the instant action under Article 50 of the Constitution on behalf of the people of Butamira and other citizens of Uganda. They were therefore not busybodies. This is a very positive ruling and helps a lot in the advancement of environmental conservation and management efforts by opening the doors of the court to any aggrieved person for redressal of their grievance.

X. Conclusion

There are mainly two key players in the conservation and management of biodiversity, *viz.* Governments and local communities, and hence, effective conservation and management efforts should have these key stakeholders in mind.

There is a need to enact a specific law covering the subject of biodiversity conservation in Uganda. Currently, as discussed above, there is no self-contained law on the subject, and most of the enactments have tended to focus on environmental protection; others just deal with specific components of the biodiversity universe. This causes lack of focus, and it is a bit disappointing that this has not been done following the ratification of the Convention on Biological Diversity (CBD), way back in 1993. The National Biodiversity Strategy and Action Plan

(NBSP) should actually be concretized by enacting it in the form of a statute so as to have more efficacy.

All conservation and management effort can succeed to a large extent in a watershed of unwavering government support and non-interference in the work of conservation agencies. There is no benefit to having agencies and adequate legislation in place, yet at the same time stifling their work. In Uganda, this is true of the work of the National Environment Management Authority (NEMA) and the National Forestry Authority (NFA), which have presided over the destruction of vast habitats in the form of wetlands and forests, right under their noses, to enable industries to be established.

Local communities can be good guardians of the environment, and traditionally in East Africa, many of these communities have treated these natural resources as their source of livelihood.[55] In Uganda in 2007, at least three people were killed during a demonstration of about 1,000 for the protection of the Mabira Forest. There were also riots against Asians, since the Mehta Group is Indian-owned. SCOUL plantations were set on fire, and e-mails and SMS messages calling for the boycott of SCOUL's Lugazi sugar circulated.

President Museveni defended the deforestation plans, saying that he shall "not be deterred by people who don't see where the future of Africa lies". According to him, the Save Mabira activists "don't understand that the future of all countries lies in processing". In May 2007,[56] the Ugandan environmental minister announced that the deforestation plans were suspended and that the government was trying to find alternative land for the Mehta Group.

With a rapid increase in population, what is required is coordination and engagement with the communities in conservation efforts, otherwise the conservation costs are higher.

Analysis of the approach taken by oil companies in stakeholder–community engagement, using Arnstein's "Ladder of Participation", reveals that they are operating at the level of "compliance" and limited "consultation".[57]

In most cases, issues of biodiversity conservation or environmental protection have been brought to the fore by the actions of spirited individuals and civil society organizations, sometimes against the compromised conservation agencies for complacency and arguably presiding over biodiversity loss through environmental destruction, following abuse of their permits or by being powerless to effectively regulate the activities of industries. The conservation agencies, being creatures of the Governments, have been undermined by the same Governments to promote industrial activity in a haphazard and dangerous manner.

Courts should continue to be receptive and supportive to efforts such as Public Interest Litigation.

Related to this is the issue of costs of suit. Courts have been slow and most times totally opposed to granting of costs in Public Interest Litigations. This also negatively impacts on the conservation efforts as the court process is expensive; and if, after a protracted case, no costs are awarded, it serves to demoralize the conservation enthusiasts in bringing such cases to court.

To sum up, it may be stated that biodiversity conservation and management is a matter of global concern that should be envisioned by every Government, Agencies, Civil Society Organizations and citizens as the imperative for the preservation of the human race in the ultimate, by undertaking rigorous, consistent and sustained efforts to minimize the environmental footprint of developmental projects. It is, however, very tragic when governments gang with corporations against efforts targeted at environmental conservation and management, as witnessed in Nigeria involving the activities of Royal Dutch Petroleum (Shell)[58] and the ongoing subtle destruction of forests like Mabira in Uganda with government consent through timber permits. It is hoped that the suggestions made above, if implemented, would go a long way in supporting biodiversity conservation and management efforts in Uganda and East Africa in general.

Notes

1 National Environmental Management Authority (NEMA), *State of Environment Report for Uganda 2010*, Kampala: NEMA, 2010, p. 97.
2 International Union for Conservation of Nature (IUCN) 2016, 'The IUCN Red List of Threatened Species; Version 2016–2', www.iucnredlist.org/search (accessed on 25 September 2016).
3 National Environmental Management Authority (NEMA), *National Biodiversity Strategy and Action Plan (NBSP) 2015–2025*, Kampala: NEMA, 2015, pp. 73–162.
4 NEMA, *State of Environment Report for Uganda 2010*, p. 97.
5 *Ibid.*, p. 105.
6 *Ibid.*, p. 106.
7 Nile Basin Initiative (NBI), 'The Environmental Resources of the Nile', p. 71, http://nileis.nilebasin.org/system/files/Nile%20SoB%20Report%20Chapter%203%20-%20%20Environment.pdf (accessed on 30 September 2016).
8 NEMA, *National Biodiversity Strategy and Action Plan (NBSAP) 2015–2016*, p. 18.
9 NEMA, *NBSAP (2015–2025)*, p. 19.
10 *Ibid.*, p. 20.
11 *Ibid.*

12 *Ibid.*
13 *Ibid.*, p. 21.
14 *Ibid.*
15 'National Forestry Authority (NFA)', Annual Report 2011, Kampala, www.nfa.org.ug/index.php/downloads1#annual-reports (accessed on 30 September 2016).
16 National Wetlands Programme, 'Wetlands Inventory Report for Kampala, Kampala', p. 1, www.mwe.go.ug/index.php?option=com_docman& task=cat_view&gid=46&Itemid=223 (accessed on 30 September 2016).
17 R. Ridderhof, 'Shell and Ogoni People: (S)oil Pollution in the Niger Delta', www.peacepalacelibrary.nl/2013/02/shell-and-ogoni-people-soil-pollu tion-in-the-niger delta (accessed on 30 September 2016).
18 National Wildlife Federation, 'Bird Habitats Threatened by Oil Spill', 2010, www.nwf.org/News-and-Magazines/National-Wildlife/Birds/Arc hives/2010/Oil-Spill-Birds.aspx (accessed on 30 September 2016).
19 Economic Policy Research Centre (EPRC), *National Resource Management in the Albertine Graben Region of Uganda: Baseline Survey Report*, Kampala: EPRC, 2015, pp. 34–35.
20 R. M. Pringle, 'The Nile Perch in Lake Victoria: Local Responses and Adaptations', *Africa: Journal of the International African Institute*, 2005-01-01, 75(4): 510–538.
21 Nile Basin Initiative (NBI), 'Nile Basin Trans-boundary Wetlands', *Project on Biodiversity Conservation and Sustainable Utilisation of Ecosystem Services of Wetlands on Trans-Boundary Relevance in the Nile Basin*, Kampala: NBI, 2015, p. 2.
22 National Wetlands Programme, 'Wetlands Inventory Report for Kampala', 2000, p. 15, www.mwe.go.ug/index.php?option=com_docman&task=cat_ view&gid=46&Itemid=223 (accessed on 30 September 2016).
23 Government of the Republic of Uganda (GoU), *The Constitution of the Republic of Uganda 1995*, National Objectives and Directives of State Policy, Objective XXVII: The Environment, 1995.
24 GoU, Constitution, Article 245.
25 GoU, Constitution, Article 39 and Article 50.
26 GoU, 'The National Environment Act', 1995, Section 4.
27 GoU, 'The National Environment Act', 1995, Section 6.
28 GoU, 'The National Environment Act', 1995, Sections 7–10.
29 GoU, 'The Agricultural Seeds and Plants Act', 2006, Section 3.
30 GoU, 'The Agricultural Seeds and Plants Act', 2006, Section 6.
31 GoU, 'The Agricultural Seeds and Plants Act', 2006, Sections 15–21.
32 GoU, 'The Plant Protection and Health Act', 2015, Preamble.
33 GoU, 'The Plant Protection and Health Act', 2015, Sections 9–11.
34 GoU, 'The Plant Protection and Health Act', 2015, Sections 27–28.
35 'The Uganda Wildlife Act 1996', Section 2.
36 'The Uganda Wildlife Act 1996', Section 4.
37 'The Uganda Wildlife Act 1996', Section 17.
38 'The Uganda Wildlife Act 1996', Section 18.
39 'The Uganda Wildlife Act 1996', Section 38.
40 'The Uganda Wildlife Act 1996', Section 22(1).
41 'The Uganda Wildlife Act 1996', Section 74.
42 'The Uganda Wildlife Act 1996', Section 29(1).

43 'The Uganda Wildlife Act 1996', Section 41.
44 GoU, 'The Water Act', 1997, Section 9.
45 GoU, 'The Water Act', 1997, Section 10.
46 GoU, 'The Water Act, 1997', Section 39.
47 NEMA, *NBSAP (2015–2025)*, 2015, pp. 73–162.
48 *Ibid.*
49 *Ibid.*, p. 74.
50 *Ibid.*, p. 77.
51 *Advocates Coalition for Development and Environment (Acode)* v. *Attorney General*, Misc. Cause No. 0100 Of 2004.
52 *Greenwatch* v. *Attorney General and Another*, Misc. Cause No. 140/2002.
53 Misc. Application No. 39/2001.
54 High Court Civil Application No. 27/2003.
55 *International Alert: Governance and Livelihoods in Uganda's Oil-Rich Albertine Graben*, Kampala: I.A, 2013, pp. 17–23, www.internationalalert.org/sites/default/files/publications/Uganda_2013_OilAndLiveli hoods_EN.pdf (accessed on 1 October 2016).
56 'Deforestation Plans', Uganda, https://en.wikipedia.org/wiki/Mabira_Forest, (accessed on 1 May 2015).
57 S. R. Arnstein, 'A Ladder of Citizen's Participation', *Journal of the American Institute of Planners*, 1969, 35(4): 216–224.
58 'The Case Against Shell: *Landmark Human Rights Trial (Wiwa* v. *Shell)*', https://ccrjustice.org/home/get-involved/tools-resources/videos/caseagainst-shell-landmark-human-rights-trial-wiwa-v-shell (accessed on 30 May 2015).

Part III

CONVENTION ON BIOLOGICAL DIVERSITY, ABS AND TRIPS AGREEMENT

5

IMPLEMENTING THE NAGOYA PROTOCOL ON ACCESS AND BENEFIT SHARING IN INDIA

Pushpa Kumar Lakshmanan

The Nagoya Protocol on Access to Genetic Resources and the Fair and Equitable Sharing of Benefits Arising From Their Utilization to the Convention on Biological Diversity, 2010 (the Nagoya Protocol)[1] sets the global norms for regulating access to genetic resources and traditional knowledge and their sustainable use. The Nagoya Protocol strives to address all three objectives of the Convention on Biological Diversity (CBD),[2] namely, conservation of biodiversity; utilization of biological resources in a sustainable manner; and fair and equitable sharing of benefits arising out of the utilization of genetic resources and traditional knowledge.

Prior to the Nagoya Protocol, the CBD prescribed that access to genetic resources should be based on mutually agreed terms and the prior informed consent of the member country providing the genetic resources.[3] In order to implement the CBD, India has introduced domestic regulatory mechanisms by enacting the Biological Diversity Act, 2002[4] (the BD Act) and the Protection of Plant Varieties and Farmers' Rights Act, 2001.[5] The National Biodiversity Authority has been regulating access and benefit sharing (ABS) with the help of State Biodiversity Boards and the local level Biodiversity Management Committees.[6] Even before the coming into force of the Biological Diversity Act, the concept of ABS pertaining to genetic resources and traditional knowledge was implemented in India. The Kani-Tropical Botanic Garden and Research Institute (TBGRI) model of benefit sharing stands as a testimony to this.[7]

India ratified the Nagoya Protocol on 9 October 2012 and committed itself to implementing the Protocol at the national level. The provisions of the Nagoya Protocol require a fair and non-arbitrary regulatory mechanism within the jurisdictions of the Contracting Parties with proper access rules, benefit-sharing procedures and compliance mechanisms based on prior informed consent and mutually agreed terms (MATs).

Taking advantage of the BD Act, India is trying to implement the Nagoya Protocol under this domestic law. It is pertinent to examine how far the BD Act is meeting the requirements of the Nagoya Protocol and what needs to be done in giving effect to the provisions of the Protocol. This chapter critically looks at the meeting points and gaps between the international and domestic legal systems on ABS and suggests the measures to comply with the Protocol.

I. The Nagoya Protocol on access and benefit sharing

Adoption of the Nagoya Protocol is seen as a landmark achievement for the international community, particularly the developing countries. The biodiversity-rich developing countries were facing several difficulties in protecting their biological resources in the wake of biopiracy.[8] The populations living in these countries heavily depend on the biological resources for their food, medicine and other livelihood opportunities. The same resources were misappropriated by foreign pharmaceutical and cosmetic industries to produce modern medicines and other valuable products, and these industries tried to protect the resources under different intellectual property systems – to the disadvantage of the native communities, who have conserved the resources and shared their traditional knowledge.

Even though the principle of ABS was entrenched into the CBD itself in 1992, the Nagoya Protocol tries to give fresh life to the ABS provisions of the CBD within the scope of its Article 15, which regulates access to genetic resources with the help of legally binding mandates for implementation of ABS. The Nagoya Protocol also applies to traditional knowledge associated with genetic resources within the scope of the CBD and to the benefits arising from the utilization of such knowledge for research and commercial purposes.[9] The Nagoya Protocol requires the Contracting Parties to introduce appropriate national legislative, administrative or policy measures for ABS for genetic resources and the associated traditional knowledge.

II. Access to genetic resources and traditional knowledge

A Contracting Party may choose to regulate access to genetic resources without Prior Informed Consent (PIC) based on domestic legislation.

If the Contracting Party decides to regulate access to genetic resources through PIC, they have to follow Articles 5 and 6 of the Nagoya Protocol that insist on application of PIC and mutually agreed terms (MAT) for regulating ABS. If a Party does not want to regulate access to genetic resources based on PIC, it need not enact such legislation. Access to genetic resources for their utilization shall be subject to the PIC of the Party providing such resources, that is the country of origin of such resources or a Party that has acquired the genetic resources in accordance with the CBD, unless otherwise determined by that Party.[10]

Article 5 of the Nagoya Protocol provides that, in accordance with Article 15, paragraphs 3 and 7 of the CBD, the benefits arising from the utilization of genetic resources as well as subsequent applications and commercialization shall be shared in a fair and equitable manner with the Party providing such resources, that is, the country of origin of such resources or a Party that has acquired the genetic resources in accordance with the CBD.

The Nagoya Protocol mandates every State Party to take suitable legislative, administrative or policy measures for the purpose of ensuring that the benefits arising out of the utilization of genetic resources are held by indigenous and local communities. This has to be done in accordance with domestic legislation regarding the established rights of indigenous and local communities over genetic resources. The domestic legislation should provide mechanisms for sharing of benefits in a fair and equitable manner with the communities based on MAT.[11]

According to Article 7 of the Protocol, each Party shall take national measures with the aim of ensuring that traditional knowledge associated with genetic resources held by indigenous and local communities are accessed with the PIC or approval and involvement of indigenous and local communities. In addition to PIC, the access should be based on MAT. The terms, national measures 'in accordance with domestic law' and 'as appropriate', provide sufficient flexibility to the Parties to choose the level of PIC and MAT as per their convenience.

The domestic law for the PIC or approval system, with the help of a Competent National Authority, should provide for a mechanism that ensures legal certainty, clarity and transparency of the domestic access and benefit-sharing legislation or regulatory requirements. The domestic measures should also provide for fair and non-arbitrary rules and procedures on accessing genetic resources. The procedures should clarify the criteria, processes for obtaining prior informed consent or approval, and clear rules for establishing MAT. The terms should set out a dispute settlement clause; details of benefit sharing, including in relation to intellectual property rights (IPR); terms on subsequent third-party use, if any; and the terms on changes of intent, if applicable.[12]

Where the indigenous and local communities have established the right to grant access to genetic resources, their involvement is required for access to genetic resources. The law should provide for the criteria and/or processes for involvement of indigenous and local communities for access to genetic resources. The domestic law should provide for a written decision or permit by a Competent National Authority, in a cost-effective manner and within a reasonable period of time, as evidence of the decision to grant PIC and MAT.[13] The Competent National Authority should notify the Access and Benefit-sharing Clearing-house of all such details.[14]

In order to ensure proper compliance with the domestic legislation on ABS, the Protocol obligates the Parties to designate National Focal Points, Competent National Authorities and checkpoints. It also provides for special considerations for certain research activities and global multilateral benefit-sharing mechanisms.[15] The Annex to the Protocol lists ten monetary benefits and 17 forms of non-monetary benefits, but does not limit the scope of benefits with those mentioned in the Annex. The Parties are at liberty to apply any other form of benefit sharing.[16] The Protocol has given life to the provisions of the CBD pertaining to ABS through an internationally legally binding instrument. The success of this Protocol will depend on how the State Parties and different stakeholders give life to its provisions, emphasizing effective compliance and implementation.

III. Utilizing the BD Act to implement the Nagoya Protocol

In India, the BD Act was enacted in 2002 to implement the CBD. The Biodiversity Rules (the BD Rules) 2004[17] supplement the BD Act and provide elaborate details for implementation. The BD Act aims at conserving the biological diversity present in the country, sustainable use of its components, and fair and equitable sharing of benefits arising out of use of the biological resources and the associated knowledge. As the BD Act was enacted in the background of numerous biopiracy cases, the Act prescribes strict rules for non-Indian entities to access the biological resources and traditional knowledge. The National Biodiversity Authority (NBA) is the highest-level body in the three-tier system created by the BD Act.[18] The NBA grants approval for access to the biological resources and the associated knowledge for persons who are non-Indians.[19] It also grants approvals to applications connected with IPR involving biological resources or traditional knowledge.[20] The NBA advises the Central Government on matters concerning conservation of biological diversity, sustainable use of its components and equitable sharing of benefits arising out of the utilization of biological resources.[21] It also advises the State Government in the selection of areas of biodiversity importance to be notified as heritage sites and in devising strategies for their management.[22]

The NBA coordinates with the State Biodiversity Boards (SBBs) and Biodiversity Management Committees (BMCs) in implementing the BD Act. While approving access and determining benefit-sharing terms, the NBA consults with the local bodies.[23] The NBA is also empowered to take necessary measures to oppose the granting of IPR in any country outside India on any biological resource obtained from India or knowledge associated with any biological resource derived from India.[24]

The SBBs advise the State Government on matters relating to the conservation of biodiversity, its sustainable use and benefit-sharing aspects.[25] It grants approval to the request for commercial utilization or bio-survey or bio-utilization of any biological resource by Indian applicants. An Indian citizen or any corporate body, association or organization which is registered in India should give prior intimation to the SBB concerned before obtaining any biological resource for commercial utilization, or any bio-survey and bio-utilization for commercial utilization.[26] This provision will not apply to local people and communities of the area, including growers and cultivators of biodiversity, and *vaids* and *hakims*,[27] who have been practicing indigenous medicine.

In every State, the local bodies constitute the BMCs for the purpose of promoting conservation, sustainable use and documentation of biological diversity.[28] The BMCs also engage in the preservation of natural habitats; the conservation of land races,[29] folk varieties and cultivars,[30] domesticated stock and breeds of animals, and microorganisms; and the documentation of knowledge associated with biodiversity. The BMCs prepare, maintain and validate People's Biodiversity Registers (PBR) in consultation with the local people. The BMC maintains a Register giving information about the details of access to biological resources and traditional knowledge, the details of the collection fee imposed by the BMC, the details of benefits derived from each case and the mode of sharing. The BMCs also advise on any matter referred to it by the SBB or NBA.[31]

The BD Act gives power to the NBA to determine benefit sharing based on six broad ways, namely (a) grant of joint ownership of

IPR to the National Biodiversity Authority, or where benefit claimers are identified, to such benefit claimers; (b) transfer of technology; (c) location of production, research and development units in such areas which will facilitate better living standards to the benefit claimers; (d) association of Indian scientists, benefit claimers and the local people with research and development in biological resources and bio-survey and bio-utilization; (e) setting up of venture capital funds for aiding the cause of benefit claimers; and (f) payment of monetary compensation and non-monetary benefits to the benefit claimers.[32] The ABS Guidelines as discussed in the following section elaborates the monetary and non-monetary forms of benefit sharing.

A. ABS Guidelines

The Guidelines on Access to Biological Resources and Associated Knowledge and Benefits Sharing Regulations, 2014[33] (hereafter ABS Guidelines) issued under the BD Act provide detailed procedures for implementing ABS in the country. The NBA accords approval to the applicants by entering into a benefit-sharing agreement which is deemed to be the grant of approval for access to biological resources for research.[34]

The ABS Guidelines prescribe different parameters of benefit sharing for different purposes. If the *biological resource has high economic value*, the agreement may contain a clause to the effect that benefit sharing shall include upfront payment of an identified amount, as agreed between the NBA and the applicant. In cases of biological resources having high economic value such as sandalwood, red sanders, etc. and their derivatives, the benefit sharing may include an upfront payment of not less than 5.0 per cent on the proceeds of the auction or sale amount as decided by the NBA or SBB.[35]

Where an applicant or trader or manufacturer has not entered into any prior benefit-sharing negotiations with the joint forest management committee or a forest dweller, tribal cultivator or the Gram Shaba for the purchase of any biological resource directly from these persons, the traders shall share the benefit in the range of 1.0 to 3.0 per cent of the purchase price of the biological resources.[36] The benefit-sharing obligations on the manufacturers shall be in the range of 3.0 to 5.0 per cent of the purchase price of the biological resources.[37]

When the biological resources are accessed for commercial utilization, or bio-survey and bio-utilization that lead to commercial utilization, the applicant has an option of paying the benefit sharing ranging from 0.1 to 0.5 per cent based on the annual gross ex-factory sale of the product.[38] If the ex-factory sale is up to INR 10 million, the applicant has to pay 0.1 per cent of the annual gross ex-factory sale amount as the benefit-sharing component.[39] If the sale is more than 10 million and less than INR 30 million, the benefit-sharing component is 0.2 per cent. If the sale amount exceeds 30 million, the benefit-sharing component is 0.5 per cent.[40] The collection fees, if any, levied by the BMC for accessing or collecting any biological resources for a commercial purpose from areas falling within the jurisdiction of the BMC shall be in addition to the benefit sharing payable to the NBA or SBBs.[41]

If anyone intends *to obtain any intellectual property rights* in or outside India for any invention based on research or information on any biological resources obtained from India, prior approval of the NBA is required and the applicant has to pay such monetary or non-monetary benefits as agreed between the applicant and the NBA.[42] If the applicant commercializes the process or product or innovation, the monetary sharing shall be in the range of 0.2 to 1 per cent based on the sectoral approach, which shall be worked out on the basis of the annual gross ex-factory sale minus government taxes.[43] When the applicant assigns or licenses the process or product or innovation to a third party for commercialization, the applicant shall pay to the NBA 3.0 to 5.0 per cent of the fee received and 2.0 to 5.0 per cent of the royalty amount received annually from the assignee or licensee based on the sectoral approach.[44]

In the case of *transfer of results of research*, the applicant shall pay to the NBA such monetary and/or non-monetary benefits as agreed to between the applicant and the NBA. If the applicant receives any monetary benefit on such transfer, the applicant shall pay to the NBA 3.0 to 5.0 per cent of monetary considerations.[45]

If there is a *transfer of accessed biological resource and/or associated knowledge* to a third party for research or commercial utilization, the applicant shall pay to the NBA 2.0 to 5.0 per cent of any amount or royalty received from the transfer as benefit sharing throughout the term of the agreement.[46]

Where the benefit sharing has been determined by the NBA while according to approval of any type, 5.0 per cent of the accrued benefits will go to the NBA, out of which half of the amount shall be retained by the NBA, and the other half may be passed on to their concerned SBBs for administrating charges. The rest of the 95 per cent of the accrued benefits shall go to concerned BMCs and/or benefit claimers. If the biological resource and knowledge is sourced from an individual or group of individuals or an organization, the amount received will directly go to such individual or group of individuals or organization.

If the benefit claimers are not identified, such funds shall be used to support the conservation and sustainable use of biological resources and to promote the livelihood of local people from where biological resources are accessed.[47]

The above discussion shows that, as far as India is concerned, the BD Act and the ABS Guidelines take care of most of the provisions of the Nagoya Protocol and the country need not enact any new law to implement the Protocol. Instead, what is required is to introduce certain provisions within its existing legal framework to fruitfully implement the Protocol.

IV. Issues and challenges of implementing the Nagoya Protocol

When the implementation of the BD Act is considered in India, it can certainly be said that the NBA has taken several measures to give effect to the Act and has taken the enforcement of law to the length and breadth of the country. All the States have been asked to set up SBBs with huge efforts. Enormous efforts have also been taken by the NBA and SBBs to establish BMCs at the local level in every state. It is thought that unless the local people are aware of the ABS process and unless the local bodies set up the BMCs and Local Biodiversity Fund, it would be very difficult to provide access to locally available biological resources. Whenever biological resources are accessed, it is the BMC which fixes the charges in the form of collection fees or access fees. Hence, it is important for the local bodies to constitute BMCs. The money collected by way of collection fees or access fees is deposited in the Local Biodiversity Fund and spent for conservation purposes within the jurisdiction of the local body and for the benefit of the community.[48] The major challenge for the NBA and the SBBs, in this process, is convincing the local bodies and constituting BMCs. It has been noted that 37,769 BMCs have been constituted in 29 states as of 2 September 2015.[49] Secondly, choosing the right people for the BMCs and training them to handle ABS issues is another issue. For most of the local people, the system of ABS itself is an alien concept to them. It is a great challenge to build their capacities and involve them in the ABS process.

A. Dealing with biopiracy through the Patent Office

The Biological Diversity Act, 2002 states that no person shall apply for any intellectual property right within or outside India for any invention based on any research or information on a biological resource obtained from India without obtaining the previous approval of the NBA.[50] However, if a person applies for a patent, permission of the NBA may be obtained after the acceptance of the patent but before the sealing of the patent by the patent authority concerned. The NBA has to dispose of any such application within 90 days from the date of receipt of the application.[51] Under the BD Act, the NBA has the option to impose benefit sharing by way of a fee, royalty or both.[52] While granting approval for IPR-based applications, the NBA can also impose conditions as to the sharing of financial benefits arising out of commercial utilization of such proprietary rights.[53] The BD Act mentions only about an 'invention based on any research or information on a biological resource obtained from India.'[54] This refers to the 'invention based on any research involving biological resource or traditional knowledge obtained from India.' Otherwise, the term 'information' could mean anything and may not necessarily imply traditional knowledge.

The Patents Act, 1970[55] does not contain any such provision requiring the applicant to first apply to the NBA for prior approval before finally deciding the patent or other IPR applications. Two amendments were made to the Patents Act since 2002, one in 2002 itself and the other in 2005; still, the Patents Act did not incorporate a bridging provision with the BD Act to give effect to its Section 6. Nevertheless, the 2005 Amendments to the Patents Act introduced a provision to the effect that a pre-grant and post-grant opposition could be made to stop or revoke the patent that involves traditional knowledge and without any inventive step.[56] Again, the patent applicant was obligated to fully disclose the source and geographical origin of the biological material in the specification whenever such biological material is used in an invention.[57] Now, Form-1 under the Patent Rules 2003 requires the patent applicant to mandatorily declare that the invention as disclosed in the specification uses the biological material from India and that the applicant will submit the approval from NBA once it is obtained.[58]

In 2013, the Office of the Controller General of Patents, Designs and Trade Marks issued detailed guidelines for patent applicants when the patent application deals with an invention based on genetic material.[59] Full disclosure has to be made at the time of filing of the patent application that includes the use of genetic or biological material in the invention process and traditional knowledge. Failure to comply with these provisions would result in cancellation or revocation of the patent. In this way, the Patent Office is functioning as one of the checkpoints to monitor the utilization and flow of the biological materials at the time of intellectual property protection.

Similarly, Sub Section 3 of Section 6 of the BD Act states that prior approval for IPR-based applications is not required for any person making an application for any right under the Protection of Plant Varieties and Farmers' Rights Act, 2001[60] (PPVFR Act) because such applications will be dealt with by the Plant Varieties and Farmers' Rights Authority under the PPVFR Act while deciding benefit sharing. Whenever any intellectual property right is granted by the Plant Varieties and Farmers' Rights Authority, it has to endorse a copy of such document granting the right to the NBA as specified in Sub Section 4 of Section 6 of the BD Act. However, no such corresponding provision is found in the PPVFR Act to ensure such compliance by the Plant Varieties and Farmers' Rights Authority.

B. Need for a provision to ensure compliance mechanism

The Nagoya Protocol demands that the user countries should enact a law to ensure that the users of genetic resources or traditional knowledge within their jurisdiction have complied with the regulatory requirements of the provider country while obtaining the genetic resources or traditional knowledge.[61] The BD Act contains provisions to regulate and check the flow of genetic resources as a provider country, but it does not have any provision for the user country to check the compliance of another provider country's rules in India. The BD Act should incorporate adequate provisions to check that the genetic resources and traditional knowledge brought from other countries into India for research or commercial purposes have complied with the regulatory requirements of those countries. The Norwegian law user country measures are a good model to refer to in this context.[62]

C. Regulating access to traditional knowledge by Indians

As a provider country, a country needs to set out its criteria for PIC in relation to access to genetic resources and traditional knowledge. With regard to the regulation of genetic resources, the Indian BD Act contains sufficient provisions to meet the requirements of the Nagoya Protocol for Indian and non-Indian applicants.[63] As far as access to traditional knowledge by Indian users is concerned, there is no clearcut provision or procedure to regulate it under the BD Act. The SBBs are expressly given powers to deal with biological resources.[64] However, as of now, the SBBs regulate both access to genetic resources and traditional knowledge without a proper legal mandate. It would be better to clarify this aspect in the BD Act and introduce express provisions to deal with access to traditional knowledge by Indian citizens. After all, when it comes to commercial exploitation of traditional knowledge, citizenship of nationality does not make any difference.[65]

D. Involving local communities in obtaining PIC

Initially, as the local communities did not have adequate knowledge about the ABS process, the BD Act sought to regulate PIC through the NBA and local bodies with representatives from the local communities.[66] But the real spirit of PIC, as reflected in the Nagoya Protocol, is to involve the local and indigenous communities in the process directly wherever they have recognized rights under the domestic law.[67] The local communities in India should be given an opportunity to participate in the ABS process, and it is advisable to make provisions to engage them with the scope of PIC under the BD Act. Suitable changes in the BD Act to explicitly involve the local communities in the ABS process, particularly in obtaining PIC to access their traditional knowledge, will bring transparency and justice in the ABS mechanism.

The provisions of the Scheduled Tribes and other Traditional Forest Dwellers (Recognition of Forest Rights) Act, 2006[68] have significant relevance in the discussion on access and benefit sharing of biological resources, as this law recognizes the rights of access to biodiversity and community rights to intellectual property and traditional knowledge associated with biodiversity and cultural diversity of the forest-dwelling communities in India. With the help of this enactment, the forest-dwelling communities can effectively regulate or resist access to biological resources if any such access adversely affects the wild animals, forest and biodiversity.[69] One major concern in biodiversity governance is the linkage and coordination among different legal regimes in regulating access to biological resources.

E. The gap between the biodiversity law, wildlife laws and forest laws

The forest laws, wildlife laws and the BD Act do not work in tandem. The conservation mandate is not coordinated well amongst these laws. The Indian Forest Act, 1927[70] is the primary law to govern the forest areas. This law creates forest authorities to govern different protected areas in the form of reserved forests, protected forests and village forests. This law protects forests from external forces, such as encroachments, illegal forest felling and protection of wildlife. The

Forest (Conservation) Act, 1980[71] was enacted primarily for the purpose of conservation of forests and preventing conversion of forest lands for non-forest purposes. For this purpose, this law introduced a regulatory mechanism for forest clearance.[72] The Wildlife (Protection) Act, 1972 takes care of protection of wild animals and prevents illegal poaching. This law also enables the creation of protected areas to conserve wild ecosystems for the benefit of wild animals. But most of the biological resources are locked in forest areas. None of these laws have any concept of ABS, as that was not in vogue at the time these laws were enacted. The BD Act contains provisions to regulate ABS throughout the country. If a biological material found in the forest area is to be accessed for commercial utilization, how anyone would access it is a tricky issue. The failure of India's first ABS case, the Kani-TBGRI case, is due to the reason that the forest laws did not allow the cultivation and supply of 'arogyappacha' (Trichopus zeylanicus travancoricus) on a regular basis and large scale.[73] If India wants to implement ABS and get benefits out of it, sincere efforts have to be made to revisit the forest and wildlife laws to enable ABS to function smoothly.

The Wildlife (Protection) Act, 1972[74] contains some provisions for the purposes of conservation and sustainable use of biological resources. Access to biological resources and their derivatives are regulated in this Act through licenses issued by the Chief Wildlife Warden. Nevertheless, it does not contain any provision pertaining to ABS as obligated by the CBD or Nagoya Protocol on ABS. Neither the Act nor its makers can be blamed for the absence of ABS provisions in the Act because that was not the objective of the Wildlife (Protection) Act and at the time when the Act was enacted, ABS was not an issue.

However, now, in the light of the Nagoya Protocol on ABS, the Wildlife (Protection) Act requires amendments to regulate access to wildlife, birds, plants and their derivatives for different purposes that include research and other forms of commercial utilization. When the Chief Wildlife Warden issues a permit for hunting of any wild animal for the purposes of 'research'[75] or 'collection of specimens'[76] or 'derivation, collection or preparation of snake-venom for the manufacture of life saving drugs';[77] and when the Chief Wildlife Warden grants permission under Section 17B to any person to pick, uproot, acquire or collect from a forest land or the area specified under Section 17A or transport any specified plant for the purpose of scientific research, the same may be intimated by the Chief Wildlife Warden to the NBA. A provision to this effect has to be inserted in Sections 12 and 17B of the Wildlife (Protection) Act.

F. Making the BD ACT work for conservation

The BD Act does not provide ample provisions for the conservation and sustainable use of biological resources but focuses rather on providing access to biological resources and associated knowledge.[78] Nevertheless, the Act contains certain provisions to protect biodiversity by declaring heritage sites, and empowers the authorities to refuse approval for access to biological resources if such activity has the potential of causing adverse impacts on biodiversity or contrary to the objectives of conservation and sustainable use of biodiversity or equitable sharing of benefits arising out of such activity.

Even access to biological resources is not linked with any kind of environmental or social impact study. It only provides certain duties for the Central as well as State Governments to take some steps for conservation of biodiversity without providing a clear set of rules as to how to do so. As per the provisions of the BD Act, the State Governments have notified only seven Biodiversity Heritage Sites so far.[79] However, considering the number of sacred groves and heritage sites available in different states, this is less than a minuscule number of sites that have been recognized by the States. The States may identify many more ecologically important sites and sacred groves and declare them as heritage sites. This will give legal status to the sites and recognize community efforts.

G. Challenges for BMCs in documenting People's Biodiversity Registers

One of the mandates of the BMCs is to document the details of biological resources and traditional knowledge available within the jurisdiction. About 2,485 People's Biodiversity Registers (PBR) have been prepared so far to document the biodiversity and traditional knowledge in 17 States.[80] There are many issues in documentation. If the documentation is not done properly, it will not have any utility in the future. Documentation of biological resources and traditional knowledge at the BMC levels should be done by sufficiently trained people with the involvement of ecologists, academia and scientists available in the nearby higher educational institutions. Participation of local people in deciding access and determining benefit sharing should be encouraged even though the BMCs consist of representatives of local people. Such direct participation of the local people will enhance their sense of conservation and community decision-making in cases that affect biodiversity, ecological balance and livelihood security of the people.

Insufficient funding flow from the benefit-sharing process and the difficulties of actual collection of monetary benefits, etc., have to be looked at from the institutional level. A trend has also been seen in the current ABS procedure in the country that the benefit that is being agreed in various ABS agreements is largely monetary in nature. This undermines the potential of non-monetary benefits. The cost-benefit analysis of the current ABS system shows that India has invested more in building the framework for ABS than the benefits it has reaped out of it. Therefore, suitable changes are required to be introduced in the current ABS legal framework, especially in the manner in which it is being implemented in the country, to maximize the utility for India.

H. Meagre proportion of benefit sharing and limited sharing of benefits with the community

Since the coming into force of the BD Act, the NBA has given 313 ABS approvals in all as of 30 June 2016. This includes 182 approvals for obtaining IPR, 79 approvals for access to biological resources for research and commercial purposes, 28 approvals for third-party transfer, and 14 approvals for transfer of research results. These 313 approvals have been granted out of 1,261 applications. Furthermore, about 232 applications are under process at different levels.[81]

According to the available data, the monetary benefit received by the NBA amounts to INR 4.3 million as of April 2014. This amount seems to be too small when compared to the number of ABS agreements India has entered into and also when compared to the cost that India has incurred in forming a legal and institutional setup for the process of ABS. Out of this amount, the NBA has shared the royalty amount of INR 20,000 with Amarchinta BMC in Andhra Pradesh for the export of neem. The NBA is working out the modalities for distributing the royalties to the benefit claimers.

I. Joint ownership of intellectual property rights

One of the modes of benefit sharing is to opt for joint ownership of IPR between the government/communities contributing biological resources or associated knowledge and the party who is accessing them for research or commercial purposes as envisaged in the BD Act. The trend of benefit-sharing agreements shows that there has been no instance of joint ownership of IPR with the communities so far. The NBA can take some steps to encourage adopting joint ownership of IPR. This will bring long-term benefits to the Government or communities.

J. Checkpoints

India needs to designate appropriate checkpoints, ensure efficient functioning and monitoring of the flow of biological resources, and ensure compliance of the national legal and regulatory framework as per the Nagoya Protocol.[82] In a country like India with vast genetic resources, a single check point will not be able to check the movement of genetic resources. It may not be feasible for the Ministry of Environment Forests & Climate Change to function as a check point as it requires a great number of administrative staff and technical experts. The viable option would be to designate the NBA as the Apex Checkpoint with other designated checkpoints in different departments/offices, such as the Protection of Plant Varieties and Farmers' Rights Authority, the Patent Office, Customs offices in Ports and Airports, the Drugs Controller General of India (DCGI), the Genetic Engineering Approval Committee (GEAC), and the Chief Wildlife Warden in the Forest Department. If required, the National Bureau of Plant Genetic Resources (NBPGR) and the Indian Council of Medical Research (ICMR) could be considered to be designated as checkpoints if there is a need to monitor the utilization of genetic resources through their approvals. As the ABS system is at its budding stage, the need for a large number of checkpoints seems unnecessary. However, when the Nagoya Protocol is implemented widely world over, this need will be felt strongly.

The Patent Office is the vital agency that can check the utilization of genetic resources or traditional knowledge in inventions. All grants of patents from the Patent Office involving genetic resources or materials have to be monitored with vigil and intimated to the NBA before the granting of a patent or sealing of a patent.

The Customs department is the agency that can sit in the points of entry or exit of goods to monitor the flow of genetic resources. Every port/harbour or airport in the country should be equipped with sufficient technical experts to monitor the movement of biological materials and should be directed to collect the information related to the utilization of genetic resources.

In the same manner, the NBPGR and the ICMR should be posted with technical experts to monitor the export/import of genetic materials or utilization of genetic materials for any research or development purposes.

The Drugs and Cosmetics Act, 1940[83] as amended in 1995 deals with regulation of the import, manufacture, marketing approval and sale of drugs and cosmetics that includes traditional medicines such as Ayurvedic, Unani and Siddha products containing ingredients of medicinal plants. In the modern context, new drugs are being developed with the application of biotechnology or genetic engineering that involves genetic materials and traditional knowledge. Biological resources are also utilized in large scale for manufacturing cosmetic products. Even though access to genetic materials such as medicinal plants and microorganisms is regulated under the BD Act, their utilization as drugs and cosmetics needs to be monitored at the stage of licensing and market approval. The Drugs Controller General of India at the central level and the appropriate authorities at the state level should monitor the flow and utilization of biological resources and traditional knowledge. In addition, adequate linkages should be established with the SSBs and BMCs to deal with access-related issues and fixation of access fees.

As the access and benefit-sharing regime is at its nascent state in India, multiple checkpoints have not been designated in the country. As of now, only the Patent Office has been acting as a checkpoint while granting patents.

V. Conclusion

The Nagoya Protocol is an instrument of global justice. Massive populations in developing countries suffer from poverty and hunger, and still they are the living repositories of traditional knowledge and contribute to biodiversity conservation. There have been attempts to misappropriate traditional knowledge and biological resources to make new inventions. Such inventions are patented in many countries and used to make enormous profits. However, the populations that offered knowledge or their services to conserve the biological resources have not been adequately compensated. The Nagoya Protocol is an attempt to bridge this gap. The Nagoya Protocol tries to prevent misappropriation of traditional knowledge and genetic resources. It provides a mechanism to regulate access to genetic resources and traditional knowledge amongst the member countries for research, development and other commercial utilization. In the process, the Nagoya Protocol ensures fair and equitable benefit sharing. This Protocol will undoubtedly provide economic benefits and livelihood development for the local communities. If implemented well, the Protocol will also contribute to biodiversity conservation significantly.

Considering the rich biodiversity and biological resources available in the country, India should make use of the Nagoya Protocol and convert it as a tool of opportunity to help local communities and conservation efforts. The BD Act already provides a solid base for implementation of the Nagoya Protocol. However, the BD Act needs to be fine-tuned to incorporate compliance measures, checkpoints and the involvement of local communities in the ABS process in a big way. India should also consider strengthening its institutional capacities with regard to ABS, focusing on capacity building, training, conducting awareness programs for the stakeholders, including indigenous communities, and developing education materials for ensuring better implementation of the ABS laws in the country. The ABS Agreements should significantly balance the interests of the providers and users with the objective of the conservation and sustainable use of biodiversity as guiding principles.

Notes

1 The Nagoya Protocol on Access to Genetic Resources and the Fair and Equitable Sharing of Benefits Arising From Their Utilization to the Convention on Biological Diversity was adopted on 29 October 2010 in Nagoya, Japan, under the Convention on Biological Diversity (CBD), www.cbd.int/abs/text/ (accessed on 20 September 2016).
2 The Convention on Biological Diversity (CBD) was adopted on 22 May 1992 and released for signature on 5 June 1992 at the United Nations Conference on Environment and Development at Rio de Janeiro. The CBD entered into force on 29 December 1993, www.cbd.int/conven tion/text/ (accessed on 20 September 2016).
3 Article 15 of the CBD insisted on the establishment of a regulatory system at the national levels to regulate access to genetic resources and traditional knowledge and to provide equitable sharing of benefits arising out of their utilization. Later, the Sixth Session of the Conference of Parties to the CBD (COP VI), vide Decision VI/24, adopted the non-binding Bonn Guidelines on Access and Benefit Sharing in 2002 that prescribed procedures for prior informed consent and mutually agreed terms in order to facilitate access to genetic resources, traditional knowledge and equitable benefit sharing, www.cbd.int/doc/publications/cbd-bonn-gdls-en.pdf (accessed on 28 January 2016). See also S. Tully, 'The Bonn Guidelines on Access to Genetic Resources and Benefit Sharing', *RECIEL*, 2003, 12(1): 84–98.
4 'The Biological Diversity Act', 2002 (hereinafter the BD Act). The Act came into force on 5 February 2003, http://nbaindia.org/content/25/19/1/act.html (accessed on 18 September 2016).
5 'The Protection of Plant Varieties and Farmers' Rights Act', 2001 (hereinafter the Plant Varieties Act). The Act came into force in 2005, http://lawmin.nic.in/ld/P–ACT/2001/The%20Protection%20of%20Plant%20Varieties%20and%20Farmers'%20Rights%20Act%202001.pd. (accessed on 18 September 2016).
6 See Chapter VI (ss. 22 to 26) and Chapter X (s. 41) of the BD Act, 2002.
7 For details of the TBGRI case, see P. K. Lakshmanan, 'India and the Patent Regime: Grandmom Shackled', *Combat Law: The Human Rights Magazine*, 2005, 4(2): 61.

8 Though there are many instances like the *Hoodia case* (South Africa) and *Enola Bean case* (Mexico) in other jurisdictions, India itself had to pay a heavy price in fighting against cases involving biological resources as well as the traditional knowledge of local communities that were misappropriated and patented in foreign countries without any permission from Indian authorities or local communities. The *Neem* and *Turmeric cases* stand as a testimony to such misappropriation. The European Patent Office (EPO) granted a patent (No. 436,257 dated 14 September 1994) to W. R. Grace, New York, and the U.S. Department of Agriculture for a method of controlling fungi on plants by the aid of a hydrophobic extracted neem oil without acknowledging the fact that the fungicidal effect of hydrophobic extracts of neem seeds was known and widely used in India for centuries in Ayurvedic Medicine and in agriculture for controlling fungal infections in plants. After establishing the traditional knowledge of Indians, the EPO revoked the patent on 10 May 2000 on the grounds there was no inventive step. In appeal, reiterating the prior art and lack of novelty, the patent was once and for all revoked by the EPO on 8 March 2005. Similarly, two US-based Indian nationals obtained the US Patent (Patent No. 5,401,504 dated 28 March 1995) for the use of turmeric in wound healing. The patent was assigned to the University of Mississippi Medical Center, USA. When the Government of India challenged the patent with about 32 historical references of turmeric in Indian texts used as medicine, the US Patent and Trade Office revoked the patent on 21 April 1998 on the grounds of the alleged invention being anticipated and obvious in India.

There are many instances of biopiracy such as bitter gourd, neem, jamun, turmeric and brinjal patent cases wherein the genetic resources and traditional knowledge have been misappropriated by non-native researchers. These experiences prompted the thinking of many countries towards having a regulatory system to control access and use of genetic resources and the related knowledge. See also G. R. Rattray, 'The Enola Bean Patent Controversy: Biopiracy, Novelty and Fish-and-Chips', *Duke Law & Technology Review*, 2002: 1–8.

9 This shows that the scope of the Nagoya Protocol attempts to realize the contents of Articles 15 and 8(j) of the CBD.

10 The Nagoya Protocol, Article 6.

11 The Nagoya Protocol, Article 5.2.

12 The Nagoya Protocol, Article 6.3(g).

13 The Nagoya Protocol, Article 14.2(C).

14 As per Article 14.1 of the Nagoya Protocol, an ABS Clearing-House was established as a part of the clearing-house mechanism under Article 18.3 of the CBD. The ABS Clearing-House will act as an information-sharing mechanism.

15 The Nagoya Protocol, Article 13.

16 See the Annex to the Nagoya Protocol.

17 The Biological Diversity Rules, 2004 issued by the Ministry of Environment, Forests and Climate Change vide G.S.R. 261 (E) dated 15 April 2004.

18 The BD Act, Section 8.

19 See the BD Act, Sections 3, 4 and 6.

20 The BD Act, Section 19.

21 The BD Act, Section 18(3)(a).

22 The BD Act, Section 18(3)(b).
23 The BD Act, Section 21 (1) and Rule 14 (3) of the Biological Diversity Rules, 2004.
24 The BD Act, Section 18(4).
25 The BD Act, Section 22.
26 The BD Act, Section 7.
27 *Vaids* and *hakims* are medical practitioners using traditional medicines.
28 The BD Act, Section 41.
29 Land races are the primitive plant varieties grown by ancient farmers and their successors.
30 The term 'cultivars' refers to a variety of plant that is specially bred for the purpose of cultivation.
31 The BD Act, Section 41.
32 The BD Act, Section 21(2).
33 'The Guidelines on Access to Biological Resources and Associated Knowledge and Benefits Sharing Regulations, 2014', issued by the NBA on 21 November 2014 by a Gazette notification, http://nbaindia.org/uploaded/pdf/Gazette_Notification_of_ABS_Guidlines.pdf (accessed on 21 September 2016).
34 The ABS Guidelines, Rule 1 (2).
35 The ABS Guidelines, Rule 3 (3).
36 *Ibid.*
37 *Ibid.*
38 The ABS Guidelines, Rule 4.
39 *Ibid.*
40 *Ibid.*
41 The ABS Guidelines, Rule 5.
42 The ABS Guidelines, Rules 8 and 9 (1).
43 The ABS Guidelines, Rule 9 (2).
44 The ABS Guidelines, Rule 9 (3).
45 The ABS Guidelines, Rule 7.
46 The ABS Guidelines, Rule 12 (2).
47 The ABS Guidelines, Rule 15.
48 The BD Act, Sections 41, 43 and 44.
49 See 'Biodiversity Management Committees', http://nbaindia.org/content/20/35/1/bmc.html (accessed on 22 September 2016).
50 The BD Act, Section 6 (1).
51 *Ibid.*
52 The BD Act, Section 6 (2).
53 *Ibid.*
54 The BD Act, Section 6 (1).
55 'The Patents Act, 1970 as amended in 2002 and 2005', http://ipindia.nic.in/resources.htm#acts (accessed on 20 September 2016).
56 The Patents Act, 1970 as amended in 2005, Section 25.
57 The Patents Act, Section 10(4) (ii) (D).
58 Form I is the Application for Grant of Patent:

> *9.DECLARATIONS:*
> *(iii) Declaration by the applicant(s):*
> *I/We, the applicant hereby declare(s) that:–*

> . . .
> *The invention as disclosed in the specification uses the biological material from India and the necessary permission from the competent authority shall be submitted by me/us before the grant of patent to us.*

59 Guidelines for Examination of Biotechnology Applications for Patent, March 2013, www.ipindia.nic.in/whats_new/biotech_Guidelines_25March 2013.pdf (accessed on 20 September 2016).
60 The Plant Varieties Act.
61 The Nagoya Protocol, Articles 15 and 16.
62 'The Norwegian Nature Diversity Act', 2009, www.wipo.int/wipolex/en/text.jsp?file_id=179529 (accessed on 15 September 2016). Section 60 states as follows:

> **Genetic material from other countries:**
> *The import for utilisation in Norway of genetic material from a state that requires consent for collection or export of such material may only take place in accordance with such consent. The person that has control of the material is bound by the conditions that have been set for consent. The state may enforce the conditions by bringing legal action on behalf of the person that set them.*
> *When genetic material from another country is utilised in Norway for research or commercial purposes, it shall be accompanied by information regarding the country from which the genetic material has been received (provider country). If national law in the provider country requires consent for the collection of biological material, it shall be accompanied by information to the effect that such consent has been obtained.*
> *If the provider country is a country other than the country of origin of the genetic material, the country of origin shall also be stated. The country of origin means the country in which the material was collected from in situ sources. If national law in the country of origin requires consent for the collection of genetic material, information as to whether such consent has been obtained shall be provided. If the information under this paragraph is not known, this shall be stated.*
> *The King may make regulations prescribing that if utilisation involves use of the traditional knowledge of local communities or indigenous peoples, the genetic material shall be accompanied by information to that effect.*
> *When genetic material covered by the International Treaty on Plant Genetic Resources for Food and Agriculture of 3 November 2001 is utilised in Norway for research or commercial purposes, it shall be accompanied by information to the effect that the material has been acquired in accordance with the Standard Material Transfer Agreement established under the treaty.*

63 See the BD Act, Section 3, and Rule 14 of the BD Rules, 2004.
64 See the BD Act, Sections 7 and 24. Section 7 of the BD Act covers only access to biological resource for the purposes of commercial utilization,

bio-survey or bio-utilization for commercial utilization. It does not deal with traditional knowledge.

65 It is pertinent to note here that one of the earliest cases involving misappropriation of traditional knowledge was by two Indians settled in the United States. See the *Turmeric patent case* (US Patent No. 5,401,504 March 28, 1995) on the use of turmeric in wound healing, which was revoked on 21 April 1998 after an intervention by the Government of India, http://patft. uspto.gov/netacgi/nph-Parser?Sect1=PTO1&Sect2=HITOFF&d=PALL& p=1&u=/netahtml/PTO/srchnum.htm&r=1&f=G&l=50&s1=5,401,504. PN.&OS=PN/5,401,504&RS=PN/5,401,504 (accessed on 15 September 2016).

66 The BD Act, Section 21, and Rule 14 (3) of the BD Rules, 2004.

67 The Nagoya Protocol, Article 7.

68 The Scheduled Tribes and other Traditional Forest Dwellers (Recognition of Forest Rights) Act, 2006 (Act 2 of 2007) (hereinafter as the Recognition of Forest Rights Act), http://tribal.nic.in/WriteReadData/CMS/Documen ts/201306070147440275455NotificationMargewith1Link.pdf (accessed on 15 September 2016).

69 The Recognition of Forest Rights Act Sections 3 and 5 and its Rules 21 and 24,

70 'The Indian Forest Act, 1927 (Act XVI of 1927)', http://forest.and.nic.in/ ActsNRules%5CIFA-1927.pdf (accessed on 15 September 2016).

71 'The Forest (Conservation) Act, 1980 (Act 6 of 1980)', http://faolex.fao. org/docs/html/ind3172.htm (accessed on 15 September 2016).

72 The Forest (Conservation) Act, 1980, Section 2.

73 The Kani-TBGRI model of benefit sharing could not be sustained in the long run due to several reasons, one of the reasons being the Kanis could not supply an uninterrupted supply of biological materials of the arogyapacha plant as the forest laws of India did not permit large-scale cultivation of the plant in the reserved forest land for commercial purposes and the forest department officials prevented it. For a detailed account, see R. V. Anuradha, *Sharing the Benefits of Biodiversity: The Kani-TBGRI Deal in Kerala, India*, Pune: Kalpavriksh, 2000, p. 24. See also R. V. Anuradha, 'Sharing the Benefits of Biodiversity: the Kani TBGRI Deal in Kerala, India', *Journal of International Wildlife Law & Policy*, 2000, 3(2): 125–151.

74 'The Wildlife (Protection) Act, 1972 (Act 53 of 1972)', http://nbaindia.org/ uploaded/Biodiversityindia/Legal/15.%20Wildlife%20(Protection)%20 Act,%201972.pdf (accessed on 15 September 2016).

75 The Wildlife (Protection) Act, 1972, Section 12 (b).

76 The Wildlife (Protection) Act, 1972, Section 12 (c).

77 The Wildlife (Protection) Act, 1972, Section 12 (d).

78 Barring certain provisions such as Sections 24, 36, 37 and 38 of the BD Act. However, conservation is not standing out as the prime focus of the BD Act.

79 These include four sites in Karnataka (Nallur Tamarind grove, Devnahalli; Hogrekal, Kadur, Chikmagalur; University of Agricultural Sciences, GKVK Campus; and Ambaraguda, Shimoga); two sites in West Bengal (Tonglu and Dhotrey BHS under the Darjeeling forest division); and one in Maharashtra (Glory of Allapalli, Gadchiroli). National Biodiversity Authority,

Biodiversity Heritage Site, http://nbaindia.org/content/106/29/1/bhs.html (accessed on 22 January 2016).

80 'People's Biodiversity Register, National Biodiversity Authority', http://nbaindia.org/content/105/30/1/pbr.html (accessed on 22 September 2016).

81 'Approval granted to the Applicants, National Biodiversity Authority', http://nbaindia.org/content/683/61/1/approvals.html (accessed on 22 September 2016).

82 The Nagoya Protocol, Article 17.

83 'The Drugs and Cosmetics Act, 1940 (23 of 1940) as amended on 30th June, 2005', www.cdsco.nic.in/writereaddata/Drugs&CosmeticAct.pdf (accessed on 22 September 2016).

6

AGREEMENT ON TRADE-RELATED ASPECTS OF INTELLECTUAL PROPERTY RIGHTS VERSUS CONVENTION ON BIOLOGICAL DIVERSITY

A call for harmonization

Vandana Mahalwar

Traditionally, intellectual property rights were not applied to life-forms[1] merely because life-forms are not manmade, and cannot be termed as 'inventions'. Hence, a discovery of what is already in existence in nature was not considered to be patentable. Yet, there were some exceptions, as in 1873, a patent was granted to Louis Pasteur, a French microbiologist, for isolating yeast, a living organism.[2] The ownership of living organisms has always fuelled debate, especially in the intellectual property regime. In the age of globalization and technological advancement, the world continues with its struggle in dealing with the patentability of life.[3] This whole debate came to an end when Chief Justice Warren E. Burger stated in *Diamond* v. *Chakraborty*[4] that 'anything under the sun is made by man' and is eligible for patent protection. After the dictum, technology-rich industrialized countries acted swiftly to obtain patents over biotechnology and genetically engineered[5] crops, food and microorganisms comprising life-forms. This intellectual property breakthrough just added to the whole argument over the subject matter.

The regulatory principle underlying international instruments, e.g. the World Intellectual Property Organization (WIPO) and the Trade-Related Aspects of Intellectual Property Rights (TRIPS)Agreement, has been referred to as 'maximalist rights culture' to refer to the people who believe that more intellectual property would serve and promote better innovation, education and trade.[6] Intellectual Property Maximalists support strong or even further enhanced intellectual property (IP) laws. Intellectual Property Maximalists do not think that the present protective levels are adequate and therefore advocate for longer terms, broader protection and enhanced punishment for violations of intellectual property rights. Secondly, maximalists align themselves with the argument that providing economic incentives through an IP system is a must for the development of the society. They also hold the view that without the tight protection, people would stop creating. Any reduction in the current level of protection therefore will lead to less creativity. Thirdly, maximalists, at the very least, desire to conserve the *status quo* when it comes to IP protection and therefore resist efforts to enhance access to creative work that may weaken the present IP laws.[7] But this notion of protection is progressively being challenged through the introduction of other normative and ethical considerations regarding the conservation of biodiversity.

The significance of intellectual property rights in biotechnology-related innovations on the one hand, and control over genetic resources, crucial for such innovations, on the other hand is becoming more apparent in international economic activities. Some scholars also argue that stronger protection of intellectual property rights will advantage the states producing almost all the intellectual property, which are mostly the developed countries with effective channels to transfer research from educational and research facilities to industries.[8] On the other hand, developing countries like India stand to lose the most from strong intellectual property protection, as with strong intellectual property protection they drop the access to affordable drugs, crop chemicals and educational matters.[9]

Developing countries worry that patenting biological resources gives the world's most precious assets over to large corporations of the wealthy and industrialized states. Other developed states are greatly advantaged from patenting biotechnology and assert that patent protection is essential for the advancement of science, technology and global economic development. The strain between the two positions has developed considerably as developing countries claim their resources are unfairly taken under acts of bio-piracy, where corporations and industrialized nations allegedly take and commercialize genetic resources of other biologically varied nations.[10] The protection of intellectual property rights at apposite concentrations can benefit both developed and developing countries.[11] Intellectual property can assist indigenous cultures in protecting their craftsmanship from overseas misuse and protect their traditional way of life.[12]

In the middle of the whole debate relating to bio-piracy are the two major international agreements that aim to solve the concerns of both sides, but in some ways have only broadened the gap between them. The Agreement on Trade-Related Aspects of Intellectual Property (TRIPS) and the Convention on Biological Diversity (CBD) depict the separating lines between the biodiversity-rich developing countries and the technology-rich industrialized countries.[13] While TRIPS backs stronger patent protection, the CBD promotes fair and equitable sharing of biological resources. In an attempt to unite the two agreements, developing countries have proposed an amendment to TRIPS that requires disclosure of genetic source and origin in the patent applications. This chapter examines the relationship between TRIPS and the CBD and discusses the amendments proposed by developing countries to TRIPS.[14]

I. Meaning of biological diversity and bio-piracy

Biological diversity has been explained as 'the foundation upon which human civilizations are built'.[15] Biological diversity is 'an umbrella term for the degree of nature's variety' and also includes 'the multitude of plant and animal species and the ecological complexes in which they occur.'[16] Biodiversity means the variety of life in all its forms and at all levels of organization. 'In all its forms' suggests that biological diversity includes plants, invertebrate animals, fungi, bacteria and other microorganisms, as well as the vertebrates that garner most of the attention. 'All levels of organization' means that biological diversity refers to the diversity of genes and ecosystems, as well as species diversity.[17] The Wildlife Society (1993) defines biodiversity as 'the richness, abundance, and variability of plant and animal species and communities and the ecological processes that link them with one another and with soil, air, and water'.[18]

Biodiversity is the underpinning of the ecosystem upon which all human life depends. Genes, ecosystems and species are said to be its three fundamental and hierarchically related levels. In legal terms, biodiversity has been defined as 'the variability among living organisms from all sources including, *inter alia*, terrestrial, marine and other aquatic ecosystems and the ecological complexes of which they are a part; this includes diversity within species, between species and of ecosystems.'[19]

Human life depends on biodiversity in our daily lives, in ways that are not always apparent or appreciated. Human health ultimately depends upon ecosystem products and services such as availability of fresh water, food and fuel sources, which are essential for fine human health and productive livelihoods.[20] Hence, loss of biodiversity may have major impacts on human life if ecosystem services are no longer ample enough to meet social needs. Changes in ecosystem services have indirect impacts also, as they affect livelihoods, income, local migration and, on occasion, may even cause political conflict.[21]

India is one of the 17 mega biodiverse countries in the world.[22] The country is divided into 10 bio-geographic regions having rich biological resources.[23] An ancient civilization with written history dating to over 5,000 years, India has a rich repertoire of traditional knowledge in the use of biological resources and practices that sustain life amidst the various ecosystems in a healthy and sustainable manner.

Traditional knowledge associated with biological diversity is in itself a great component. Traditional knowledge has the potential of being translated into monetary benefits by providing leads for the development of useful products and processes.[24] The commercial potential involved in biological resources and associated traditional knowledge has assumed enormous magnitude in the last few decades with the tremendous proliferation of the biotechnology industry. The world's biodiversity-rich countries are mostly developing countries and these nations could have been in a better position to benefit substantially by trading in such resources. The multinational corporations (MNCs) of the developed countries appropriate the biological resources and the associated traditional knowledge without the prior informed consent of the legitimate holders of the biological resources. The country of origin of the biological resources and associated traditional knowledge remains deprived of its legitimate share of the profits generated out of them. This practice is called 'bio-piracy'.

The term 'bio-piracy' was coined in 1993 by Pat Mooney, president of the Rural Advancement Foundation International (RAFI, now the ETC Group), and refers specifically to

> the use of intellectual property systems to legitimize the exclusive ownership and control of biological resources and knowledge, without recognition, compensation or protection for contributions from indigenous and rural communities . . . thus bio-prospecting cannot be considered anything but bio-piracy.[25]

According to Vandana Shiva, the term 'bio-piracy' means the appropriation of biological resources and traditional knowledge of farmers or local communities by patents without their permission, i.e. prior informed consent, or payment, i.e. benefit sharing, by multinational corporations in order to use the patent for their own purpose of investigation, production and marketing activities.[26]

Daniel F. Robinson has mentioned three types of bio-piracy:

- Patent-based bio-piracy: This means the patenting of inventions based on biological resources and/or traditional knowledge that are extracted, without any authorization and benefit-sharing, from developing countries or indigenous or local communities.

- Non-patent bio-piracy: This refers to other intellectual property control (through plant-variety protection or deceptive trademarks) based on biological resources and/or traditional knowledge that have been extracted, without adequate authorization and benefit-sharing, from other (usually developing) countries or indigenous or local communities.

- Misappropriation: The unauthorized appropriation of biological resources and/or traditional knowledge from other developing countries or indigenous or local communities, without sharing of adequate benefits with them.[27]

It is such a practice where developed countries take a free ride or unfair advantage of the genetic resources and traditional knowledge of developing countries which are rich in biodiversity. Biological materials which were earlier treated as a common resource for humankind, now are viewed as a form of property. Bio-piracy is a negative term for the exploitation of legal rights over indigenous biomedical knowledge without giving any compensation to indigenous groups who actually originally developed such knowledge. The practice of bio-piracy enhances the inequalities between developing and less-developed countries rich in biodiversity, and developed countries.[28]

Bio-prospecting precedes bio-piracy and is a search for and further development of new sources of chemical compounds, genes, micro- and macro-organisms and other valuable bio-products.[29] Bio-prospecting is the exploration of biological diversity for new biological resources of huge social and economic value.[30] It denotes the search and collection of biological materials to be used for commercial uses, and it places a reward on the natural resources of countries rich in biological diversity.[31] Developed countries which are not rich in biogenetic resources, but are better equipped in research and development, find it really profitable to search and experiment on natural resources such as soil samples, marine waters, insects, tropical plants and animals in developing countries.[32]

II. Bio-piracy cases in India

There are several cases of bio-piracy of traditional knowledge in India, a few of which are cited here:

A. Turmeric case

Turmeric is an herb that is grown in East India. Turmeric powder has a unique deep yellow colour and bitter taste that is used for different purposes such as for dye; the rhizomes of turmeric are used as a spice for flavouring cuisines, and as a litmus in a chemical test; and it has medicinal value also as it has been used for centuries for treating wounds and rashes.[33] In 1955, the US Patent and Trademark Office (USPTO) granted a patent (Patent No. 5,401,504) on the 'use of turmeric for wound healing' to two US scientists from the University of Mississippi. Two years later, a complaint was filed by the Council of Scientific and Industrial Research (CSIR) challenging the patent on the grounds of prior art. The CSIR opposed on the grounds that turmeric had already been in use for centuries for healing wounds and rashes, and hence, it was not a novel invention. The CSIR also adduced the documentary evidences in its support. The USPTO, after examining the validity of patent, revoked the patent in 1997, maintaining that there was no novelty in it. The turmeric case is one of the landmark cases, as in this case for the first time a patent based on the traditional knowledge of a developing country was successfully challenged.

B. Neem case

In India, the neem tree (*Azadirachta indica*) has been used from the pre-Vedic times for different purposes. It was also written in Indian texts 2,000 years ago. *Azadirachtin* is an active compound present in the bark, leaves, flowers and seeds of the neem tree.[34] This active component has been in use for centuries by indigenous people in agriculture as a potent insecticide and pesticide, and for medicinal purposes as it is useful in treating diabetes, ulcers and skin disorders.

In 1971, a timber company in the United States reckoned on the *Azadirachta indica*'s utility as a pesticide and started planting *Azadirachta indica* trees. In 1983, Terumo Corporation obtained the first US patent on therapeutic preparation from neem bark.[35]

In 1985, Robert Larson from the United States Department of Agriculture (USDA) was granted a patent for his preparation of neem seed extract and the Environmental Protection Agency approved this product for use in the US market. In 1988, Robert Larson sold the patent on an extraction process to the US company W.R. Grace. In 1992, W.R. Grace secured its rights to the formula that used the emulsion from the neem tree's seeds to make a powerful pesticide, and it also began suing Indian companies for making the emulsion.[36] The whole campaign was led by the EU Parliament's Green Party, the India-based Research Foundation for Science, Technology and Ecology (RFSTE) and the International Federation of Organic Agriculture Movements (IFOAM), against the US multinational corporation W.R. Grace. The main argument made by RFSTE was that the fungicidal qualities of the neem tree and its uses had already been known in India for more than 2,000 years.[37] After a dispute of more than 15 years, the European Patent Office revoked the patent on March 8, 2005.

C. Basmati case

Basmati is a staple crop that is grown in Punjab, Pakistan, and North India. Basmati, a long-grained rice with a fine texture and a unique aroma, is the costliest rice in the world. On September 2, 1997, a Texas-based company, RiceTec Inc., obtained a patent (No. 5,663,484) from the USPTO on a rice that was similar to the traditional Indian basmati rice. On learning about the same, the government of India reacted immediately, contending that it would request the USPTO to re-examine the patent.[38] After filling of the re-examination request, RiceTec chose to withdraw its crucial claims on the basmati patent.

According to Dr Vandana Shiva, the only purpose of obtaining the patent by RiceTec Inc. was to deceive consumers by leading them to believe there is no variation between spurious basmati and real basmati. She states that

> theft involved in the Basmati patent is, therefore, threefold: a theft of collective intellectual and biodiversity heritage on Indian farmers, a theft from Indian traders and exporters whose markets are being stolen by RiceTec Inc., and finally a deception of consumers since RiceTec is using a stolen name Basmati for rice which are derived from Indian rice but not grown in India, and hence are not the same quality.[39]

III. The Convention on Biological Diversity

The Convention on Biological Diversity (CBD) is the first multilateral treaty that focuses on the issue of preserving the world's biological resources.[40] At the United Nations Environment Programme (UNEP), in 1987, the United States proposed that UNEP 'establish an "umbrella" convention' to make the conservation agreements throughout the world compatible with one another.[41] After two years, an Ad Hoc Working Group of Experts was created to draft a harmonized document for the conservation and sustainable use of biological diversity, while considering 'the need to share costs and benefits between the developed and developing countries and the ways and means to support innovation by local people.'[42] There are 196 parties to the Convention and 168 signatories.

The Preamble of the CBD outlines the three key objectives of the Convention: (1) conservation of biological diversity; (2) sustainable use of biological diversity; and (3) fair and equitable sharing of benefits arising out of the utilization of genetic resources.[43] The CBD endeavours to regulate biodiversity and also the use of biological resources. The Convention aims to promote the conservation of biodiversity by providing for

> national monitoring of biological diversity, the development of national strategies, plans and programs for conserving biological diversity, national in situ and ex situ conservation measures, environmental impact assessments of projects for adverse effects on biological diversity, and national reports from parties on measures taken to implement the convention and the effectiveness of these measures.[44]

Article 1 of the CBD states that 'equitable sharing of benefits' includes access to genetic resources and 'the appropriate transfer of relevant technologies'.[45]

The CBD also confirms that states have sovereign rights over their own biological resources and that states are responsible for conserving their biological resources in a sustainable manner.[46] It also asserts that genetic resources are the 'common heritage of mankind'. In Article 15.7, the CBD obligates that use of biological resources should be 'fair and equitable':

> Each Contracting Party shall take legislative, administrative or policy measures, as appropriate, and in accordance with Articles 16 and 19 and, where necessary, through the financial mechanism established by Articles 20 and 21 with the aim of sharing in a fair and equitable way the results of research and

development and the benefits arising from the commercial and other utilization of genetic resources with the Contracting Party providing such resources. Such sharing shall be upon mutually agreed terms.[47]

Though this article does not explicitly refer to intellectual property rights, Article 16 of the CBD further requires that 'access and transfer shall be provided on terms which recognize and are consistent with the adequate and effective protection of intellectual property rights'.[48] Articles 15 and 16 together are at the core of the controversy between the developed and developing countries, and hence need to be discussed further.

In Article 16.3, countries of origin, especially developing countries, are given access to technology that incorporate the use of that country's biological resources. This article also signifies that the transfer of technology shall be consistent with the adequate and effective protection of intellectual property rights. It further provides that access to and transfer of the benefits derived by biotechnology are required to meet the objectives of the CBD.

As it is difficult for the least developed countries to take substantive steps to conserve biological diversity, the Convention has vested in developing states the right to exclude nationals of foreign states from accessing the biological organisms of their own country.[49] Article 16.5 recognizes the significance of intellectual property rights but appears to give more importance to the transfer of technology.[50]

There was much of controversy on CBD over the international response voiced by the developed and developing countries. The United States took a very different view on CBD and also refused to sign the same, reasoning that intellectual property and technology transfer were not dealt with uniformly in the CBD.[51] The United States found the CBD was forcing the developed states to transfer the technology to developing countries while it pleased the developing countries by relieving them of the burden to recognize the patent protection for US biotechnology corporations.[52] The US contended that Article 16 treated intellectual property rights as a limitation on technology transfer rather than a precondition.[53] It also revealed its concern at how the CBD would affect the competitiveness of biotechnology firms.[54] Other developed nations also expressed similar apprehensions and desired a tough international intellectual property rights protection so as to generate incentives for technological development.[55] Developing states, on the other hand, conveyed their concern for the protection of their right to control the access to their own nation's biological resources. The US biotech companies who initially opposed the CBD feared that a refusal to sign the Convention could be more detrimental than participating in the agreement.[56] Ultimately, the US signed the CBD.

IV. The Agreement on Trade-Related Aspects of Intellectual Property Rights

The most significant international agreement regarding intellectual property rights is the World Trade Organization's TRIPS Agreement. In 1994, after efforts to bring uniformity in international intellectual property rights protection, the Uruguay Round under the General Agreements of Tariffs and Trade (GATT) resulted in the foundation of the World Trade Organization (WTO) and the Trade-Related Aspects of Intellectual Property Rights (TRIPS) Agreement, which is administered by the World Trade Organization.[57] This Agreement, which was concluded during the Uruguay Round of trade negotiations, is binding on all the members and it lays down uniform and minimum standards for the protection and enforcement of intellectual property rights by all WTO member states.[58] Before the TRIPS Agreement, issues of international intellectual property rights were dealt with by the World Intellectual Property Organization (WIPO) treaties, bilateral agreements and the GATT.[59]

The TRIPS Agreement came into effect on January 1, 1995, and is the most detailed multilateral agreement on intellectual property. The preamble to the TRIPS Agreement provides that it was designed for: (1) the reduction of distortion and impediments to international trade; (2) the promotion of effective and adequate protection of intellectual property rights; and (3) the assurance that the measures and procedures used to enforce intellectual property rights do not themselves become barriers to legitimate trade.

According to its objectives, included in Article 7, TRIPS seeks to promote technological innovation and transfer in a manner 'conducive to social and economic welfare, and to a balance of rights and obligations'.[60] The Agreement comprises various intellectual property rights, including copyright, trademarks, geographical indications, trade secrets, industrial designs and patents. It requires its members to provide intellectual property rights to the nationals of other members as they provide to their own nationals[61] and to extend the same favourable terms they grant to the national of any member to the nationals of every other member.[62]

TRIPS is a minimum-standards agreement that allows its member states to give greater protection to intellectual property rights. Article 1.1 of TRIPS provides as follows:

> Members shall give effect to the provisions of this Agreement. Members may, but shall not be obliged to, implement in their law more extensive protection than is required by this Agreement, provided that such protection does not contravene the provisions of this Agreement. Members shall be free to determine the appropriate method of implementing the provisions of this Agreement within their own legal system and practice.

One such standard is that patents must be awarded in all fields of technology, including products and processes.[63] For an invention to be recognized as a patent, it must 'involve an inventive step' and have 'industrial application'. There are certain important exceptions to this requirement which are significant to the successful implementation of the purposes of the CBD. First, members may exclude inventions from patentability where it is

> necessary to protect *ordre* public or morality, including to protect human, animal or plant life or health or to avoid serious prejudice to the environment provided that such exclusion is not made merely because the exploitation is prohibited by their law.[64]

Second, members may also exclude from patent ability plants and animals other than micro-organisms, and essentially biological processes for the production of plants or animals other than non-biological and microbiological processes.[65] Third, members are also allowed to provide limited exceptions to the exclusive rights conferred by patents only if such exceptions do not unreasonably conflict with a normal exploitation of the patent and do not unreasonably prejudice the legitimate interests of the patent owner.[66]

The exception given in Article 27.3(b) also talks of providing *sui generis* protection to plant varieties. Article 27.3(b) provides that members shall provide protection to plant varieties, either by patents or an effective *sui generis* system or by a combination of both. A *sui generis* system is a system of its own kind that is chosen by the member itself and is planned to suit the particular requirements of that country. Members are also allowed to select a combination of patents and *sui generis* protection.

Moreover, considering the complexities of the obligations provided in the Agreement, developing countries were given a transition period of five years for implementation until January 1, 2000.[67] From that date, developing nations were given an additional five years to extend patent protection over products in those areas of technology for which they were offered no protection when the WTO Agreement entered into force.[68] The least developed states were given a buffer of 10 years to implement the agreement.[69]

Developed member states are required to give incentives to corporations of their own country to transfer technology to other developing member states.[70] TRIPS also obligates the developed countries to provide, on request and on mutually agreed terms and conditions, technical and financial cooperation to developing and least developed member countries in implementing a legal infrastructure for intellectual property rights protection.[71]

V. Relationship between CBD and TRIPS

The CBD and TRIPS stand in direct opposition to each other, as one speaks of conserving biological diversity and advancing sustainable development while the other talks about protection of intellectual property rights.[72] Many states which are rich in genetic resources face huge challenges in the combined implementation of TRIPS and CBD that results in conflict over the use of their genetic resources.[73] At the centre of the whole debate is Article 27 of TRIPS which required members to review its provisions in 1999.[74]

The Doha Declaration adopted in November 2001 mandates further review of Article 27[75]:

> We instruct the Council for TRIPS, in pursuing its work programme including under the review of Article 27.3(b), the review of the implementation of the TRIPS Agreement under Article 71.1 and the work foreseen pursuant to paragraph 12 of this declaration, to examine, inter alia, the relationship between the TRIPS Agreement and the Convention on Biological Diversity, the protection of traditional knowledge and folklore, and other relevant new developments raised by members pursuant to Article 71.1. In undertaking this work, the TRIPS Council shall be guided by the objectives and principles set out in Articles 7 and 8 of the TRIPS Agreement and shall take fully into account the development dimension.

The developing countries expressed the concern regarding the need to harmonize the provisions of TRIPS with the CBD. Ethiopia was one of the first member states to the CBD to propose the examination of the relationship between TRIPS and the CBD.[76] In 1996, India formally proposed to the WTO that the Committee on Trade and the Environment (CTE) review the consonance between the CBD and TRIPS.[77] In the next few years, various communications were made by the like-minded developing countries to the TRIPS Council of the WTO, emphasizing the recognition of rights regarding benefit sharing arising out of the innovation which is based on their knowledge.[78] A proposal was made by India that the CBD and TRIPS Agreement needed to be reconciled by amending TRIPS through incorporation of genetic resource disclosure requirement in patent applications.[79]

VI. Proposed TRIPS amendment

The amendments have been proposed by the developing countries that are rich in biological resources. These proposals from developing countries refer to the problem of bio-piracy. The pretence of the demand for intellectual property protections is dual: it is not just that the developing countries have to pay a high premium for the patented products that are reintroduced in their country, but that developing countries are failing to use the intellectual property laws against the piracy relating to their own indigenous and local resources and knowledge.[80] Brazil, being rich in biological resources, became the first signatory to the CBD and has been a keen proponent of the amendment to TRIPS.[81] The African Group, the Andean Community, Bolivia, Brazil, China, Columbia, Cuba, Ecuador, India, Indonesia, Kenya, Pakistan, Peru, Thailand, Venezuela and Zimbabwe together made the proposal for amendment to the TRIPS Agreement.

These countries proposed three types of disclosure requirements to be incorporated in TRIPS:

(1) disclosure of source and country of origin of the genetic materials and associated traditional knowledge used in developing the invention claimed in the patent application; (2) disclosure of the evidence of prior informed consent; and (3) disclosure of the evidence of a benefit–sharing agreement.[82]

The proposal suggests that the main aim of the disclosure requirement is to avoid the grant of bad patents, as revocation of an erroneously granted patent is more costlier and troublesome than the mandatory disclosure requirements. The mandatory disclosure requirement would function as a crucial factor in the ascertainment of the patentability of biotechnological inventions.

The proposal was to add an exception to Article 27 as:

> Members may also exclude from patentability '(c) products or processes which directly or indirectly include genetic resources or traditional knowledge obtained in the absence of compliance with international and national legislation on the subject, including failure to obtain the prior informed consent of the country of origin or the community concerned and failure to reach agreement on conditions for the fair and equitable sharing of benefits arising from their use. Nothing in TRIPS shall prevent Members from adopting enforcement measures in their domestic legislation, in accordance with the principles and obligations enshrined in the Convention on Biological Diversity.'[83]

The proposal was vehemently opposed by some developed countries.[84] Some developed countries expressed that the mandatory disclosure requirement would result in uncertainty in the patent system and would also complicate the implementation of benefit sharing.[85] Moreover, in 2001, the United States submitted one of its first papers to the World Trade Organization expressing that it finds no conflict between the TRIPS Agreement and the CBD at all. The US also stated that the CBD and TRIPS provisions are not conflicting, but rather are complementary to each other. It also mentioned that the provisions relating to theft and misappropriation of genetic resources don't lie within the purview of TRIPS, but rather 'are appropriately the domain of a separate regulatory system'.[86] The US asserts that the mandatory 'disclosure requirements may upset the careful balance created by the patent system to promote innovation'.[87] Japan also maintained that a disclosure requirement may infringe various provisions of the TRIPS Agreement.

On April 19, 2011, a communication from Brazil, China, Colombia, Ecuador, India, Indonesia, Peru, Thailand, the ACP Group and the African Group was made to amend the TRIPS Agreement by inserting a new Article 29*bis* in it.[88] The proposed article requires the disclosure of the origin of genetic resources and/or associated traditional knowledge.[89] Paragraph 5 of the proposed article also provides for administrative sanctions, criminal sanctions, fines and an adequate amount for

144

damages if the applicant fails to disclose the information or submitted false or fraudulent information.

Some states have already adopted laws regarding disclosure of genetic resources in national legislation. In 1998, the European Community adopted a directive that deals with the disclosure information regarding the geographical origin of the genetic material in an invention. This directive does not mandate to disclose but is merely directive, which encourages the applicant to mention the geographical origin of biological material in the patent application.[90]

VII. Conclusion

The misappropriation of genetic resources and the relinquished profits derived from their use continues to bring forth serious worries to the biodiversity community and indigenous peoples. The TRIPS Agreement and the CBD are aiming to strive a correct balance between the interest of biodiversity-rich developing countries and the industrialized developed nations. But, in their efforts to create a way to the solution, the Convention on Biological Diversity and the Agreement on Trade-Related Aspects of Intellectual Property Rights intersected each other in a sphere of great controversy. Despite some conflicts, both of the international agreements are endeavouring to advance economic development by acknowledging the role of intellectual property rights while aiming to protect the interests of the owners of genetic resources. The Convention on Biological Diversity and the Agreement on Trade-Related Aspects of Intellectual Property Rights reflect various conflicting interests of industrialized developed countries and developing countries. Developed countries state that the CBD fails to create a correct equilibrium between the sovereign rights of states over their biodiversity and intellectual property rights. On the other hand, the developing nations view TRIPS as liberally permitting the industrialized nations to access their genetic resources without sharing in the benefits derived from them. In this backdrop, there is a need to bring harmony in these two conflicting international agreements. A mandatory disclosure requirement, however, needs to be introduced in the TRIPS Agreement to realize any progress in protecting the developing countries' interests in conserving biological diversity. Requirements for the mandatory disclosure of the origin of genetic resources and traditional knowledge in intellectual property applications presently appears to be the most transparent and appropriate solution to generate mutual support for the access and benefit-sharing system being bargained under the CBD and TRIPS Agreement. An amendment incorporating a genetic resource disclosure requirement may take much time as a vast majority is required to bring such a change in TRIPS, but it would be a significant step towards the protection of the owners of genetic resources.

Notes

1 M. Kruger, 'Harmonizing TRIPs and the CBD: A Proposal From India', *Minnesota Journal of Global Trade*, 2001, 10: 169–207.
2 'Bioethics and Patent Law: The Relaxin Case', *WIPO Magazine*, April 2006, www.wipo.int/wipo_magazine/en/2006/02/article_0009.html (accessed on 5 October 2016).
3 J. Carr, 'Agreements That Divide: TRIPS vs. CBD and Proposals for Mandatory Disclosure of Source and Origin of Genetic Resources in Patent Applications', *Florida State University Journal of Transnational Law Policy*, 2008, 18(1): 132–154.
4 *Diamond v. Chakraborty* 447 U.S. 303 (1980) (the case involved an issue whether genetically engineered micro-organisms are excluded from patent protection.) The Supreme Court of the United States pronounced by a majority of five to four that a new strain of bacteria produced artificially by bacterial recombination was a patentable invention. It contended that the bacteria had utility as it could disperse oil slicks. The decision was of much importance. The Supreme Court of the United States stated that 'anything under the sun made by man' was patentable subject matter.
5 Genetic engineering is a method of manipulating genes in plants, animals and micro-organisms to enhance the desirable qualities or eliminate undesirable qualities. This includes insertion of a gene from one species into another.
6 J. Boyle, 'A Manifesto on WIPO and the Future of Intellectual Property', *Duke Law and Technology Review*, 2004, 9(10): 1–13.
7 D. Halbert 'The Politics of IP Maximalism', *The WIPO Journal*, 2011, 3(1): 81–93.
8 J. H. Reichman, 'Intellectual Property in International Trade: Opportunities and Risks of a GATT Connection', *Vanderbilt Journal of Transnational Law*, 1989, 22(4): 747–892.
9 D. Brenner-Beck, 'Do as I Say, Not as I Did', *UCLA Pacific Basin Law Journal*, 1992, 11(1): 84–118.
10 C. R. McManis, 'Intellectual Property, Genetic Resources and Traditional Knowledge Protection: Thinking Globally, Acting Locally', *Cardozo Journal of International and Comparative Law*, 2003, 11(2): 547–584.
11 'Winning the War Against Bio-Colonisation', *The Hindu*, 17 May 2000.
12 R. J. Coombe, 'Intellectual Property Human Rights & Sovereignty: New Dilemmas in International Law Posed by the Recognition of Indigenous Knowledge and the Conservation of Biodiversity', *Indiana Journal of Global Legal Studies*, 1998, 6(1): 59–116.
13 McManis, 'Thinking Globally, Acting Locally'.
14 Coombe, 'Intellectual Property Human Rights & Sovereignty'.
15 A. K. Sharma, 'The Global Loss of Biodiversity: A Perspective in the Context of the Controversy over Intellectual Property Rights', *University of Baltimore Intellectual Property Law Journal*, 1995, 4(1): 1–32, 1.

16 J. L. Dunoff, 'Reconciling International Trade With Preservation of the Global Commons: Can We Prosper and Protect?' *Washington and Lee Law Review*, 1992, 49(4): 1407–1454, 1431.
17 M. L. Hunter, Jr. and J. Gibbs, *Fundamentals of Conservation Biology*, Oxford: Blackwell Publishing, 2007.
18 *Ibid.*, p. 24.
19 G. K. Venbrux, 'When Two Worlds Collide: Ownership of Genetic Resources Under the Convention on Biological Diversity and the Agreement on Trade-Related Aspects of Intellectual Property Rights', *Pittsburgh Journal of Technology Law & Policy*, 2005, 6: 1–36, 3.
20 'Climate Change and Human Health', World Health Organization, www.who.int/globalchange/ecosystems/biodiversity/en/ (accessed on 26 September 2016).
21 *Ibid.*
22 P. Pushpangadan and K. Narayanan Nair, 'Value Addition and Commercialization of Biodiversity and Associated Traditional Knowledge in the Context of the Intellectual Property Regime', *Journal of Intellectual Property Rights*, 2005, 10: 441–453.
23 C. Ganguly, 'Fast Facts on India's Biodiversity Part 1 – Biogeographic Zones', *Core Sector Communique*, 2012, http://corecommunique.com/new–post–6/ (accessed on 30 January, 2016).
24 V. Shiva, *Protect or Plunder? Understanding Intellectual Property Rights*, London: Zed Books Ltd., 2001.
25 G. C. Delgado, 'Biopiracy and Intellectual Property as the Basis for Biotechnological Development: The Case of Mexico', *International Journal of Politics, Culture and Society*, 2002, 16(2): 297–318.
26 V. Shiva, *Biopiracy: The Plunder of Nature and Knowledge*, Cambridge, MA: South End Press, 1997.
27 D. F. Robinson, *Confronting Biopiracy: Challenges, Cases and International Debates*, London: Earthscan, 2010.
28 P. S. Sudha, 'Combating Biopiracy of Indian Traditional Knowledge, (TK) – A Legal Perspective', *Bharati Law Review*, 2014, 3(2): 42–62.
29 R. D. Singh et al., 'Pharmaceutical Biopiracy and Protection of Traditional Knowledge', *International Journal of Research and Development in Pharmacy and Life Sciences*, 2014, 3(2): 866–871.
30 *Ibid.*
31 D. A. Posey and G. Dutfield, *Beyond Intellectual Property: Toward Traditional Resource Rights for Indigenous Peoples and Local Communities*, Ottawa, Canada: IDRC, 1996.
32 Singh et al., 'Pharmaceutical Biopiracy and Protection of Traditional Knowledge'.
33 A. Slack, 'Turmeric', *TED Case Studies*, 770, 2004, www1.american.edu/ted/turmeric.htm (accessed on 25 September 2016).
34 S. Hasan, 'The Neem Tree, Environment, Culture, and Intellectual Property', *TED Case Studies*, 665, 2002, www1.american.edu/ted/neemtree.htm (accessed on 28 September 2016).
35 A. Shaukat, 'Neem It Up!' *Spicy IP*, 27 November 2007, http://spicyip.com/2007/11/neem-it-up.html (accessed on 28 September 2016).
36 O. Singh et al., 'Neem (*Azadirachta indica*) in Context of Intellectual Property Rights', *Recent Research in Science and Technology*, 2011, 3(6): 80–84.

37 P. Pani and D. Nigam, 'Traditional Knowledge', *IPPRO*, 2008: 1–12, www.ipproinc.com/admin/files/upload/7fbc4e0da416089922a31b566b0 8378c.pdf (accessed on 10 September 2016).

38 J. Adewumi, 'Basmati', *TED Case Studies*, June 1998, www1.american. edu/TED/basmati.htm (accessed on 10 September 2016).

39 A. Mukherjee and P. Shastri, 'The Battle for Basmati', *Outlook Magazine*, 2 March 1998, www.outlookindia.com/magazine/story/the–battle–for-basmati/205154 (accessed on 12 September 2016).

40 A. Hubbard, 'The Convention on Biological Diversity's Fifth Anniversary: A General Overview of the Convention – Where Has It Been and Where Is It Going?' *Tulane Environmental Law Journal*, 1997, 10(2): 413–444.

41 D. Keating, 'Access to Genetic Resources and Equitable Benefit Sharing Through a New Disclosure Requirement in the Patent System: An Issue in Search of a Forum', *Journal of the Patent and Trademark Office Society*, 2005, 87(7): 525–547.

42 *Ibid.*, p. 528.

43 Convention on Biological Diversity (hereinafter as the CBD), Article 1 defines the objectives of the Convention as:

> the conservation of biological diversity, the sustainable use of its components and the fair and equitable sharing of the benefits arising out of the utilization of genetic resources, including by appropriate access to genetic resources and by appropriate transfer of relevant technologies, taking into account all rights over those resources and to technologies, and by appropriate funding.

44 E. B. Weiss, 'Introductory Note to the Convention on Biological Diversity', *International Legal Material*, 1992, 31: 814–817, 814.

45 Venbrux, 'When Two Worlds Collide', p. 10.

46 The CBD, Article 3.

47 The CBD, Article 15, entitled 'Access to Genetic Resources', provides as follows:

> 1 Recognizing the sovereign rights of States over their natural resources, the authority to determine access to genetic resources rests with the national governments and is subject to national legislation.
>
> 2 Each Contracting Party shall endeavour to create conditions to facilitate access to genetic resources for environmentally sound uses by other Contracting Parties and not to impose restrictions that run counter to the objectives of this Convention.
>
> 3 For the purpose of this Convention, the genetic resources being provided by a Contracting Party, as referred to in this Article and Articles 16 and 19, are only those that are provided by Contracting Parties that are countries of origin of such resources or by the Parties that have acquired the genetic resources in accordance with this Convention.
>
> 4 Access, where granted, shall be on mutually agreed terms and subject to the provisions of this Article.
>
> 5 Access to genetic resources shall be subject to prior informed consent of the Contracting Party providing such resources, unless otherwise determined by that Party.

6 Each Contracting Party shall endeavour to develop and carry out scientific research based on genetic resources provided by other Contracting Parties with the full participation of, and where possible in, such Contracting Parties.

7 Each Contracting Party shall take legislative, administrative or policy measures, as appropriate, and in accordance with Articles 16 and 19 and, where necessary, through the financial mechanism established by Articles 20 and 21 with the aim of sharing in a fair and equitable way the results of research and development and the benefits arising from the commercial and other utilization of genetic resources with the Contracting Party providing such resources. Such sharing shall be upon mutually agreed terms.

48 The CBD, Article 16.2, provides as follows:

Access to and transfer of technology referred to in paragraph 1 above to developing countries shall be provided and/or facilitated under fair and most favourable terms, including on concessional and preferential terms where mutually agreed, and, where necessary, in accordance with the financial mechanism established by Articles 20 and 21. In the case of technology subject to patents and other intellectual property rights, such access and transfer shall be provided on terms which recognize and are consistent with the adequate and effective protection of intellectual property rights.

49 M. D. Coughlin, 'Using the Merck–INBio Agreement to Clarify the Convention on Biological Diversity', *Columbia Journal of Transnational Law*, 1993, 31(2): 337–376.

50 The CBD, Article 16.5, provides as follows:

The Contracting Parties, recognizing that patents and other intellectual property rights may have an influence on the implementation of this Convention, shall cooperate in this regard subject to national legislation and international law in order to ensure that such rights are supportive of and do not run counter to its objectives.

51 Venbrux, 'When Two Worlds Collide'.

52 Coughlin, 'Using the Merck–INBio Agreement'.

53 J. Chen, 'Diversity and Deadlock: Transcending Conventional Wisdom on the Relationship between Biological Diversity and Intellectual Property', *Environment Law Reporter*, 2001, 31(6): 10625–10644.

54 Coughlin, 'Using the Merck–INBio Agreement'.

55 Ibid.

56 Venbrux, 'When Two Worlds Collide'.

57 Keating, 'New Disclosure Requirement'.

58 C. Monagle, 'Biodiversity & Intellectual Property Rights: Reviewing Intellectual Property Rights in Light of the Objectives of the Convention on Biological Diversity', *WWF International & CIEL, Joint Discussion Paper*, March 2001, 1–34.

59 Keating, 'New Disclosure Requirement'.

60 The Agreement on Trade-Related Aspects of Intellectual Property Rights (hereinafter TRIPS), Article 7, provides as follows:

> The protection and enforcement of intellectual property rights should contribute to the promotion of technological innovation and to the transfer and dissemination of technology, to the mutual advantage of producers and users of technological knowledge and in a manner conducive to social and economic welfare, and to a balance of rights and obligations.

61 TRIPS, Article 3 on 'National Treatment', provides as follows:

> Each Member shall accord to the nationals of other Members treatment no less favourable than that it accords to its own nationals with regard to the protection (3) of intellectual property, subject to the exceptions already provided in, respectively, the Paris Convention (1967), the Berne Convention (1971), the Rome Convention or the Treaty on Intellectual Property in Respect of Integrated Circuits.

62 TRIPS, Article 4 on 'Most-Favoured-Nation Treatment', provides as follows:

> With regard to the protection of intellectual property, any advantage, favour, privilege or immunity granted by a Member to the nationals of any other country shall be accorded immediately and unconditionally to the nationals of all other Members.

63 The Agreement on Trade-Related Aspects of Intellectual Property Rights, Article 27.1, provides that:

> Subject to the provisions of paragraphs 2 and 3, patents shall be available for any inventions, whether products or processes, in all fields of technology, provided that they are new, involve an inventive step and are capable of industrial application.

64 TRIPS, Article 27.2.
65 TRIPS, Article 27.3.
66 TRIPS, Article 30.
67 TRIPS, Article 65.2.
68 TRIPS, Article 65.4.
69 TRIPS, Article 66.1.
70 TRIPS, Article 66.
71 TRIPS, Article 67.
72 Carr, 'Agreements That Divide'.
73 L. Laxman and A. H. Ansari, 'The Interface Between TRIPS and CBD: Efforts Towards Harmonisation', *Journal of International Trade Law and Policy*, 2012, 11(2): 108–132.
74 TRIPS: Members may also exclude from patentability, from Article 27.3(b):

> plants and animals other than micro–organisms, and essentially biological processes for the production of plants or animals other than non–biological and microbiological processes. However,

> Members shall provide for the protection of plant varieties either by patents or by an effective sui *generis* system or by any combination thereof. The provisions of this subparagraph shall be reviewed four years after the date of entry into force of the WTO Agreement.

75 'World Trade Organization, Ministerial Declaration of 14 November 2001, article 19, WT/MIN(01)/DEC/1', 41 *I.L.M*, 2002, 41: 746–749.

76 Ethiopia made the recommendation to the Secretariat of the CBD:

> To request the WTO/TRIPS Council to take into account and accommodate the concerns of the Contracting Parties to the [CBD] before taking any decisions or measures in relation with the TRIPS Agreement that may affect biological diversity and the protection of knowledge, innovations, and practices of local and indigenous communities.

77 Keating, 'New Disclosure Requirement'.

78 P. Kumar, 'Biopiracy, GM Seeds and Rural India', *Global Research*, 2 June 2009, www.globalresearch.ca/biopiracy–gm–seeds–and–rural–india/ 13820 (accessed on 30 September 2016).

79 Keating, 'New Disclosure Requirement'.

80 K. Aoki, 'Neocolonialism, Anticommons Property, and Biopiracy in the (Not-So-Brave) New World Order of International Intellectual Property Protection', *Indiana Journal of Global Legal Studies*, 1998, 6(1): 11–58.

81 Council for Trade-Related Aspects of Intellectual Property Rights (hereinafter Council for TRIPS), 'Communication from Brazil: Review of the Provisions of Article 27.3(b)', *IP/C/W/164*, 29 October 1999.

82 Council for TRIPS, 'Communication From Bolivia, Brazil, Colombia, Cuba, India and Pakistan: The Relationship between the TRIPS Agreement and the Convention on Biological Diversity (CBD) and the Protection of Traditional Knowledge', *IP/C/W/459*, 18 November 2005.

83 Council for TRIPS, 'Communication from Peru: Article 27.3(B), Relationship between the TRIPS Agreement and the CBD and Protection of Traditional Knowledge and Folklore', *VII, IP/C/W/447*, 8 June 2005.

84 Council for TRIPS, 'Communication from the United States: Views of the United States on the Relationship Between the Convention on Biological Diversity and the TRIPS Agreement', *IP/C/W/257*, 13 June 2001.

85 C. Saez, 'WIPO Members Debate Disclosure of Origin for Genetic Resources in Patents', *Intellectual Property Watch*, 17 February 2016, www.ip–watch. org/2016/02/17/wipo–members–debate–disclosure–of–origin–for–genetic–resources–in–patents/ (accessed on 28 September 2016).

86 Council for TRIPS, 'Communication from the United States: Article 27.3(B), Relationship between the TRIPS Agreement and the CBD, and the Protection of Traditional Knowledge and Folklore', para. 4, IP/C/W/469 (13 March 2006).

87 *Id*. at 35.

88 Trade Negotiations Committee, World Trade Organization, 'Draft Decision to Enhance Mutual Supportiveness Between the Trips Agreement and the Convention on Biological Diversity', *TN/C/W/59*, 19 April 2011, http://commerce.nic.in/trade/wtopdfs/tn-c-w-59.pdf (accessed on 30 September 2016).

89 Paragraph 2 of the proposed Article 29*bis.* provides as follows:

> Where the subject matter of a patent application involves utilization of genetic resources and/or associated traditional knowledge, Members shall require applicants to disclose: (i) the country providing such resources, that is, the country of origin of such resources or a country that has acquired the genetic resources and/or associated traditional knowledge in accordance with the CBD; and,
> (ii) the source in the country providing the genetic resources and/or associated traditional knowledge.

90 World Trade Organization, Council for TRIPS, 'Minutes of Meeting', para 127, *IP/C/M/49*, 31 January 2006.

Part IV

GENETIC ENGINEERING
Problems and prospects

7

GENETICALLY MODIFIED CROPS

Neither poison nor panacea

Parikshet Sirohi

Genetically Modified Organisms (GMOs) can be defined as organisms, whether plants, animals or microorganisms, in which the basic genetic material (DNA) has been altered in a manner that does not occur naturally by mating and/or natural recombination. This technology is often called 'modern biotechnology'[1] or 'gene technology',[2] and is sometimes also referred to as 'recombinant DNA technology'[3] or 'genetic engineering'.[4] It allows selected individual genes to be transferred from one organism to another, and also allows such transfer amongst organisms belonging to non-related species. Foods produced from or using GMOs are often referred to as Genetically Modified (GM) foods.[5] The resulting plant is said to be 'genetically modified' although in reality, all crops have been genetically modified from their original wild state by domestication, selection and controlled breeding over long periods of time.[6]

Doing something that does not occur in nature, unsurprisingly, makes some people nervous. Globally, a record 17.3 million farmers grew GM crops in the year 2012, up from 16.7 million farmers in 2011. Worldwide, 170.3 million hectares were planted with GM crops in 28 countries, which is a hundred-fold increase since the time they were first introduced in the year 1996. The area under GM crop cultivation today is about the same size as the territories of Spain, Germany, France and the UK combined.[7]

Traditionally, farmers and plant breeders have attempted to produce plants with desired traits by undertaking the process of exchange of genes between two species of plants. This is done by transferring the male (pollen) of one plant to the female organ of another. This cross-breeding, however, is limited to exchanges between the same or very

155

closely related species. It may also take an extremely long period of time to achieve the desired results and frequently, characteristics of interest do not exist in any related species.[8] GM technology has enabled plant breeders to bring together, in one plant, useful genes from a wide range of living sources, not just from within the crop species or from closely related plants, but from other organisms as well. This powerful tool allows plant breeders to do, in a faster and more effective manner, something that they have been doing for years, namely generation of superior plant varieties.[9] The additional advantage of this technology is that it significantly expands these possibilities as it enables scientists and breeders to go beyond the limits imposed by conventional plant breeding.[10]

Initially, GM seed developers wanted their products to be accepted by producers and thus concentrated on bringing forth innovations that brought about a direct benefit to farmers in particular, and the overall food industry in general.[11] One of the objectives for developing plants based on GM organisms is to improve crop protection. The GM crops currently in the market, especially in the developed world, mainly aim at achieving an increased level of crop protection through the introduction of resistance against plant diseases caused by insects or viruses or through increased tolerance towards herbicides.[12, 13]

Resistance against insects is achieved by incorporating into the food plant the gene for toxin production from the bacterium *Bacillus thuringiensis* (or *Bt*).[14] This toxin is currently used as a conventional insecticide in agriculture and is safe for human consumption. GM crops that inherently produce this toxin have been shown to require lower quantities of insecticides in certain specific situations, for instance, in places where pest pressure is high.[15] Virus resistance is achieved through the introduction of a gene from certain viruses which cause disease in plants. This makes the plants less susceptible to diseases caused by such viruses, consequently resulting in higher crop yields.[16] Herbicide tolerance is achieved through the introduction of a gene from a bacterium conveying resistance to some herbicides. In situations where the weed population is high, the use of such crops has resulted in a reduction in the quantity of the herbicides used.[17]

For many in the First World, GM crops have become the latest incarnation of evil biotechnology, which sacrifices humans and the environment at the altar of revenues and shareholder value. On one side of this heated debate are people who firmly believe that GM crops pose a threat to human health and biodiversity, and on the other are scientists and breeders who are convinced that genetic engineering represents a technology which has an enormous potential to increase food production in an environmentally benign manner.[18] The opposition to GM crops is in part due to the fact that a lot of consumers in the First World have not yet seen for themselves any direct advantages of products derived from this new technology – be it in the form of lower prices or improved nutritional quality or 'better' products. Given this apparent lack of benefit, many consumer associations and environmental groups feel that it is unjustified to accept any possible risk to the environment that might arise due to the use of GM crops.[19] Many of these critics do not trust either the industry or the regulatory agencies, and regard them as allies of the chemical industry and biotechnology companies.[20]

Most scientists, on the other hand, consider transgenic crops as safe as, or even safer, than comparable products obtained through traditional breeding. However, some scientific journals have published negative reports about the safety of GM crops,[21] such as the potentially harmful effects of pollen from insect-resistant corn on the larvae of the monarch butterfly.[22] Publications like these have increased public pressure on the regulatory authorities of various countries to either prohibit or delay the use of GM crops. While environmental and consumer activists in the First World fight against the worldwide use of GM crops, hundreds of millions of people in the Third World continue to remain malnourished. Any serious attempts to discuss the future of GM plants and long-term plans regarding their usage must therefore take into account the conditions that exist in the case of poor countries like India, which have, so far, remained largely ignored by opponents of this technology.

I. Growth of population and need for increased food production

Human population is constantly growing and the rate of growth has been faster than what was previously anticipated. In 2015, the UN published its latest world population estimates, according to which the world's population is likely to be around 9.35 billion in the year 2050, which is around 450 million more than what was previously estimated.[23] In order to feed all of these people and prevent famine, upheavals and civil wars, we need to produce more quantities of food and even better quality of food. This will not only benefit the economies of the food-importing countries, but will also reduce their dependence on the industrialised world.

The Food and Agriculture Organisation (FAO) estimates that we need to grow 70 percent more food by the year 2050.[24] Either we do this on the same land we have today, or we chop down forests so as to be able to create farms and pastures to meet that demand, something that no one wants to do. Renowned environmental scientist Dr Jon Foley[25] quite rightly points out that it is meat consumption and not population, that is really driving the global food demand.[26] Unfortunately, consumption of meat has roughly quadrupled in the last 50 years or so,[27] primarily driven by increasing wealth in the developing world, and shows no signs of abating in the immediate future. Practical methods to reduce meat consumption worldwide are very welcome, but until then, we have to keep finding newer ways to keep boosting food production. Another way to feed the world is to close the 'yield gap' between farms in the rich and poor worlds. The per-acre food production in the US is today twice the world average.[28] This is largely because farmers in that country can typically afford better farm equipment, fuel, fertiliser and pesticides as compared to their counterparts in the developing world, who simply cannot afford many of these inputs.[29] Some of this gap, undoubtedly, will be closed as poverty drops around the world; but it is unrealistic to assume that all of it will.[30]

Farmers, in general, are neither opponents nor proponents of GM crops. They adopt whatever technologies promise them lower production costs or increased productivity. Indeed, GM crops have been successfully used not only in the US but also in other countries like Argentina, China, Brazil and Mexico, showing that farmers in developing countries can also benefit from their cultivation.[31] Another important issue is the question of how to ensure that these new technologies would actually help people in developing countries. Some people argue that GM technology is controlled by large multinational companies and thus can never really be used for the benefit of small farmers.[32] It is for these reasons that proponents of GM technology argue that instead of condemning and blocking GM crop technology, government-funded institutions and Non-Governmental Organisations (NGOs) should find ways to ensure that this knowledge is transferred to developing countries.[33]

Since Third World countries lack access to GM technology, the only alternative that appears plausible for them in order to achieve their target of increased foodgrain production is the increased usage of chemical fertilisers, insecticides and herbicides – something that is certainly inimical to the environment. Furthermore, most farmers in poor countries simply cannot afford these chemicals that have been developed for large mechanised farms in the First World.[34] GM technology has already demonstrated that it has the potential to increase food production while decreasing production costs.[35] Scientific consensus appears to be that GMOs are as safe to eat as any other food – they reduce soil-damaging tillage, carbon emissions, insecticide use and usage of toxic herbicides in favour of far milder ones. Despite these obvious advantages, GM crops tend to have certain limitations and some of their benefits are threatened by the increased incidence of pesticide resistance.[36]

The future is easy to discount, so let us come back to the present, and in particular, the present reality for the six billion people who live outside of the rich world. Until recently, the majority of the acres of GM farmland in the world have been in the rich nations. This means that when we look at how GM crops perform, we tend to focus on how they do in countries where farmers have better access to farm inputs.[37] It must be said that in countries like the US, Brazil and Argentina, we see a real but modest benefit; however, the situation is markedly different in many parts of the developing world.

II. Criticism of GM crops

Concerns about GM crops can be put into three main categories *viz.* health, environment, and the manner in which GM seeds are brought to the market. The last category, namely the commercialisation of GM technology, is not so much a criticism of the technology, but more a concern as to who benefits from it'll three issues are complex and deserve to be deliberated by various stakeholders. Concerns about the possible impact of GM foods on the health of the consumers often point to the risk of GM products creating new allergic responses in people, and the possible toxic or reduced nutritional benefits of GM foods. Allergic responses may occur due to inserting material from a food – such as peanuts, wheat or egg – that already causes allergic reactions in some people. Toxins are already present in many widely consumed foods, but GM plant breeding to develop a particular characteristic (such as pest resistance) could actually increase the toxicity of a plant or reduce its nutritional value.[38]

The foremost criticism against the GM system all over the world is that seed companies make it illegal to save seed. In pursuance of this goal, seed companies have even sought to sue farmers in North America whose crop was cross-pollinated by GM pollen! About 80 percent of small-scale farmers in African and Asian countries save their seed. Questions of how these farmers are supposed to protect from GM contamination the varieties of seeds developed by them through generations have been left unanswered by proponents of this regime. GM cotton has been widely blamed for an epidemic of suicides among

Indian farmers, who have been plunged into debt due to high seed and pesticide costs and failing crops.[39] These crops are claimed to be drought-tolerant, but this claim is yet to be effectively tested. Instead of waiting for expensive GM solutions that may never arrive, critics argue that governments are better off working with communities who were able to produce surplus food in times of drought by returning to their traditional varieties. The only way to ensure real food security is to revive seed diversity and healthy soil ecology.

As far as the commercialisation of GM technology is concerned, GM companies are often criticised for having enormous market power that allows them to keep the price of GM seeds unrealistically high.[40] Further, as farmers are legally prevented from holding back seeds to plant in the following year, they have to perforce buy new seeds each season.[41] Beyond the issue of the market power of particular companies, there is the concern that the ownership of agricultural genetic resources has shifted from the public domain to the hands of a select few; this is something that policy makers need to urgently redress.[42]

GM crops have been controversially linked to cancer and a number of other diseases. However, a study suggesting that GM maize might be the cause of cancer in rats was later discredited by the European Food Safety Authority (EFSA).[43] In its summary, the report said that no differences were found to implicate a higher risk to human health from these foods than from their non-GM counterparts.[44] Three US National Academies of the Sciences, Engineering and Medicine studied the impact of using GM maize, soybean and cotton instead of convention crops[45] and cautioned in their 388-page report that 'any new kind of food – whether GM or non GM – may have some subtle favourable or adverse health effects that are not detected even with careful scrutiny and that health effects can develop over time'.[46] The report acknowledged that there was a pressing need to test all new GM products before they were consumed by people. Cancer rates had changed over time in the US and Canada, but they were 'generally similar' to those in the UK and western Europe 'where diets contain much lower amounts' of GM food.[47] There was absolutely no suggestion that eating GM food could increase the risk of diabetes, kidney disease, obesity, autism or any other food allergies.[48]

The argument that GM plants are harmful to biodiversity needs careful consideration. The term 'biodiversity' is actually a contraction of 'biological diversity' and refers to the variety and variability of life on our planet. One of the most widely used definitions defines it in terms of the variability within species, between species and between ecosystems.[49] This can refer to genetic variation, ecosystem variation, or species variation (number of species) within an area, biome, or planet. Regions home to many different species are high in biodiversity. Ecosystems with high biodiversity are characterised by complex interactions between different species, which can help the ecosystem remain intact and healthy in the face of disturbance and environmental change.[50] It is for these reasons that studying biodiversity is a good parameter to assess the overall health of an ecosystem.

Producing crops, by its very nature, means getting rid of wild plants on the farm. There are many management strategies used by farmers to reduce the growth of weedy wild plants. If other plants are permitted to grow alongside crops, the crops will have to compete for valuable resources such as light, water and nutrients. This would have huge negative effects on crop productivity. However, research has shown that presence of certain wild plants on the field can be helpful for sustainable agricultural practices.[51] Well thought-out agricultural practices, thus, leave a place for wild plants and still provide crops with good growing conditions.

There are also widespread fears that GM organisms may have an adverse impact upon the environment. For instance, herbicide-resistant GM crops could result in more herbicide-resistant weeds and pesticide-resistant insects. Furthermore, it is feared that GM crops and products will inevitably spread to areas where they are not wanted.[52] Till date, there does not seem to be any substantial evidence of GM plants exacerbating weed growth as cross-pollination rates are very low. Similarly, there is no evidence of a large-scale increase in pesticide resistance in insects due to GM, although Monsanto recently reported that a common cotton pest in India appeared to have become more resistant to GM pest-resistant cotton.[53] Critics argue that non-GM agriculture still has considerable potential to reduce food insecurity in the world and increase overall agricultural productivity.[54] While there appears to have been no large-scale adverse effect of GM crops on the environment, this should not be taken as given because of the fact that the effects of this technology may manifest themselves in the years to come, and hence the environmental impact of GM technology is something that needs to be continuously monitored.[55]

The arguments against GM crops discussed in this section also find favour in other corners of the world. African farmers and civil society have repeatedly rejected GM crops and asked their governments to ban them. Critics of the system are quick to point out that farmers in Africa already have effective approaches to seed and agriculture, which are far more environmentally suitable and farmer-friendly than their GM counterparts. African farmers have a vast amount of ecological knowledge and are more capable of producing newer varieties of seeds that are more suited to their local environment.[56] Having many different types of seeds which are bred for their flavours and better nutrition, and which have evolved over a period of time with local pests and diseases, is a far better strategy of resilience than developing a single crop that is bound to fail in the face of climate change.[57] Critics aver that by pushing just a few varieties of seed that need fertilisers and pesticides, agri-business has eroded indigenous crop diversity. It is not a solution to the challenges posed by hunger and malnutrition, but a cause of these problems. If northern governments genuinely wish to help African agriculture, they should support the revival of seed-saving practices and ensure that there is diversity in farmers' hands.[58]

III. The Indian perspective

India allows only one GM crop *viz.* GM cotton with the *Bt* trait. This variety of cotton is naturally resistant to insects and reduces the need to spray insecticides. In the US, there is today a broad consensus that *Bt* corn has reduced insecticide spraying but there is less evidence regarding increased productivity of such corn crops. In India, where quite a large number of farmers are not in a position to readily afford pesticides and farm equipment is often lacking, the situation is dramatically different. For the decade between 1991 and 2001, cotton yields in India were flat, at around 300 kilograms per hectare.[59] In 2002, *Bt* cotton was introduced in the country for the very first time. Farmers were quick to adopt it, and yields of cotton soared by two-thirds in just a few years to more than 500 kilograms per hectare.[60] Between 1975 and 2009, researchers found that *Bt* cotton produced around 19 percent of India's yield growth, despite the fact that it had been in the market for only eight of these 24 years.[61]

There are seven million cotton farmers in India. Peer-reviewed studies have revealed that because *Bt* cotton increases the farm produce of these farmers, it raises their farm profits by as much as 50 percent,[62] helps lift them out of poverty and reduces their risk of falling into hunger.[63] By reducing the amount of insecticides used (which, in India, are mostly sprayed by hand), *Bt* cotton has also significantly reduced cases of insecticide poisoning to farm workers to the tune of some 2.4 million cases per year.[64] In China, similar impacts of *Bt* cotton have been seen, with multiple studies showing that *Bt* cotton has increased yields, boosted the incomes of four million small and marginal farmers and also reduced the incidence of pesticide poisoning among them.[65]

Studies showed that *Bt* brinjal was safe, could cut pesticide use by half and could nearly double yields by reducing crop loss due to insects.[66] While regulators in India approved the planting and sale of this crop, activists cried out against it and prompted the government to declare an indefinite moratorium on it. Similar things have taken place in other jurisdictions too. The very same *Bt* brinjal crop was supported by regulators in the Philippines,[67] who were convinced by the available data regarding enhanced productivity, but the move was blocked by the court on grounds that reflected not specific concerns, but general, metaphorical and emotional arguments.[68] The same arguments that kept *Bt* eggplant (*Bt* brinjal) out of the Philippines have also been used, often by Western groups, to keep GM crops out of virtually all of Africa, as has been documented by Robert Paarlberg[69] in his powerful work *Starved For Science*.[70]

On 4 December 2014, the then Union Minister for Environment, Forest and Climate Change, Prakash Javadekar, made a statement in the *Rajya Sabha*[71] which was widely seen as a major shot in the arm for the GM movement in India: 'There is no scientific evidence to prove that GM crops harm the soil, human health and the environment.' This statement was made apropos the controversial decision of the Genetic Engineering Appraisal Committee (GEAC)[72] granting approval for open field trials of 12 GM crops.[73] Until around six years ago, India was a hostile place for researchers testing GM crops. The Union Government barred the commercial planting of *Bt* brinjal after protests from anti-GM activists. Further, it gave state governments the power to veto field trials of transgenic crops. The result of these measures was an effective moratorium on such trials. 'We felt as if we had come up against a brick wall, and might as well chuck it in and do something else', says molecular biologist Bharat Char.[74] Under the present government, India has quietly changed its course with regards to field testing of GM crops. During the period 2014–15, as many as eight Indian states have approved field trials of GM crops and allowed tests that include transgenic rice, cotton, maize (corn), mustard, brinjal and chickpea.[75] The relaxed attitude to GM crop trials is not only reviving the enthusiasm of Indian biotech researchers but is also being keenly watched around the world because the Indian situation perfectly epitomises the tensions that surround the use of GM technology in developing nations. According to Dominic Glover,[76] 'India's attitude towards transgenic crops has a symbolic importance beyond its borders'. On the one hand, India must improve its agricultural productivity to feed its rapidly growing population, and on the other hand is the problem of rising population. The country should thus embrace

GM efforts to develop higher-yield crops that are resistant to pests or grow well in droughts or harsh environments such as salty soil', says eminent biochemist Prof. Govindarajan Padmanabhan.[77] We also have more than 100 million farmers who are concerned that if GM crops were to become prevalent, their livelihoods and the nation's food supply would increasingly rely on expensive, rapidly changing and proprietary seed technologies owned by large corporations. These tensions came to the fore for the first time in the year 2010, when farmers and anti-GM groups organised huge public protests that led to the banning of *Bt* brinjal in the country. They regularly flare up in criticisms of India's 2002 adoption of GM cotton, which contains genes to ward off certain insects. Despite *Bt* cotton being the country's only permitted commercial GM crop, the fact that it is grown in such large quantities makes India the world's fourth-biggest GM crop producer behind the United States, Brazil and Argentina.[78]

The new lenience with regards to field trials has not reached all of India though – nearly 20 States and Union Territories are still exercising their vetoes as of the date of writing this chapter. In meetings between March and July 2016, the GEAC granted permission to over 80 field-trial applications, but state government vetoes meant that many of these trials never got to see the light of day,[79] and activists like Dr Vandana Shiva[80] have gone on record to state that 'We must end Monsanto's[81] colonization and its enslavement of farmers.'

These local bottlenecks have mainly hurt biotech researchers in India's universities and public-sector institutions, feels Padmanaban. 'They unnecessarily retard the progress of domestic technology, whereas multinational firms such as Monsanto can test GM crops elsewhere.' Leading geneticist Prof. Akshay Pradhan[82] says that his team had its technology ready as far back as 2002 but resumed field trials in 2015 after a two-yearlong hiatus. Agricultural scientists want faster progress in these areas.[83]

Like in other parts of the world, activists and NGOs argue that India should be wary of welcoming transgenic crops. They frequently raise concerns that such crops may be unsafe for the environment or human health, and that Indian regulators have conflicts of interest and have not put in place sufficient mechanisms to carefully monitor field trials. Similar criticisms were also raised in 2012 by a technical committee convened by the Supreme Court of India. The court is still considering the issue of placing a complete moratorium on the planting of GM crops in the country (including field trials). This is something that became amply clear during the course of the hearing of a petition that was filed by NGOs and activists over a decade ago.

One issue that critics and scientists agree on is the need for legislation to improve regulations in the area of biotechnology. The creation of a National Biotechnology Regulatory Authority of India (BRAI) was first suggested under the Draft National Biotechnology Authority bill[84] prepared by the Department of Biotechnology in the year 2008.[85] The bill, which aimed to set up the BRAI as an independent authority to regulate organisms and products of modern biotechnology, failed to get Parliamentary approval in the year 2013 and is now being revised, but it could take up to two years or more for the bill to be passed, says Dr Sunkeswari Raghavendra Rao.[86] Certain key features of the bill in its present form are as follows:

- The BRAI will regulate the research, transport, import, containment, environmental release, manufacture and use of biotechnology products in India.
- Regulatory approval by the BRAI will be granted through a multilevel process of assessment, which shall be undertaken by scientific experts.
- The BRAI will certify that the product developed is safe for its intended use. All other laws governing the product will continue to apply; thus, the BRAI will coexist with other legislations like the Patents Act[87] and the Protection of Plant Varieties and Farmers' Rights Act,[88] that currently operate in this area.
- A Biotechnology Regulatory Appellate Tribunal will hear civil cases that involve a substantial question relating to modern biotechnology and hear appeals arising from the decisions and orders of the BRAI.
- Penalties are specified in the bill for providing false information to the BRAI, conducting unapproved field trials, obstructing or impersonating an officer of the BRAI and for contravening any other provisions contained in the bill.[89]

India ratified the Cartagena Protocol,[90] which protects biodiversity from potential risks posed by GMOs that are the products of modern biotechnology. The Protocol requires setting up of a regulatory body. Currently, the GEAC is responsible for approval of GM products in India. If the bill is passed, this responsibility will be taken over by the Environment Appraisal Panel, which is a sub-division of the BRAI.

The bill has faced opposition from farmer groups and anti-GMO activists alike. Dr Suman Sahai[91] feels that the bill is flawed. According to her, the bill proposes to set up new institutions without clearly defining their powers and responsibilities. She has also stated that the bill was introduced without consulting the people who would be affected the most by it.[92] Distinguished molecular biologist and administrator Dr P. M. Bhargava[93] has been scathing in his commentary on the bill and has called it unconstitutional, since agricultural policy is the domain of state governments. He avers that the bill proposes formation of several subdivisions which will be staffed with bureaucrats who possess neither scientific knowledge nor domain expertise. He has alleged conflict of interest on the part of the Department of Biotechnology, which will be involved in the process of selection of members of the BRAI, since the Department also plays the role of a promoter of genetic technology in India. He contends that the broadly defined term 'confidential commercial information' has been kept outside the purview of the Right to Information Act.[94] Perhaps the most stringent criticism of the bill by Dr Bhargava is that the bill has used vague wordings which would criminalise sequencing or isolation of DNA and Polymerase Chain Reaction (PCR)[95] techniques, requiring approvals for each usage, thus hindering education and research. He argues that the bill has no provision for mandatory labeling of GM foods and has criticised the grant of powers to the BRAI to punish parties making false or misleading statements regarding GM crops, calling it unprecedented.[96]

In September 2010, Jairam Ramesh[97] pointed out that the BRAI was only meant to deal with the issues of safety and efficacy of GM products. The issue of commercialisation had, in fact, been left unaddressed, and therefore decisions regarding commercialisation could fall under the purview of either the Ministry of Environment and Forests, the Ministry of Health, the Ministry of Agriculture, or the Ministry of Science and Technology. On the other hand, the Association of Biotechnology Led Enterprises (ABLE)[98] has supported the bill.

It needs to be kept in mind that the Indian government has treaded more cautiously on the issue of commercial cultivation of GM crops as compared to the issue of field trials. Farmers in neighbouring Bangladesh began cultivating GM brinjal in 2014 and such transgenic brinjal, or its seeds, will eventually make its way into India anyway, thanks to the porous Indo-Bangladesh border. The government has appeared reluctant to engage in transparent debates about the pros and cons of pushing forward the use of GM technology in India and has also fallen shy of sharing details regarding field trials of GM crops that are currently being allowed in various parts of the country.

Activist Aruna Rodrigues[99] argues that there is a clear conflict of interest in the role played by various governmental agencies that are currently dealing with the issue of GM crops in India. She says that the Union Ministry of Agriculture is a vendor of GM crops and has no mandate for regulating GMOs. The same Ministry had lobbied and fought to include an additional member on the Technical Experts Committee (TEC)on GM Crops[100] after its interim report had been submitted. Further, the Indian Council of Agriculture Research (ICAR) promotes Public–Private Partnerships (PPPs) with the biotechnology industry. It does this with the active backing of the Ministry of Science and Technology. Critics argue that multinational seed companies like Monsanto have been given access to our agri-research public institutions through the PPP route and this has enabled these organisations to be in a position to seriously influence agri-policy in the country.[101]

Rodgrigues' is, however, not the sole voice in India against the use of GM crops. Opposition has come from several quarters – both on the left and the right of the political spectrum. The *Swadeshi Jagaran Manch* (SJM), which is the economic wing of the *Rashtriya Swayamsevak Sangh* (RSS), has termed the decision of the Union Government to allow field trials of GM crops as a 'betrayal of peoples' trust'. The Coalition for a GM-Free India[102] says:

> When most countries around the world are not adopting this risky technology which has a large number of attendant risks to health, environment, and livelihoods, and when several credible official bodies in India have asked for a stopping of field trials, it is extremely irresponsible that our apex biotechnology regulator has thrown such caution to the winds to approve open air field trials.[103]

The TEC appointed by the Supreme Court had, in the year 2013, recommended an indefinite moratorium on field trials of GM crops till the time the government fixed regulatory and safety aspects in these areas. It also recommended a ban on the introduction of GM varieties in regions of their origin. The final report that was submitted to the court contained a dissenting note from Dr R. S. Paroda.[104] The committee recommended imposition of four conditions for 'meaningfully' considering allowing trials of GM crops for commercial release. The conditions that were suggested by the committee were as follows:

- Setting up of a Secretariat of experts to fix gaps in bio-safety protocol.
- Housing the new biotechnology regulatory in either the Ministry of Environment and Forest (as it then was) or the Ministry of

Health and Family Welfare and not under the Ministry of Science and Technology, as was originally envisaged.

- Identification of specific sites for conducting field tests of GM crops.
- Mandatory civil society participation as part of risk management/ risk mitigation strategies.

Once these conditions were met, the TEC had suggested that the trials should be allowed only on land owned by GM crop proponents and not on leased land, which is the present position. The TEC did not find any 'compelling' reasons for allowing the commercial release of *Bt* for food crops such as rice and brinjal in India, and gave global examples of countries where transgenics such as soybean, corn and canola that are in use are meant primarily for the purpose of oil extraction or cattle feed, and that too after necessary processing. The report also said that allowing GM crops in the area of origin would impact India's food export, especially in the case of rice where exports are to the tune of ₹12,000 crore (₹120 billion) every year. This was based on the submission of the Department of Agriculture and Cooperation (DAC) that India does not have a system to ensure proper labeling of GM and non-GM foods. The committee also disallowed herbicide-tolerant crops on the grounds that they would exert a highly adverse impact on sustainable agriculture, rural livelihood and the local ecology and environment in the long run.[105]

IV. Conclusion

Elimination of hunger is the first and foremost priority in the process of development. Food security at various levels is the *sine qua non* for ensuring an equitable society. Poverty and hunger pose the greatest challenges to mankind in the twenty-first century. This century has seen several social, political and technological advancements, but the nature of development seen in this era has left many people hungry and has often been criticised by scholars and policy makers as being lopsided. The agricultural sector has always enjoyed top priority in independent India and farmers have always been supported by the State in times of need. Seeds are the most vital input in agriculture. Timely availability of good quality seeds of the right varieties is essential to ensure the health of our agriculture-based economy and also to fulfill the food requirements of our burgeoning population.

The Green Revolution of the 1960s took place due to the introduction of several high-quality seeds, which led to a phenomenal increase in farm productivity across Punjab, Haryana and western Uttar Pradesh. However, there has been a policy shift in the post-1991 era which has also impacted the seed sector – seed development and distribution is today no longer a small-scale enterprise, but has become highly institutionalised. Today, the private sector is an important player in this vital area, and the advent of biotechnology has made it possible for the private sector to claim monopoly over the production and sale of certain types of seeds. There are apprehensions in several quarters that seeds may soon become a patentable commodity, and hence the farmer would lose his freedom and would no longer be in a position to decide what to grow, how much to grow, whom to sell his produce to, etc. Activists and NGOs are apprehensive regarding the extensive use of GM seeds in Indian agriculture and their consequent impact on human and environmental health. The experiment with GM crops in parts of Africa also does not inspire much confidence amongst people working in these sectors. The introduction of Genetic Use Restriction Technology (GURT),[106] popularly known as 'suicide seeds' or 'terminator seeds', has caused further complications in the seed sector.

The establishment of the new economic order under the auspices of the World Trade Organisation (WTO) has led to a drastic realignment in national priorities, cultures and economic systems. Sectors like agriculture are now within the purview of trade regulation. The Agreement on Agriculture (AoA) and the Trade Related Aspects of Intellectual Property Rights (TRIPS) Agreement have added to the complexities as GM seeds have become patentable in some jurisdictions.[107]

Proponents of GM technology argue that most of the perceived ills of GM crops are either illusory or far smaller than believed. Data that are available in this regard seem to suggest that the benefits of this technology, while modest in the rich world today, might be quite substantial in the foreseeable future. Such benefits are already much larger in the parts of the world where the battle over GM approval is raging most actively.

GM crops are thus neither poison nor panacea. They can act as a toolkit, albeit a varied one, with real benefits to the environment and also to millions of people. This author believes that the response to any adverse economic aspects of GM technology should not be to ban GM products altogether, but rather to use trade practices and legislation to reduce the market power of GM seed companies and increase the power of farmers and consumers, both in developing and developed countries. The bottom line is not a choice between conventional crops and GM crops, but between sustainable and unsustainable farming practices. The experience with GM techniques till date indicates that the challenge of feeding the world cannot be solved by any one single approach or technology. In an increasingly hungry world, GM technology should not be ruled out. The government should, therefore, be encouraging a diversity of approaches in agriculture, entailing the use of both GM technology and the more traditional processes that have stood the test of time. This appears to be the need of the hour, especially at a time when nature has taught us that the best of technologies can never really match up to the wisdom of an innate evolutionary process.

Notes

1 Term adopted by international convention to refer to biotechnological techniques for the manipulation of genetic material and the fusion of cells beyond normal breeding barriers. The most obvious example is genetic engineering to create genetically modified/engineered organisms (GMOs/GEOs) through transgenic technology involving the insertion or deletion of genes, www.grida.no/graphicslib/detail/biotechnology–and–modern-biotechnology–defined_b9d8 (accessed on 4 October 2016).

2 Term given to a range of activities concerned with understanding gene expression, taking advantage of natural genetic variation, modifying genes and transferring genes to new hosts, www.csiro.au/en/Research/Farming–food/Innovation–and–technology–for–the–future/Gene–technology/Overview (accessed on 5 October 2016).

3 Recombinant DNA (rDNA) molecules are DNA molecules formed by laboratory methods of genetic recombination (such as molecular cloning) so as to bring together genetic material from multiple sources, creating sequences that would not otherwise be found in the genome.

P. Eggleston, 'The Control of Insect-Borne Disease through Recombinant DNA Technology', The 1990 Balfour Lecture given on 4 April 1990 at the 213th Meeting of the Genetical Society, University of Birmingham, www.researchgate.net/publication/247650827_The_control_of_insect–borne_disease_through_recombinant_DNA_technology (accessed on 9 October 2016).

4 Term given to direct manipulation of an organism's genome using biotechnology. It is a set of technologies used to change the genetic makeup of cells, including the transfer of genes within and across species' boundaries to produce improved or novel organisms.

M. Vert et al., 'Terminology for Biorelated Polymers and Applications (IUPAC Recommendations 2012)', Pure and Applied Chemistry, 2012, 84(2): 377–410.

5 Genetically Engineered Crops in the United State: Report (eds.), J. Fernandez-Cornejo et al., Washington, DC: Economic Research Service of USDA, February 2014.

6 B. E. Tabashnik et al., 'Insect Resistance to Bt Crops: Lessons from the First Billion Acres', Nature Biotechnology, 2013, 31(6): 510–521.

7 C. James, 'Global Status of Commercialized Biotech/GM Crops: 2012', ISAAA Brief No. 44, ISAAA, 2012, www.isaaa.org/resources/

publications/briefs/44/download/isaaa-brief-44-2012.pdf (accessed on 28 September 2016).

8 N. du Rand, 'Isolation of Entomopathogenic Gram Positive Spore Forming Bacteria Effective Against Coleoptera', Ph.D. Thesis, University of KwaZulu-Natal Pietermaritzburg, South Africa, July 2009.

9 A. Bravo, S. S. Gill and M. Soberon, 'Mode of Action of Bacillus Thuringiensis Cry and Cyt Toxins and Their Potential for Insect Control', *Toxicon*, 15 March 2007, 49(4): 423–435.

10 G. H. Toenniessen, J. O'Toole and J. De Vries, 'Advances in Plant Biotechnology and Its Option in Developing Countries', *Current Opinion*, 2003, 6: 191–198.

11 C. James, 'Global Status of Commercialized Biotech/GM Crops'.

12 D. M. Suter, M. Dubois-Dauphin and K.-H. Krause, 'Genetic Engineering of Embryonic Stem Cells', *Swiss Medical Weekly*, 2006, 136(27–28): 413–415.

13 S. P. Moose and R. H. Mumm, 'Molecular Plant Breeding as the Foundation for 21st Century Crop Improvement', *Plant Physiology – Journal of the American Society of Plant Biologists*, July 2008, 147(3): 969–977, www.plantphysiol.org/content/147/3/969.short (accessed on 1 October 2016).

14 Gram-positive, soil-dwelling bacterium which is commonly used as a biological pesticide. It is found to occur naturally in the gut of caterpillars of various types of moths and butterflies. It can also be found on leaf surfaces, aquatic environments, animal faeces, insect-rich environments and inside flour mills and grain-storage facilities. During the process of sporulation, certain *Bt* strains produce crystal proteins (proteinaceous inclusions), called δ-endotoxins, that have insecticidal action. This has led to their use as insecticides, and more recently in their use as genetically modified crops using *Bt* genes, such as *Bt* corn, *Bt* brinjal, etc.

15 P. A. Kumar, V. S. Malik and R. P. Sharma, 'Insecticidal Proteins of Bacillus Thuringiensis', *Advances in Applied Microbiology*, 1996, 42: 1–43.

16 M. T. Madigan and J. M. Martinko (eds.), *Brock Biology of Microorganisms*, Upper Saddle River, NJ: Prentice Hall, 2005, 11th ed.

17 J. Y. Roh et al., 'Bacillus Thuringiensis as a Specific, Safe, and Effective Tool for Insect Pest Control', *Journal of Microbiology and Biotechnology*, 2007, 17(4): 547–559.

18 L. Herrera–Estrella and A. Alvarez-Morales, 'Genetically Modified Crops: Hope for Developing Countries?' *EMBO Reports*, April 2001, 2(4): 256–258, www.ncbi.nlm.nih.gov/pmc/articles/PMC34204/ (accessed on 28 September 2016).

19 F. Duchin, 'Sustainable Consumption of Food: A Framework for Analyzing Scenarios About Changes in Diets', *Journal of Industrial Ecology*, 2005, 9: 99–114.

20 P. P. Jauhar and G. S. Khush, 'Importance of Biotechnology in Global Food Security', in R. Lal et al. (eds.), *Food Security and Environmental Quality in the Developing World*, Boca Raton: Lewis Publishers, 2003, pp. 107–128.

21 B. Lomborg, *The Skeptical Environmentalist*, Cambridge: Cambridge University Press, 2001.

22 J. E. Losey, L. S. Rayor and M. E. Carter, 'Transgenic Pollen Harms Monarch Larvae', *Nature*, 1999: 214–221.

23 The 2015 Revision of World Population Prospects is the 24th round of official United Nations population estimates and projections that have been prepared by the Population Division of the Department of Economic and Social Affairs of the United Nations Secretariat, http://esa.un.org/unpd/wpp/ (accessed on 7 October 2016.)

24 *2050: A Third More Mouths to Feed: Report*, Rome: Food and Agriculture Organization of the United Nations, 23 September 2009.

25 Currently serves as the Executive Director of the California Academy of Sciences, where he is also the William R. and Gretchen B. Kimball Chair. In this role, he leads the greenest museum on the planet and one of the most future-focused scientific institutions in the world. His work focuses mainly on the sustainability of our planet and the ecosystems and natural resources that we depend upon. He is a trusted advisor to governments, environmental groups, foundations, NGOs and business leaders around the world. He has published over 130 scientific articles, including many highly cited works in *Science, Nature* and the *Proceedings of the National Academy of Sciences*. In 2014, Thomson Reuters named him a Highly Cited Researcher in ecology and environmental science, placing him among the top 1 percent of the most cited global scientists.

26 J. Foley, 'Changing the Global Food Narrative', 12 November 2013, http://ensia.com/voices/changing-the-global-food-narrative/ (accessed on 1 October 2016).

27 B. Stallard, 'Chickens Have Quadrupled in Size in the Last 50 Years', *Nature World News*, 14 October 2014, www.natureworldnews.com/articles/9587/20141014/chickens-quadrupled-size-last-50-years.htm (accessed on 1 October 2016).

28 *Wheat in the World: Report* (ed.) B. C. Curtis, Rome: FAO Corporate Document Repository, 2012.

29 N. Johnson, 'Soil Proprietor: Do GMOs Promote Dirt Conservation?' *Grist*, 22 October 2013.

30 R. J. Wright, V. C. Baligar and R. Paul Murrmann (eds.), *Plant–Soil Interactions at Low pH: Proceedings of the Second International Symposium on Plant-Soil Interactions at Low pH, 24–29 June 1990*, Beckley: Springer–Science+Business Media, B.V., 1991.

31 R. M. Solomon–Blackburn and H. Barker, 'Breeding Virus Resistant Potatoes (*Solanum tuberosum*): A Review of Traditional and Molecular Approaches', *Heredity*, January 2001, 86(1): 17–35.

32 C. Ford Runge et al., *Ending Hunger in Our Lifetime: Food Security and Globalization*, Baltimore: The John Hopkins University Press, 2003.

33 M. Qaim, 'Transgenic Virus Resistant Potatoes in Mexico: Potential Socioeconomic Implications of North-South Biotechnology Transfer', *ISAAA Briefs No. 7*, ISAAA, 1998.

34 C. Marris, 'Public Views on GMOs: Deconstructing the Myths', *EMBO Report*, 2001, 2(7): 545–548.

35 R. Sederhoff, 'Regulatory Science in Forest Biotechnology', *Tree Genetics and Genomes*, April 2007, 3(2): 71–74.

36 R. Naam, *The Infinite Resource: The Power of Ideas on a Finite Planet*, Lebanon: University Press of New England, 2013.

37 K. Kloor, 'Why Organic Advocates Should Love GMOs', *Collide-a-Scape*, 12 April 2013, http://blogs.discovermagazine.com/collideascape/2013/

04/12/why-organic-advocates-should-love-gmos/ (accessed on 30 September 2016).

38 *Supra* Note 7.

39 B. Ahmed, 'Behind India's "Epidemic" of Farmer Suicides', *ThinkProgress*, 17 April 2015, https://thinkprogress.org/behind-indias-epidemic-of-farmer-suicides fa820ad674f3#.fti9pxctt (accessed on 10 October 2016).

40 I. Ahmed, 'Killer Seeds: The Devastating Impacts of Monsanto's Genetically Modified Seeds in India', *Global Research*, 2012, www.globalresearch.ca/killer-seeds-the-devastating-impacts-of-monsanto-s-genetically-modified-seeds-in-india/(accessed on 10 October 2016).

41 *Ibid.*

42 J. E. Carpenter, 'Peer-Reviewed Surveys Indicate Positive Impact of Commercialized GM Crops', *Nature Biotechnology*, 2010, 28(4): 319–321.

43 Most of EFSA's work is undertaken in response to requests for scientific advice from the European Commission, the European Parliament and EU Member States. The organisation also carries out scientific work on its own initiative.

44 I. Johnston, 'GM Crops Are Neither "a Panacea" Nor "Monsters" But Are "Pretty Much Just Crops" Expert Says: Major Review by Three US National Academies Finds Genetically Modified Crops are Probably Safe to Eat but don't Seem to Increase Yields', *The Independent*, 17 May 2016, www.independent.co.uk/news/science/gm-genetically-modified-genetically-engineered-cancer-autism-health-environment-yields-a7034206.html (accessed on 11 October 2016).

45 Crops grown using fertilisers and pesticides which allow for higher yield, out of season growth, greater resistance, greater longevity and a generally greater mass. The term 'convention(al) crops' is usually used in juxtaposition with the term 'organically grown crops'.

46 www.biotecnika.org/2016/05/neither-a-panacea-nor-monsters-expert-says-gm-crops-are-just-crops/ (accessed on 11 October 2016).

47 Board on Agriculture and Natural Resources (Division on Earth and Life Studies, National Academies of Science, Engineering and Medicine), *Genetically Engineered Crops: Experiences and Prospects*, Washington, DC: The National Academies Press, 2016, pp. 16–17.

48 www.biotecnika.org/2016/05/neither-a-panacea-nor-monsters-expert-says-gm-crops-are-just-crops/ (accessed on 11 October 2016).

49 K. J. Gaston, 'Global Patterns in Biodiversity', *Nature*, 11 May 2000, 405(6783): 220–227.

50 D. P. Tittensor et al., 'Global Patterns and Predictors of Marine Biodiversity Across Taxa', *Nature*, 28 July 2010, 466(7310): 1098–1101.

51 N. Myers et al., 'Biodiversity Hotspots for Conservation Priorities', *Nature*, 24 February 2000, 403(6772): 853–858.

52 D. Umali-Deininger and S. Shapouri, 'Food Security: Is India at Risk?' in Rattan Lal et al. (eds.), *Food Security and Environmental Quality in the Developing World*, Lewis Publishers: Boca Raton, Florida 2002, pp. 31–52.

53 S.-E. Jacobsen, 'Feeding the World: Genetically Modified Crops Versus Agricultural Biodiversity', *Agronomy for Sustainable Development*, October 2013, 33(4): 651–662.

54 P.M.T. Hansen and J. A. Narvhus, 'Storage and Processing of Agricultural Products', in Rattan Lal et al. (eds.), *Food Security and Environmental Quality in the Developing World*, pp. 277–286.

55 D. Alnwick, 'Significance of Micronutrient Deficiencies in Developing and Industrialized Countries', in G. F. Combs et al. (eds.), *Food–Based Approaches to Preventing Micronutrient Malnutrition: An International Research Agenda*, Ithaca: Cornell University Press, 1996.

56 R. Tirado and J. Cotter, 'Ecological Farming: Drought-Resistant Agriculture', *Greenpeace International*, April 2010: 21–22.

57 K. Bomba and D. Glickman, 'The Ethiopian Approach to Food Security', *Stanford Social Innovation Review*, 31 January 2014: 3–5.

58 H. Dempewolf et al., 'Food Security: Crop Species Diversity', *Science*, 2010, 328: 169–170.

59 One hectare is approximately 2.5 acres.

60 G. D. Stone, 'Constructing Facts: *Bt* Cotton Narratives in India', *Economic & Political Weekly*, 22 September 2012, XLVII(38): 62–70.

61 G. P. Gruere and Y. Sun, 'Measuring the Contribution of *Bt* Cotton Adoption to India's Cotton Yields Leap', IFPRI Discussion Paper 2012 presented at the International Food Policy Research Institute, www.ifpri. org/publication/measuring–contribution–bt–cotton–adoption–indias–cotton–yields–leap (accessed on 27 September 2016).

62 J. Kathage and M. Qaim, 'Economic Impacts and Impact Dynamics of *Bt* (*Bacillus thuringiensis*) Cotton in India', *CrossMark*, 2012, 109(29): 11652–11656, www.pnas.org/content/109/29/11652.short (accessed on 5 October 2016).

63 M. Qaim and S. Kouser, 'Genetically Modified Crops and Food Security', *PLOS ONE*, 5 June 2013, http://journals.plos.org/plosone/article?id=10.1371/journal.pone.0064879 (accessed on 9 October 2016).

64 N. Gilbert, '*Bt* Cotton Cuts Pesticide Poisoning', *Nature Newsblog*, 29 July 2011, http://blogs.nature.com/news/2011/07/bt_cotton_cuts_pesticide_poiso.html (accessed on 10 October 2016).

65 C. E. Pray et al., 'Five Years of *Bt* Cotton in China – The Benefits Continue', *The Plant Journal*, August 2002, 31(4): 423–430.

66 S. Morse, A. M. Mannion and C. Evans, 'Location, Location, Location: Presenting Evidence for Genetically Modified Crops', *Applied Geography*, 2012, 34(2): 274–280.

67 The European NGO Network on Genetic Engineering (GENET) has been collecting and distributing information on various topics in the field of genetic engineering in agriculture, food production and health since the year 1999. With this special topic of '*Bt* eggplants in the Philippines', GENET aims at providing an overview about the current debate regarding the development and approval of *Bt* eggplants in that country, based on material that is stored in its archives. This organisation also collects and disseminates data on various other topics that are of interest to its patrons viz. Genetically Engineered Trees, Genetically Engineered Animals and the debate surrounding the cultivation of *Bt* brinjals in India, www. genet-info.org/information-services/bt-eggplants-in-the-philippines. html (accessed on 7 October 2016).

68 D. Ropeik, 'Filipino Ruling on *Bt* Eggplant', *Scientific American*, 3 June 2013, http://blogs.scientificamerican.com/guest-blog/filipino-ruling-on–bt–eggplant/ (accessed on 5 October 2016).

69 Professor at Wellesley College and Associate at the Weatherhead Center for International Affairs at Harvard University. He is the author of several

books and numerous articles. His research focuses on the international agricultural and environmental policy, and regulation of modern technology, including biotechnology.

70 R. Paarlberg, *Starved for Science: How Biotechnology Is Being Kept Out of Africa*, Cambridge: Harvard University Press, 2009.

71 The *Rajya Sabha* or the Council of States is the upper house of the Parliament of India. Members sit for staggered six-year terms, with one-third of the members retiring every two years.

72 Apex body constituted under the aegis of the Ministry of Environment and Forests, Government of India under the 'Rules for Manufacture, Use, Import, Export and Storage of Hazardous Microorganisms/Genetically Engineered Organisms or Cells 1989', under the Environment Protection Act, 1986. The Rules of 1989 also define five competent authoritics, i.e. the Institutional Bio-safety Committees (IBSC), Review Committee of Genetic Manipulation (RCGM), Genetic Engineering Approval Committee (GEAC), State Biotechnology Coordination Committee (SBCC) and District Level Committee (DLC) for handling of various aspects of these rules.

73 K. P. Prabhakaran Nair, 'Genetically Modified Crops and Indian Agriculture', *Vijayvaani.com – The complete opinions forum*, 7 January 2015, www.vijayvaani.com/ArticleDisplay.aspx?aid=3450 (accessed on 29 September 2016).

74 Employed with Maharashtra Hybrid Seeds Company (MAHYCO), a Jalna (Maharashtra) based firm that pioneered GM brinjal in India.

75 S. Kumar, 'India Eases Stance on GM Crop Trials', *Nature – International Weekly Journal of Science*, 12 May 2015, www.nature.com/news/india-eases-stance-on-gm-crop-trials-1.17529 (accessed on 1 October 2016).

76 An agricultural socio-economist at the University of Sussex in Brighton, he specialises in the study of technology and processes of socio-technical change, particularly in small-scale farming in the global South. He has more than 15 years of experience in research, policy analysis and communication on technological change, innovation, knowledge systems, governance and policy processes relating to agriculture, biotechnology and rural development.

77 Former Director of the Indian Institute of Science (IISc), Bangalore, and presently serves as Honorary Professor in the Department of Biochemistry at IISc. Source from Article published in Nature magazine. Available at www.nature.com/news/india-eases-stance-on-gm-crop-trials-1.17529 (accessed on 16 September 2017).

78 'India World's 4th in GM Crop Acreage, Well Ahead of China', *The Indian Express*, 2 February 2015.

79 V. Shiva, 'We Must End Monsanto's Colonization, Its Enslavement of Farmers', *The International Reporter*, 29 July 2015, http://theinternational reporter.org/2015/07/29/vandana–shiva–we–must–end–monsantos–colo-nization–its–enslavement–of–farmers/ (accessed on 5 October 2016).

80 Trained as a Physicist at the University of Panjab (India) and completed her Ph.D. on the topic 'Hidden Variables and Non-locality in Quantum Theory' from the University of Western Ontario, Canada. She later shifted to inter-disciplinary research in science, technology and environmental policy, which she carried out at the Indian Institute of Science and the Indian Institute of Management (IIM), Bangalore.

81 Publicly traded American multinational agrochemical and agricultural bio-technology corporation. It is a leading producer of GM seeds in the world. From an Article published in EcoWatch magazine. Available at www.ecowatch.com/vandana-shiva-we-must-end-monsantos-colonization-its-enslavement-of-fa-1882075931.html (accessed on 16 September 2017).

82 Works on transgenic mustard plants at the University of Delhi's Centre for Genetic Manipulation of Crop Plants.

83 N. E. Borlaug, 'Ending World Hunger: The Promise of Biotechnology and the Threat of Antiscience Zealotry', *Plant Physiol*, 2000, 124: 487–490.

84 www.indiaoppi.com/sites/default/files/PDF%20files/DRAFT%20 NATIONAL%20BIOTECHNOLOGY%20REGULATORY%20BILL, %202008.PDF (accessed on 13 October 2016).

85 Written reply dated 22 May 2012 of the then Union Minister of State for Agriculture and Food Processing Industries, Harish Rawat, in reply to a question in the *Lok Sabha* (House of the People). Website of the Press Information Bureau, Government of India, http://pib.nic.in/newsite/ere lease.aspx?relid=84347 (accessed on 13 October 2016).

86 Adviser to the Department of Biotechnology, Government of India.

87 Act No. 15 of 2005.

88 Act No. 53 of 2001.

89 'PRS Legislative Research, The Biotechnology Regulatory Authority of India Bill', 2013, www.prsindia.org/billtrack/the-biotechnology–regula tory–authority–of–india–bill–2013–2709/ (accessed on 13 October 2016).

90 The Cartagena Protocol on Biosafety to the Convention on Biological Diversity is an international agreement on biosafety as a supplement to the Convention on Biological Diversity, effective since 2003. The Biosafety Protocol seeks to protect biological diversity from the potential risks posed by GMOs resulting from modern biotechnology.

91 *Padma Shri*, recipient and holder of a Ph.D. from the Indian Agricultural Research Institute (IARI), Sahai has successively worked at the University of Alberta, University of Chicago and the University of Heidelberg, Germany, where she obtained her Habilitation (Post Doctorate) in human genetics. According to the *Web of Science*, Sahai has published over 40 articles, mostly on policy issues relating to GMOs, which have been cited over 150 times, giving her an h-index of 7. She serves as the founder-director of the NGO, Gene Campaign.

92 'Activists Oppose Biotechnology Regulatory Authority of India Bill', *Daily News & Analysis*, 9 July 2013, www.dnaindia.com/ahmedabad/ report–activists–oppose–biotechnology–regulatory–authority–of–india–bill–1858894 (accessed on 13 October 2016).

93 Founder-director of the Centre for Cellular and Molecular Biology, Hyderabad (India).

94 Act No. 22 of 2005.

95 Technique used in molecular biology to amplify a single copy or a few copies of a piece of DNA across several orders of magnitude, generating thousands to millions of copies of a particular DNA sequence. It is an easy and cheap tool to amplify a focussed segment of DNA, which proves to be useful in the diagnosis and monitoring of genetic diseases, identification of criminals (in the field of forensics), studying the function of targeted segment, etc.

J. M. S. Bartlett and D. Stirling, 'A Short History of the Polymerase Chain Reaction', *Methods in Molecular Biology*, 2003, 226: 3–6.

96 P. M. Bhargava, 'Unconstitutional, Unethical, Unscientific', *The Hindu*, 5 January 2012, www.thehindu.com/opinion/lead/unconstitutional–uneth ical–unscientific/article2752711.ece (accessed on 13 October 2016).

97 The then Union Minister for Environment and Forest.

98 A not-for-profit pan-India forum that acts as a representative of the biotechnology sector. It was launched in April 2003, after industry leaders felt that there was a need to form an exclusive forum to represent the interests of this sector. It has over 400 members from all across India representing different verticals of the sector like agribiotech, bio-pharma, industrial biotech, bioinformatics, investment banks and venture capital firms, leading research and academic institutes and law firms, and equipment suppliers.

99 Lead petitioner before the Supreme Court in the case for a moratorium on GM foods in India.

100 The body of experts constituted by the Supreme Court of India to aid and advise the Court on the issue of GM crops, vide its order dated 10 May 2012 in Writ Petition (Civil) No. 260 of 2005 titled *Aruna Rodrigues and Ors. v. Union of India*.

101 A. Rodrigues, 'Nip this in the Bud', *The Hindu*, 12 August 2013, www. thehindu.com/opinion/lead/nip–this–in–the–bud/article5012989.ece (accessed on 30 September 2016).

102 An informal network of organizations and individuals from across India, who have come together to campaign to keep India GM-free.

103 '21 New Varieties of Genetically Modified Crops Approved for Field Trials by Narendra Modi Government', *Daily News & Analysis*, 15 July 2014, www.dnaindia.com/india/report-21-new-varities-of-geneti cally-modified-crops-approved-for-field-trials-by-narendra-modi-government-2002487 (accessed on 3 October 2016).

104 Former Director General of the Indian Council of Agricultural Research (ICAR) and Secretary, Department of Agricultural Research and Education, Government of India. He was also the nominee of the Union Ministry of Agriculture on the Technical Experts Committee (TEC) appointed by the Supreme Court.

105 www.hindustantimes.com/India–news/NewDelhi/SC–committee–says–no–to–GM–crops–for–time–being/Article1–1096481.aspx (accessed on 27 September 2016).

106 Methods for restricting the use of genetically modified plants by causing second-generation seeds to be sterile. The technology was developed under a cooperative research and development agreement between the United States' Department of Agriculture and Delta and Pine Land Company in the 1990s.

107 P. P. Rao, Dr and A. Kumar, Dr, 'Genetically Modified Seeds and Environment Risks', in Prof. (Dr) Bimal N. Patel and Dr Ranita Nagar (eds.), *Food Security Law: Interdisciplinary Perspectives*, Lucknow: Eastern Book Company, 2014, 1st ed., pp. 51–79.

Part V

JUDICIAL RESPONSES

8

PUBLIC INTEREST
LITIGATION – AN EFFECTIVE
TOOL FOR BIODIVERSITY
CONSERVATION

An Indian experience

Mohan Parasaran and Sidharth Luthra

In a 1984 lecture, former Chief Justice of India, Justice P. N. Bhagwati, identified Public Interest Litigation or 'PIL' as it is popularly known, in its nascent stages in India. In his words,

> PILs represent a sustained effort on the part of the highest judiciary to provide access to justice for the deprived sections of Indian humanity. With a legal architecture designed for a colonial administration and a jurisprudence structured around a free-market economy, the Indian judiciary could not accomplish much in fulfilling the constitutional aspirations of the vast masses of underprivileged people during the first three decades of freedom. During the last five or six years, however, social activism has opened up a new dimension of the judicial process, and this new dimension is a direct emanation from the basic objectives and values underlying the Indian Constitution.[1]

Now, more than 30 years later, the history of PIL in India is diverse, varied and of immense relevance in society, be it the determination of the rights of citizens, or the duties of the State and various other bodies. In India, judicial activism has been central to socio-legal progress in various areas, and has been of particular relevance in the protection of rights of women and of the environment. In the context of the environment, PIL has a special role as the entire population is interested

181

and affected, albeit that the impact of environmental degradation is differentially experienced across social strata.

According to Justice Bhagwati, judicial activism is central to judicial function,[2] and the view that endorses strict separation of powers is ignorant of not only the judicial process but also of the law-making process. The history of law and legislation in India is intimately linked to the judicial activism undertaken by Judges, particularly through Public Interest Litigation.

While the concept of 'Juristic or Judicial activism' is familiar, Justice Bhagwati identifies the endeavours of Judges in achieving distributive justice as 'social activism', since '[they] have to justify their decision-making within the framework of Constitutional values'[3] and against the backdrop of Fundamental Rights. Justice Bhagwati, therefore, identifies a crucial expectation of Judges in India – that they pass Orders and render decisions that promote, propagate and instigate social justice.[4]

In *S.P. Gupta* v. *Union of India*,[5] the Supreme Court relaxed rules of 'standing' and allowed a third party to invoke writ jurisdiction in the case of violation of the Fundamental Rights of another person or 'determinate class of persons' who, for reason of poverty, helplessness, disability or social or economic disadvantage, were unable to move the Court personally for relief.[6] The said case also introduced the concept of 'epistolary jurisdiction', where Supreme Court Judges may treat letters, or 'epistles' written to them as Writ Petitions under Article 32 of the Constitution of India.

The power of the Court to treat letters written to the Chief Justice of India or any Judge as a petition was explained in *Bandhua Mukti Morcha* v. *Union of India*.[7] Justice Pathak observed that there is a good reason for maintaining the Rule that, except in special circumstances, the document petitioning the Court for relief should be supported by satisfactory verification. This requirement is all the greater, he emphasized, where petitions are received by the Court through the post.[8] However, he cautioned that it is never beyond the bound of possibility that an unverified communication received through the post by the Court may in fact have been employed mala fide, as an instrument of coercion or blackmail or other oblique motive against a person named therein who holds a position of honour and respect in society.[9] In order to be vigilant against the abuse of its process, he ruled that an appropriate verification of the allegations be supplied before issuing notice to the respondent. The requirement was held to be imperative in private law litigation and equally attracted to Public Interest Litigation. Justice Pathak further clarified that though this Court had readily acted upon letters and telegrams in the past, there is need to insist now, on an appropriate verification of the petition or other communication, before acting on it. But then again, he emphasized that there may be exceptional circumstances which may justify a waiver of the Rule.[10]

The view taken by Justice Pathak was modified in *MC Mehta and Anr. v. Union of India and Ors*,[11] wherein it was held that letters addressed to *any* Judge could be treated as a letter petition, and to insist upon an affidavit as a condition for entertaining the letter-petition would frustrate the purpose of epistolary jurisdiction.[12]

At the time, not only did the Supreme Court relax the rules of standing to allow environmental organizations to bring petitions under Article 32 by way of PIL (or Social Action Litigation, for short SAL), it also allowed petitions where respondents were private parties that do not fall within the meaning of 'State' under Article 12 of the Constitution. In *Indian Council for Enviro-Legal Action and Ors. v. Union of India and Ors.*,[13] a Social Action Litigation was filed on behalf of the villagers of Bichhri, whose right to life was invaded and seriously infringed by the respondents by establishing a chemical industry without obtaining the requisite permission and clearances. The Apex Court, in this context, framed the issue:

> if the industry is continued to be run in blatant disregard of law to the detriment of life and liberty of the citizens living in the vicinity, can it be suggested with any modicum of reasonableness that this Court has no power to intervene and protect the fundamental right to life and liberty of the citizens of this country.[14]

Dealing with the above issue, the Court held that this writ petition is not really for issuance of appropriate writ, order or directions against the respondents but is directed against the Union of India, Government of Rajasthan and RPCB to compel them to perform their statutory duties enjoined by the law on the ground that their failure to carry out their statutory duties is seriously undermining the right to life of the residents of Bichhri, guaranteed by Article 21 of the Constitution.[15] The Court further stressed that if it finds that the said authorities have not taken the action required of them by law and that their inaction is jeopardizing the right to life of the citizens of this country or of any section thereof, it is the duty of the Court to intervene. The Court categorically made it clear that if it is found that the respondents are flouting the provisions of law and the directions and orders issued by the lawful authorities, this Court can certainly make appropriate directions to ensure compliance with the law.[16]

The history of PIL (or SAL) in India is important while considering the progress on environmental issues and concerns that has been achieved through directions passed by High Courts and the Supreme Court, such as in *Indian Council for Enviro-Legal Action and Ors* discussed above. More such cases will be discussed and analysed for their contribution later on in this chapter. However, before harping on to the discussion to elaborate on the environmental changes brought about, a consideration of the impact on biodiversity conservation specifically first begs a consideration of how 'biodiversity' may be defined.

It has been asserted that the definitions of biodiversity are 'as diverse as the biological resource'[17] and definitions are employed to include or exclude life at various levels, including but not limited to human, animal, plant and cellular levels. Bruce A. Wilcox defined biological diversity as 'the variety of life forms, the ecological roles they perform and the genetic diversity they contain.'[18]

Noss and Cooperrider adopted a modified version of the definition developed during the Keystone Dialogue[19] and defined biodiversity as

the variety of life and its processes. It includes the variety of living organisms, the genetic differences among them, the communities and ecosystems in which they occur, and the ecological and evolutionary processes that keep them functioning, yet ever changing and adapting.[20]

These authors clarify that biodiversity cannot mean homogenization and that variety does not mean that external or exotic species are introduced into an ecosystem. Rather, they emphasize '*native biodiversity*', which is concerned with the distinctiveness of a habitat of ecosystem.[21] They also state that defining biodiversity is less important than identifying its components but posit a working definition that takes into consideration life at several levels of biological organization, namely, genetic, species, community/ecosystem, and landscape and regional levels.

Noss and Cooperrider exclude cultural or social diversity from their definition, although this was included in the United Nations Convention on Biological Diversity (hereinafter CBD).[22] They do so for fear that including human diversity will dilute the concept of biodiversity and make it unworkable. However, Noss and Cooperrider define cultural diversity in an oversimplified and literal manner: to preserve human cultural diversity in the context of biodiversity conservation need not mean 'maintaining Nazis, slave owners, or those who enjoy using desert tortoises for target practice.'[23] Biodiversity must take into account the diversity of races, cultures, traditions and gender because ecosystems are tailored by their inhabitants, and humans (through traditional/cultural practices) exist both symbiotically and parasitically with their environment. To discount human diversity is to paint an incomplete picture of the components of biodiversity, particularly in the culturally diverse contexts such as in India.

This chapter, however, adopts the broad-based and inclusive definition of 'biological diversity' as posited by the CBD in Article 2, wherein biological diversity is taken to mean the variability among living organisms from all sources including, inter alia, terrestrial, marine and other aquatic ecosystems and the ecological complexes of which they are part; this includes diversity within species, between species and of ecosystems.[24]

India is a party to the CBD and this definition has also been adopted in the Biological Diversity Act, 2002, which came into force on 5 February 2003.

I. Biodiversity in the Indian context

A. Biodiversity and traditional knowledge

The Preamble of the Biological Diversity Act, 2002 (hereinafter called the 'BDA') contains an acknowledgment of India's rich 'traditional and contemporary knowledge' pertaining to and contributing to an extremely diverse and varied ecological environment.

This law identifies the link between traditional knowledge and biological diversity, although it does not embody the obligation placed by the CBD to 'preserve and maintain knowledge, innovations and practices of indigenous and local communities . . . relevant for the conservation and sustainable use of biological diversity and promote their wider application.'[25]

References to 'traditional knowledge' abound in the CBD; first in the Preamble itself, then in Article 8(j) – In Situ Conservation, Article 10 – Sustainable Use of Components of Biological Diversity, Article 17 – Exchange of Information, and Article 18 – Technical and Scientific Cooperation.

The BDA, in Chapter IX, Section 36(5), places an obligation on the Central Government to 'respect and protect the knowledge of local people relating to biological diversity.'

Against the backdrop of International Conventions, the Supreme Court in *Orissa Mining Corporation Limited* v. *Ministry of Environment and Forests and Others*[26] declared in no uncertain terms that

Scheduled Tribes (STs) and other Traditional Forest Dwellers (TFDs) residing in Scheduled Areas 'have a right to maintain their distinctive spiritual relationship with their traditionally owned or otherwise occupied and used lands.'[27] The Supreme Court went on to hold that many of the STs and other TFDs are totally unaware of their rights and experience a lot of difficulties in obtaining effective access to justice because of their distinct culture and limited contact with mainstream society.[28] The Court took cognizance of the fact that many times, they do not have the financial resources to engage in any legal action against development projects undertaken in their abode or the forest in which they stay. However, the Court stressed that they have a vital role to play in environmental management and development because of their knowledge and traditional practices, and therefore the State has a duty to recognize and duly support their identity, culture and interests so that they can effectively participate in achieving Sustainable Development.[29]

The Supreme Court went on further to discuss in detail the Scheduled Tribes and Other Traditional Forest Dwellers (Recognition of Forest Rights) Act, 2006 (hereinafter 'Forest Right Act' or 'FRA'), which is focused and much clearer on the link between people and their habitat, as compared to the BDA.

According to the Supreme Court, the Statement of Objects and Reasons of the FRA 'states that forest dwelling tribal people and forests are inseparable'[30] and therefore the protection of one requires the preservation of the other and vice-versa.

In India, the link between traditional or local knowledge, practices and experiences on the one hand and biological diversity on the other is crucial to bring about the equitable division of resources, sustainable use, and preservation of life forms. The inextricable nature of 'traditional knowledge' and 'biological diversity' makes PIL a particularly poignant vehicle for formulating and enforcing rules and laws pertaining to environmental protection, particularly conservation.

As has been noted in the Introduction, the Supreme Court undertook a 'relaxation of standing' in an exercise of judicial activism, as part of its bounden duty to deliver substantive social justice, leading to the birth of PIL (SAL). Without this relaxation of standing, which allows both suo motu action and through a representative–petitioner, the holders of traditional and local knowledge or those concerned with the rights of such holders would not be able to reach out to the Courts in an attempt to preserve this environment and their culture. The constitutional Courts in India, in addition to adjudicating within the present framework of laws, have been called upon to innovate and protect those people and environments that are increasingly marginalized by the industrialization and modernization of Indian society.

The role of the Court is indisputable and indispensable given the fact that, firstly, communities may be disempowered and unable to claim rights for themselves without the existence of PIL, and secondly, because there are lacunae in treaties such as the CBD and consequent laws like the BDA. These gaps may hamper biodiversity conservation due to a limited view of the role of traditional knowledge and communities and since they promote an international legal regime that would reward traditional knowledge holders for their role in preserving biodiversity and ancient knowledge – that is, for their role in preserving the public domain. But these international legal documents do not expressly recognize the inventiveness of traditional knowledge, or the attendant intellectual property rights claimed by the world's poor as authors and inventors of new knowledge.[31]

B. Biodiversity and the constitutional framework

Early environmental cases were decided by the Supreme Court of India by relaxing the strict rules of standing and hearing PILs, which have resulted in the closure of limestone quarries in the Dehradun region,[32] the installation of safeguards at a chlorine plant in Delhi[33] and the closure of polluting tanneries on the Ganges,[34] all of which are examples of citizens' activism as much as they are examples of a judiciary alive to the concerns of the citizenry.

The State has a clear obligation under Part IV, Article 48A of the Constitution of India to protect and improve the environment and safeguard forests and wildlife. There is a similar fundamental duty of a citizen under Article 51A(g). The Supreme Court has noted these and read them along with Fundamental Rights, particularly Articles 14 and 21, to develop a Constitutional Framework of interconnected rights, obligations and duties as regards conservation and ecological protection.[35]

The right to a safe and healthy environment, as also the principle of inter-generational equity, have been held to be components of the right to life with dignity under Article 21.[36] The Supreme Court has held in *T.N. Godavarman Thirumulpad (87)*[37] that 'Conservation includes preservation, maintenance, sustainable utilisation, restoration, and enhancement of the natural environment'[38] and this accords with the primary aim of the Forest Policy, 1988. It was further held in Para 72 that non-fulfilment of the basic objectives of the Forest Policy would be violative of Articles 14 and 21 of the Constitution.

As regards inter-generational equity, it was observed, by the Supreme Court, in *Rural Litigation and Entitlement Kendra, Dehradun and Ors* v. *State of UP and Ors.*[39] that natural resources have to be tapped for the purposes of social development, but one cannot forget at the same time that tapping of resources has to be done with requisite attention and care so that the ecology and environment may not be affected in any serious way; there may not be any depletion of water resources and long-term planning must be undertaken to keep up the national wealth. The Court highlighted that it has always to be remembered that these are permanent assets of mankind and are not intended to be exhausted in one generation.[40]

While Government Policy, social understanding, and legislative change may have been slow to give the environment its due importance, the Supreme Court has clearly been the harbinger of change in this regard. It has recognized that 'any threat to the ecology can lead to violation of the right of enjoyment of healthy life guaranteed under Article 21, which is required to be protected. The Constitution enjoins upon this Court a duty to protect the environment.'[41]

C. Competing concerns: biodiversity versus development

It was observed by Hassan and Azfar that 'The ecological challenge in South Asia is distinguished by the compounding of environmental problems caused by widespread poverty, illiteracy, and lack of proper governance.'[42] Quoting *Mahesh Chander Mehta* (noted environmental lawyer and activist), the authors assert that development along Western models, which prioritize rapid growth, leads to the depletion of natural resources and serious environmental decay.[43]

Job creation, economic development, the development of infrastructure and industrialization are major concerns in developing nations, including India. This is even more so in recent years, with the Government increasingly requisitioning foreign investment and easing the path for industries with swift environmental and other clearances.

Hassan and Azfar identify a crucial roadblock in achieving environmental goals and implementing directions passed by the Courts: the lack of political will. The authors state that the

> paucity of resources, lack of general awareness, and the larger objective of attaining self-sufficiency, growth, and development have prevented the executive and legislative branches of these governments from giving priority to this area of concern. The South Asian countries in the region have unleashed

particularly ambitious liberalization and industrial policies in recent years to attract foreign investment. Thus, there is almost a 'no holds barred' approach to development in these countries. The overriding conceptual 'buzz words' are growth, development, jobs, and exports. Environmental concerns clearly come in a distant second place. It is against this unhappy background that one can find a ray of hope in the region's evolving judicial activism.[44]

The history of the *Dehradun Quarrying Case*[45] reveals that the Court was aware to environmental concerns and also to the hardship and loss caused to the inhabitants of the Doon Valley due to illegal quarrying and mining for limestone. The Court, relying on various Expert Committee Reports, ordered that quarrying operations in the Doon Valley Region be stopped completely.[46] However, keeping economic and trade conditions in mind, the Supreme Court allowed the ban to be spread over a period. Two committees, the Rehabilitation Committee and a Monitoring Committee were directed to be set up to oversee the reforestation of the area and for overseeing the running of the remaining mines.

In *Lafarge Umiam Mining (P) Ltd.* v. *Union of India*,[47] the Supreme Court, while considering the clearance of a limestone mine in land belonging to tribals near a forest area in the state of Meghalaya, held that the clearances given by the MOEF were not vitiated by non-application of mind or by suppression of material facts by Lafarge. The Court has tried to balance dichotomies by applying the administrative law principles of proportionality by observing that

> the time has come for us to apply the constitutional doctrine of proportionality to the matters concerning environment as a part of the process of judicial review in contradistinction to merit review. It cannot be gainsaid that utilization of the environment and its natural resources has to be in a way that is consistent with principles of sustainable development and intergenerational equity, but balancing of these equities may entail policy choices.

In *G. Sundarrajan* v. *Union of India*,[48] an appeal by way of special leave was filed challenging the judgment of the Madras High Court, which had given the go-ahead to the Kudankulam nuclear power plant. The grounds taken included the assertion that permission for the power plant was granted without first ensuring that critical safety features recommended by the Government Task Force were put in place, and also in violation of an undertaking given by the Atomic Energy Regulatory Board before the High Court in a previous writ petition. Dismissing the Appeals, the Supreme Court held that the economic benefit has to be viewed on a larger canvas which not only augments the economic growth but alleviates poverty and generates more employment. The Court ruled that NPCIL, while setting up the NPP at Kudankulam, had satisfied the environmental principles like Sustainable Development, corporate social responsibility, Precautionary Principle, inter-/intra-generational equity and so on to implement India's National Policy to develop, control and use atomic energy for the welfare of the people and for economic growth of the country.[49] The Court laid down that larger public interest of the community should give way to individual apprehension of the violation of human rights and right to life guaranteed under Article 21.[50]

The Indian Supreme Court and High Courts have recognized that it may be challenging to develop policies and practices that equally emphasize biodiversity conservation and developmental strategy. However, it has been held in no uncertain terms that the two 'must' be made to balance each other and that development cannot be at the cost of the environment.[51] In this regard, the Supreme Court has elucidated and adopted several principles of International Environmental Law and Policy, such as 'Sustainable Development', 'Precautionary Principle' and 'Polluter Pays Principle'.[52]

In a case concerning the relocation and protection of the Asiatic lion to Kuno in Madhya Pradesh, the Supreme Court considered the provisions of the Wildlife (Protection) Act, 1972, its relationship to biodiversity conservation generally, and the implementation of the CBD and BDA.[53] The Court observed that the 'Protection of wild animals and birds' falls under Schedule VII List III Entry 17-B, that is, in the Concurrent List, and Parliament has passed several laws for the protection and preservation of different species and the environment.[54]

The Supreme Court in *Centre for Environmental Law* discussed 'Sustainable Development' and noted the argument that Sustainable Development tends towards an anthropocentric, that is, human interest focused, bias.[55] The Court chose to instead adopt a 'nature-centred' or 'Ecocentric' approach while examining the necessity of a second home for Asiatic lions.[56] The Court elucidated and applied a 'species best interest standard', that is the best interest of the Asiatic lions, and held that 'we must focus our attention to safeguard the interest of species, as species has equal rights to exist on this earth.'[57] In doing so, the Supreme Court moved beyond protecting the environment and promoting biodiversity conservation for the sake of communities and persons, but as protection emanating from the equal right of other species to exist and the obligation on the State to preserve endangered species.

D. The Public Trust Doctrine and intergenerational equity

Some key concepts emerged from the decisions of the Supreme Court of India that have given shape to the environmental jurisprudence of the country. A significant development was the Supreme Court invoking and relying upon the *'Public Trust Doctrine'* in the case of *M.C. Mehta v. Kamal Nath & Ors.*[58] The Doctrine of Public Trust originated as part of Roman Law and is based on the principal that all natural resources are held in trust by the Sovereign for the beneficial use of the general public.[59] In the modern context, the Government of India is considered the trustee of all natural resources such as air, water, forest, lakes, rivers and wildlife and is thus responsible for their proper use and protection.[60]

In *Kamal Nath*, the Supreme Court observed that the Public Trust Doctrine primarily rests on the principle that certain resources like air, sea, waters and the forests have such a great importance to the people as a whole that it would be wholly unjustified to make them a subject of private ownership. Therefore, said resources being a gift of nature, should be made freely available to everyone irrespective of the status in life.[61] The Court laid down that the Doctrine enjoins upon the Government to protect the resources for the enjoyment of the general public rather than to permit their use for private ownership or commercial purposes.[62]

Adopting the Public Trust Doctrine, the Court placed reliance on various articles and case law from the United States of America in particular, which had adopted the Doctrine to decide cases on ecological rather than commercial considerations.[63]

The Court held that 'the public trust doctrine should ... be expanded to include all ecosystems operating in our natural resources.'[64] It is pertinent to note that the Court also relied on its earlier decision in *Vellore Citizens' Welfare Forum* (supra) and, adopting the 'Polluter Pays Principle' and the 'Precautionary Principle', and directed the polluting motel to refrain from discharging effluents into the Beas river, to pay compensation for the restitution of the environmental damage caused, and to show cause against the imposition of a pollution fine. The Supreme Court also directed the State of Himachal Pradesh

Pollution Control Board to not permit the discharge of untreated efflu-ent into the river Beas, to duly inspect the hotels, etc. in the Kullu–Manali area, and to take appropriate legal action against defaulters.[65]

The Public Trust Doctrine, therefore, restricts Government Author-ity in various ways, which were elucidated by Professor Sax[66] and quoted with approval in several judgments, including *Intellectuals Forum v. State of A.P.*,[67] in which it was held as under:

> According to Prof. Sax, whose article on this subject is con-sidered to be an authority, three types of restrictions on gov-ernmental authority are often thought to be imposed by the public trust doctrine [ibid]:
>
> 1 the property subject to the trust must not only be used for a public purpose, but it must be held available for use by the general public;
> 2 the property may not be sold, even for fair cash equivalent;
> 3 the property must be maintained for particular types of use (i) either traditional uses, or (ii) some uses particu-lar to that form of resources.[68]

In other words, the Government, State or Authority must act at par with a 'trustee' and is restricted from misusing or alienating the trust property in any manner that is not beneficial to the citizens or the public at large.

The Public Trust Doctrine developed in the context of environmen-tal jurisprudence has subsequently had a far-reaching impact on the manner in which a State is allowed to distribute natural resources. In *RIL v. RNRL*,[69] although not directly on biodiversity or the environ-ment, was significant in the way the Supreme Court used the public trust narrative to restrict the power of the Government, to distribute the natural resources of the community for the common good, and to prevent the concentration of wealth by invoking clauses (b) and (c) of Article 39 of the Constitution of India.

A key component of Sustainable Development, 'intergenerational equity' is another concept that the Supreme Court has adopted and followed in dealing with matters relating to the environment and to the distribution of natural resources. The concept has been read into Article 21 as part of the right to life itself and thus is of Constitutional importance in developing environmental policy, as well as economic policy that may impact the environment. According to E. B. Weiss, the 'theory of intergenerational equity states that we, the human species, hold the natural and cultural environment of our planet in common with other species, other people, and with past, present and future generations.'[70] The principle is broad in scope and includes within its ambit the right of present and future generations of human and other species to a safe, clean and hospitable habitat. It means that we inherit the Earth from previous generations and have an obligation to pass it on in reasonable condition to future generations.

Most cases trace *intergenerational equity* to Principles 1 and 2 of the 1972 Stockholm Declaration.[71] In *A.P. Pollution Control Board* v. *Prof. M.V. Nayudu*,[72] quoting these principles, the Supreme Court said that the principle of intergenerational equity is of recent origin and in this context, the environment is viewed more as a resource basis for the survival of the present and future generations.[73] It further stated that several international conventions and treaties have recognized the above principles and, in fact, several imaginative proposals have been submitted, including the locus standi of individuals or groups to take out actions as representatives of future generations, or appointing an ombudsman to take care of the rights of the future against the present.[74]

The idea behind not reducing the ability of future generations to meet their needs is that, although future generations might gain from economic progress, those gains might be more than offset by environmental degradation and deterioration. Most people acknowledge a moral obligation to future generations, particularly as people who are not yet born can have no say in decisions taken today that may affect them. In the context of biodiversity conservation, the right of other species to co-exist with humans, and of future generations to reside on an earth that is diverse and hospitable, is of paramount importance and goes beyond the description of intergenerational equity given in the Stockholm Principles.

One of the earliest cases where the Supreme Court dealt with the concept of intergenerational equity was in the case of *Rural Litigation and Entitlement Kendra, Dehradun* pertaining to illegal and unauthorized mining damaging and destroying the local environmental system and causing ecological imbalance. The Supreme Court held that some assets are permanent and should not be exhausted in one generation and held that environmental protection should be placed on the same footing as economic development of the country.

The Supreme Court in *M.C. Mehta* also touched upon this well-cherished goal of environmental jurisprudence, viz. intergenerational equity, when it held that the principle of Sustainable Development and intergenerational equity too presupposes the higher needs of humans and lays down that exploitation of natural resources must be equitably distributed between the present and future generations. It is both logical and legally sound to understand intergenerational equity as an indelible part of substantive equality, as is understood and contemplated in the Universal Declaration of Human Rights, as also under Articles 14 to 18 of the Constitution of India.[75]

The Supreme Court in its decision in *Glanrock Estate (P) Ltd. v. State of T.N.*[76] has discussed the development of environmental jurisprudence. This judgment is significant because it elevates key principles of environmental jurisprudence to the level of being part of the basic structure of the Constitution itself. The ratio of the judgment is as under:

> In various judgments of this Court delivered by the Forest Bench of this Court in the case of *T.N. Godavarman* v. *Union of India* [Writ Petition No. 202 of 1995], it has been held that 'inter–generational equity' is part of Article 21 of the Constitution. What is inter–generational equity? The present generation is answerable to the next generation by giving to the next generation a good environment. We are answerable to the next generation and if deforestation takes place rampantly then inter–generational equity would stand violated. The doctrine of sustainable development also forms part of Article 21 of the Constitution. The 'precautionary principle' and the 'polluter pays principle' flow from the core value in Article 21. The important point to be noted is that in this case we are concerned with vesting of forests in the State. When we talk about inter–generational equity and sustainable development, we are elevating an ordinary principle of equality to the level of overarching principle. Equality doctrine has various facets. It is in this sense that in I.R. Coelho's case this Court has read Article 21 with Article 14. The above example indicates that when it comes to preservation of forests as well as environment vis-à-vis development, one has to look at the constitutional amendment not from the point of view of formal equality or equality enshrined in Article 14 but on a much wider platform of an egalitarian equality which includes the concept of 'inclusive growth'. It is in that sense that this Court has used the expression Article 21 read with Article 14 in I.R. Coelho's case. Therefore, it is only that breach of the principle of equality which is of the character

of destroying the basic framework of the Constitution which will not be protected by Article 31B. If every breach of Article 14, however, egregious, is held to be unprotected by Article 31B, there would be no purpose in protection by Article 31B. The question can be looked at from yet another angle. Can Parliament increase its amending power by amendment of Article 368 so as to confer on itself the unlimited power of amendment and destroy and damage the fundamentals of the Constitution? The answer is obvious. Article 368 does not vest such a power in Parliament. It cannot lift all limitations/restrictions placed on the amending power or free the amending power from all limitations. This is the effect of the decision in *Kesavananda Bharati*.

This decision also indicates that the scope of the 'Golden Triangle' of rights under Articles 14, 19 and 21 can only be understood fully if the impact of Articles 48-A and 51-A are also taken into consideration.

II. Conclusion

This chapter has addressed the efforts made by Indian Courts, particularly the Supreme Court, to create, adapt and implement legislation pertaining to the environment, specifically relating to biodiversity conservation, through Public Interest Litigation. The chapter begins by tracing the history of PIL, the relaxation of rules of standing, and the initiation of action by way of a petition or suo motu by Judges. It is argued that PIL is a creative tool and is indispensable for the implementation of laws relating to the environment, given the political apathy, lack of social awareness and legislative lacunae that exist in our society.

The judgment of the Supreme Court in *T.N. Godavarman Thirumulpad* v. *Union of India*[77] expresses the conundrum faced by twenty-first-century India, where developmental considerations and environmental concerns strike an uneasy and unequal balance. It was observed by the Supreme Court that 'Progress and pollution go together'[78] and yet a clean environment is as indispensable to quality of life as nutrition, health, shelter and income. The authors have identified principles, such as 'Polluter Pays', the 'Precautionary Principle', 'Sustainable Development' and within it the 'Public Trust Doctrine', and the concept of 'intergenerational equity', which have been adopted into the law of this country. What remains is perhaps a reconceptualization of what 'environment' is and where humans reside within the Earth's ecology.

The Supreme Court in *T.N. Godavarman* performed this very intellectual exercise with some elan to state that

> the environment is everything but the point of reference itself; that it is the world, its inhabitants, its invisible and visible elements in perfect camaraderie; and that 'environment' is the Earth and its animal and vegetal beings, the air, and the light and warmth of the sum.[79]

A poignant line in the decision of the Supreme Court is that 'environment is a polycentric and multifaceted problem affecting the human existence',[80] presumably because of our inability to 'progress' without contaminating our surroundings. It is this problem, ever-magnified in the context of a developing nation with a very large population, that has led to 'the tide of judicial considerations in environmental litigation' which 'symbolizes the anxiety of courts in finding out appropriate remedies for environmental maladies.'[81] This tide of judicial considerations is particularly interesting in light of the large number of environmental laws, rules and regulations that are in force. The true harbingers of change, therefore, are clearly first the organizations, groups and individuals that move the Courts, and secondly the Courts that consider such petitions and nudge and direct the State to perform its constitutional and statutory duties.

Notes

1 P. N. Bhagwati, 'Judicial Activism and Public Interest Litigation', *Columbia Journal of Transnational Law*, 1985, 23: 561–568.
2 *Ibid.*, p. 562.
3 *Ibid.*, p. 566.
4 *Ibid.*, pp. 566–567.
5 (1981) Supp SCC 87.
6 *S.P. Gupta*, 571.
7 (1984) 3 SCC 161.
8 *Bandhua Mukti*, para 53.
9 *Ibid.*
10 *Ibid.*
11 (1987) 1 SCC 395.
12 *Ibid.*, para 5.
13 (1996) 3 SCC 212.
14 *Enviro–Legal Action*, para 54.
15 *Ibid.*
16 *Ibid.*
17 Knopf, F. L., 'Focusing Conservation of a Diverse Wildlife Resource', *Transactions of the North American Wildlife and Natural Resource Conference*, 1992, 57: 241–242.

18 B. A. Wilcox, 'In situ Conservation of Genetic Resources: Determinants of Minimum Area Requirements', *Commission on National Parks and Protected Areas*, IUCN and UNEP, 1984, pp. 639–640.
19 http://energy.gov/em/downloads/keystone–dialogue (accessed on 5 October 2016).
20 R. F. Noss and A. Y. Cooperrider, 'Biodiversity and Its Value', in *Saving Natures Legacy: Protecting and Restoring Biodiversity*, Washington, DC: Island Press, 1994, pp. 1, 5.
21 *Ibid.*, p. 4.
22 *Ibid.*, p. 15.
23 *Ibid.*, p. 14.
24 United Nations Convention on Biological Diversity (CBD) [Rio de Janeiro: 5 June 1992], Article 2.
25 CBD, Article 8(j).
26 (2013) 6 SCC 476.
27 *Orissa Mining Corporation*, para 46.
28 *Orissa Mining Corporation*, para 47.
29 *Ibid.*
30 *Orissa Mining Corporation*, para 50; see also the Statement of Objects and Reasons, Forest Rights Act, which states as follows:

> WHEREAS the recognized rights of the forest dwelling Scheduled Tribes and other traditional forest dwellers include the responsibilities and authority for sustainable use, conservation of biodiversity and maintenance of ecological balance and thereby strengthening the conservation regime of the forests while ensuring livelihood and food security of the forest dwelling Scheduled Tribes and other traditional forest dwellers.

31 M. Sundar, 'The Invention of Traditional Knowledge', *Law & Contemporary Problems*, 2007, 70: 97–103.
32 *Rural Litigation and Entitlement Kendra, Dehradun and Ors. v. State of UP and Ors.* (1985) 2 SCC 431; also see *Rural Litigation and Entitlement Kendra, Dehradun v. State of UP* (1989) Supp(1) SCC 504.
33 *M.C. Mehta and Anr. v. Union of India and Ors.* (1986) 2 SCC 176.
34 *M.C. Mehta v. Union of India and Ors.* (1987) 4 SCC 463.
35 See, for example, *T.N. Godavarman Thirumulpad v. Union of India* (2006) 1 SCC 1 and *Centre for Environmental Law, World Wide Fund–India v. Union of India and Ors.* (2013) 8 SCC 234.
36 See *Subhash Kumar v. State of Bihar and Ors.* (1991) 1 SCC 598 para 7.
37 (2006) 1 SCC 1.
38 *Godavarman* para 71.
39 (1986) Supp SCC 517.
40 *Rural Litigation and Entitlement Kendra, Dehradun and Ors v. State of UP and Ors.* (1987) Supp SCC 517 para 19.
41 *Godavarman* para 77.
42 Dr P. Hassan and A. Azfar, 'Securing Environmental Rights Through Public Interest Litigation in South Asia', *Virginia Environmental Law Journal*, 22: 215, 218.
43 *Ibid.*
44 *Ibid.*, p. 219.

45 *Rural Litigation and Entitlement Kendra, Dehradun* v. *State of UP* (1989) Supp (1) SCC 504.
46 *Rural Litigation and Entitlement Kendra, Dehradun.*
47 (2011) 7 SCC 338.
48 (2013) 6 SCC 620.
49 *G. Sundarrajan*, para 201.
50 *Ibid.*
51 *Vellore Citizens' Welfare Forum* v. *Union of India* (1996) 5 SCC 647.
52 See, for example, *Vellore Citizens' Welfare Forum* and also *Indian Council for Enviro-legal Action* and *Centre for Environmental Law, World Wide Fund–India.*
53 *Centre for Environmental Law, World Wide Fund–India.*
54 For example, the Wildlife (Protection) Act, 2002, the Forest Rights Act, the Forest (Conservation) Act, the BDA and several attendant policies and action plans for their implementation.
55 *Centre for Environmental Law, World Wide Fund–India*, paras 45–46.
56 *Centre for Environmental Law, World Wide Fund–India*, paras 46–47.
57 *Centre for Environmental Law, World Wide Fund–India*, para 47.
58 (1997) 1 SCC 388.
59 Justice T. S. Doabia, 'Chapter 7 Doctrine of Public Trust', *Environmental & Pollution Laws in India*, p. 523.
60 Justice T. S. Doabia, 'Chapter 7 Doctrine of Public Trust'.
61 *Kamal Nath*, para 25.
62 *Ibid.*
63 *Kamal Nath*, para 33.
64 *Kamal Nath*, para 33.
65 *Kamal Nath*, para 39.
66 J. L. Sax, 'The Public Trust Doctrine in Natural Resource Law: Effective Judicial Intervention', *Michigan Law Review*, January 1970, 68(3): 471–566.
67 (2006) 3 SCC 549.
68 *Intellectuals Forum*, para 76.
69 (2010) 7 SCC 1.
70 E. B. Weiss, 'In Fairness to Future Generations and Sustainable Development', *American University International Law Review*, 1992, 8(1): 19.
71 'Principle 1. – Man has the fundamental right to freedom, equality and adequate conditions of life, in an environment of quality that permits a life of dignity and well-being, and he bears a solemn responsibility to protect and improve the environment for the present and future generations. . . .
Principle 2. – The natural resources of the earth, including the air, water, lands, flora and fauna and especially representative samples of natural ecosystems, must be safeguarded for the benefit of the present and future generations through careful planning or management, as appropriate.'
72 See *A.P. Pollution Control Board* v. *Prof. M.V. Nayudu* (1999) 2 SCC 718 at page 739 and *Intellectuals Forum*, page 575.
73 *Prof. M.V. Nayudu*, para 53.
74 *Ibid.*
75 See *Glanrock Estates (P) Ltd.* v. *State of TN* (2010) 10 SCC 96, para 68.
76 (2010) 10 SCC 96.

77 (2002) 10 SCC 606.
78 *T.N. Godavarman Thirumulpad*, 25.
79 *T.N. Godavarman Thirumulpad*, paras 14, 15 and 16.
80 *T.N. Godavarman Thirumulpad*, para 15.
81 *T.N. Godavarman Thirumulpad*, para 24.

9

ASSESSING INDIA'S GREEN TRIBUNAL FOR CONSERVATION OF BIODIVERSITY

*Usha Tandon**

Habitat destruction and the loss of biodiversity are two major inevitable consequences of the unbearable pressure that rising human population[1] is putting on natural resources and that has, rightly, been recognised world over as 'the largest factor contributing to the current global extinction event'.[2] This is more relevant in the Indian context, as India is likely to become the most populous country in the world soon, overtaking China.[3] Further, India's fast-growing economy has resulted in mass urbanisation at an unprecedented large scale. Many studies have shown that, in India, natural resources are under extreme pressure due to the burgeoning human population and rapid urbanisation.[4]

Environmental litigation, unlike other matters, involves complex issues of science and technology.[5] The exponential growth in environmental litigation in India in the last 30 years led the Supreme Court of India to recommend the constitution, first, of Green Benches and then the Specialized Environmental Tribunal. Strongly backed by the Apex Court and the Law Commission of India, the National Green Tribunal (NGT) saw the light of the day in October 2010. Consisting of judicial and scientific experts, the NGT is a quasi-judicial body aimed to provide quick disposal of environmental cases. It has been empowered to entertain cases raising 'substantial questions relating to the environment'[6] which arise from the implementation of seven statutory laws, including legislations on environment protection and biodiversity. It has been bestowed Original as well as Appellate Jurisdiction along with jurisdiction to award compensation to the victims of pollution and for the restitution of damaged ecology.

This chapter examines the role of the NGT in protecting and enhancing biodiversity and a wholesome environment. It starts with a brief overview of the provisions of the National Green Tribunal Act, 2010, and then identifies and discusses some landmark judgments as well as other important judgments on biodiversity, decided by the Principal Bench and the Zonal Benches of the National Green Tribunal. The various innovative methods adopted by the Tribunal giving liberal interpretation to its parent statute are analysed in the next portion of the chapter. After having examined the sharp reactions of powerful stakeholders for developmental projects, it concludes by stating that, till date, the NGT has been successful in surviving all odds against it and argues for its bright future for the protection of biodiversity, ecology and natural resources.

I. An independent specialized green tribunal

The National Green Tribunal Act[7] was passed in June 2010 and India's National Green Tribunal was established on October 18, 2010.[8] The NGT Act repealed two existing laws viz. National Environment Tribunal Act (NETA), 1995 and National Environment Appellate Authority Act (NEAAA), 1997.[9] All the cases pending before the NEAA were to be heard by the NGT. Its closure created a judicial vacuum, as there was no forum for new cases, and the pending cases were left in limbo. Without the appointment of at least one other member besides the Chairperson, the NGT couldn't function. While the Ministry of Environment, Forest & Climate Change (MoEF&CC)[10] continued to grant regulatory approvals, there was no judicial redressal mechanism to challenge it. This situation might have continued indefinitely if it hadn't been for the Supreme Court's direction that the MoEF&CC regularly report on the progress made in establishing the new Tribunal. As a result, three judicial members and four expert members were appointed on May 5, 2011, and the NGT held its first hearing on May 25, 2011,[11] with New Delhi selected as the site for the Principal Bench, later followed by four Zonal Benches in Chennai, Pune, Bhopal and Kolkata.[12]

The objective of India's NGT Act is to provide for the establishment of a National Green Tribunal for the effective and expeditious disposal of cases relating to environmental protection and conservation of forests and other natural resources, including enforcement of any legal right relating to the environment and giving relief and compensation for damage to persons and property and for matters connected therewith.[13]

Section 14 of the Act prescribes Original Jurisdiction to the Tribunal, Section 16 provides Appellate Jurisdiction to the Tribunal, and

Sections 15 and 17 deal with the powers of the Tribunal to order for the relief and compensation to the victims of pollution and restitution of environment. The Act provides only civil jurisdiction to the Tribunal, in contrast to criminal jurisdiction. The NGT has been vested with jurisdiction and power to provide relief and compensation to the victims of pollution and other environmental damage; and for restitution of damaged property and for restitution of the degraded environment.[14] The Tribunal, while passing any order or decision or award, applies the principles of Sustainable Development, the Precautionary Principle and the Polluter Pays Principle.[15]

A key feature of the NGT is the inclusion of the scientific experts in the composition of the Tribunal. India's NGT consists of a full-time Chairperson,[16] ten to twenty full-time Judicial Members, and ten to twenty full-time Expert Members.[17] A person is qualified for appointment as an Expert Member if he has a degree in Master of Science (in physical sciences or life sciences) with a Doctorate degree or Master of Engineering or Master of Technology and has an experience of fifteen years in the relevant field, including five years of practical experience in the field of environment and forests – including pollution control, hazardous substance management, environment impact assessment, climate change management and biological diversity management – in a reputed National-level institution.[18] A person having administrative experience of fifteen years, including experience of five years in dealing with environmental matters in the Central or a State Government or in a reputed National- or State-level institution, is also eligible for being appointed as Expert Member of the Tribunal.[19]

The Central Government appoints the Chairperson, Judicial Members and Expert Members of the Tribunal.[20] The decisions of the Tribunal are taken by majority.[21] If there is a difference of opinion among the Members hearing a matter and the opinion is equally divided, then the Chairperson hears such matter and decides, provided s/he has not heard such matter earlier.[22] However, where the Chairperson himself has heard such matter along with other Members of the Tribunal, and if there is a difference of opinion among the Members and the opinion is equally divided, then in such cases the Chairperson shall refer the matter to other Members of the Tribunal, who shall hear such application or appeal and decide.[23]

II. Some landmark judgments on biodiversity

Ever since the NGT became functional in May 2011 till date, it has resolved more than 5,000 cases over the last five years. Not a single day passes without an environmental judgment delivered either by the Principal Bench or Zonal Benches in different regions directing the implementing agencies concerned to protect and improve the environment.[24] Given the constraints of time and space, a detailed account of all identified orders and judgments of the NGT on biodiversity cannot be undertaken in a chapter like this. The following two sections, therefore, present a brief analysis of some landmark and other important orders and judgments of the Principal Bench as well as the Zonal Benches of the NGT.

A. Western Ghats[25]

This case relates to the protection of 'Western Ghats'[26] that consist of high mountains, gorges and deep-cut valleys, second only to the Eastern Himalayas as a treasure trove of biological diversity in India, which have been recognised as among the several global 'hotspots' of biodiversity. Certain non-governmental organisations (NGOs) approached the NGT for anticipatory[27] interim relief,[28] directing the Respondents[29] not to issue any consent or clearance under the Environmental Laws within the Western Ghats areas, particularly the Ecological Sensitive Zone I & II and a prayer that recommendations made in the Gadgil's Report should be implemented by the Government to protect the Western Ghats. These prayers were founded on the premise that the areas of Western Ghats have been 'subjected to a rapid erosion of natural capital with the building up of manmade capital, regrettably imposing excessive accepted unnecessary environmental damage in the process, accompanied by a degradation of social capital as well'.[30] During the pendency of this Application, the MoEF&CC constituted another High-Level Working Group (HLWG) under the Chairmanship of Dr. K. Kasturirangan. This Committee submitted its report to the MoEF&CC, which in turn initially took a decision to accept the said report in principle and proposed a draft notification under Section 5 of the Environment (Protection) Act, 1986 and invited objections from all stakeholders, including the States. It may be noted that the K. Kasturirangan Committee Report excluded substantial parts of the eco-sensitive area of the Western Ghats, which have been included in the report of Dr. Gadgil.

The Tribunal asked the MoEF&CC to file an affidavit and its unambiguous and clear stand before it. The Secretary of the MoEF&CC filed an affidavit stating *inter alia* that the State Governments of the Western Ghats region may, after undertaking demarcation of ESA by physical verification, propose the exclusion or inclusion of certain areas from the already notified ESA. Such proposals of the State Governments received after physical verification would have to be examined by the Ministry before taking a view on further appropriate action, including issuing a fresh draft notification, if required, to seek objections from the public on the proposals received from the State Governments of Western Ghats. The affidavit clarified that the previous direction issued, for providing immediate protection to the Western Ghats and maintaining its environmental integrity, was in force.[31] Accepting the above affidavit, the Principal Bench, New Delhi, left the matter exclusively[32] for the MoEF to determine and decide the rival contentions.

It is interesting to note that though the Tribunal assumed contested jurisdiction, in terms of anticipatory interim relief, it did not pass any direction to prohibit clearance to any developmental activity within the Western Ghats areas in this case. Nevertheless, the Tribunal dealt with, at length, the importance of Western Ghats in terms of biodiversity, holding that

> the importance of the Western Ghats in terms of its biodiversity can be seen from the known inventory of its plant and animal groups, and the levels of endemism in these taxa. Nearly 4000 species of flowering plants or about 27% of the country's total species are known from the Ghats. Of the 645 species of evergreen trees (>10 cm dbh), about 56% is endemic to the Ghats. Among the lower plant groups, the diversity of bryophytes is impressive, with 850–1000 species are mosses with 28% endemics and 280 species are liverworts with 43% endemics. Among the invertebrate groups, about 350 (20% endemic) species of ants, 330 (11% endemic) species of butterflies, 174 (40% endemic) species of odonates (dragonflies and damselflies), and 269 (76% endemic) species of mollusks (land snails) have been described from this region. The known fish-fauna of the Ghats is 288 species with 41% of these being endemic to the region. The Western Ghats are particularly notable for its amphibian fauna with about 220 species of which 88% are endemic. Similarly, the Ghats are unique in its caecilian diversity harbouring 16 of the country's 20 known species, with all 16 species being endemic. Of the 225 described species of reptiles, 62% are endemic; special mention must be made of the primitively burrowing snakes of the family Uropeltidae that are mostly restricted to the southern hills of the Western Ghats. Over 500 species of birds and 120 species of mammals are also known from this region. The

Western Ghats region harbours the largest global populations of the Asian elephant, and possibly of other mammals such as tiger, dhole, and gaur. The Western Ghats also harbour a number of wild relatives of cultivated plants, including pepper, cardamom, mango, jackfruit and plantain. This biological wealth has paid rich dividends over the years. In fact, the tract was famous for its wild produce of pepper, cardamom, sandal and ivory.[33]

B. *Kaziranga National Park*[34]

Kaziranga National Park, located in Assam, has been declared a World Heritage Site by the United Nations Educational, Scientific and Cultural Organisation (UNESCO).[35] Many rhinoceros (especially one-horned rhino), elephants, and a wide variety of *flora* and *fauna* exist in the Kaziranga National Park.[36] The Applicant in this case approached the NGT alleging flagrant violation to the Notification[37] issued by the MoEF&CC, declaring an area of 15 km around the Numaligarh Refinery, adjoining the Kaziranga National Park, as a 'No Development Zone' (NDZ). The Applicant argued that, in flagrant violation to the aforesaid Notification, a mushroom of stone quarries was installed indiscriminately within the NDZ, thereby causing immense adverse impact on the ecology, wildlife and environment, and no action was taken by the Authorities to stop the illegal installation of quarries and crushing units, in spite of the fact that the Applicant had submitted several objections and approached the State as well as Central Authorities with a prayer to take action to stop infringement and violation of law.

Citing the Supreme Court of India in the case of *Indian Council for Enviro-legal Action* v. *Union of India and others*,[38] the Principle Bench, New Delhi, regretfully remarked that tolerating infringement of law is worse than not enacting the law at all and the case in hand is a clear example of infringement of law to the optimum.[39] After perusing the records, the Tribunal found that 10 stone crusher units, 34 brick kilns,[40] and a number of tea factories and wood-based units[41] were set up in the NDZ after issuance of the Notification in 1996. Having heard the pleadings and arguments of the parties, the Tribunal emphasised on the measures to be taken to eradicate the hazards created by the installation of industrial units in the NDZ.[42] Explaining the adverse impact on the ecology and the environment by the units situated within the NDZ;[43] lamenting on the 'tolerance of the violation of law';[44] and referring to the landmark judgments delivered by the

Supreme Court of India,[45] the Tribunal observed that certain directions are necessary to be issued for the protection of the environment, ecology, biodiversity and adverse impacts on *flora* and *fauna vis- à-vis* conservation of forests and other natural resources, and to protect the ecology in Kaziranga National Park and in its vicinity, which is highly eco-sensitive.[46] The directions passed by the Tribunal included removal of all illegal stone crushers; closing down of brick kilns operating within the NDZ; and not allowing the operation of stone crusher units existing in the vicinity of the NDZ (outside the NDZ) till necessary pollution control equipments were installed to the satisfaction of Assam Pollution Control Board and Central Pollution Control Board. The Tribunal also directed the SPCB and other Authorities to ensure that no tea processing units having a boiler that uses fossil fuel operate within the NDZ.[47]

C. *Kanha National Park*[48]

The Central Zonal Bench, Bhopal took *suo moto* cognizance of the Bhopal edition of the *Times of India* dated April 10, 2013, in which a news item was published on the front page under the caption 'Dolomite mining a threat to Tiger corridor in Kanha – Foresters want ban on mining in Mandla District'. To assume jurisdiction in wildlife protection, which otherwise is not included in its jurisdictional Statutes, the Tribunal enumerated the legal structure for protection of wildlife by citing a Supreme Court judgment[49] wherein it was observed that, for achieving the objectives of various conventions, including the Convention on Biological Diversity (CBD); and also for proper implementation of IUCN, CITES, etc., and the provisions of the Wild Life (Protection) Act, Biodiversity Act and Forest Conservation Act, the Government of India laid down various policies and action plans – such as the National Forest Policy (NFP), 1988; National Environment Policy (NEP), 2006; National Biodiversity Action Plan (NBAP), 2008; National Action Plan on Climate Change (NAPCC), 2008; and the Integrated Development of Wildlife Habitats a Centrally Sponsored Scheme framed in the year 2009 – and integrated development of the National Wildlife Action Plan (NWAP) 2002–2016. Thereafter, the Tribunal took note of the fact that the dolomite mined from the mines in Mandla District was of superior quality and was in good demand in the market. Also, it was observed that this superior-quality mineral was not found elsewhere in the country. However, the Tribunal, tilting towards protection of the environment, held that mining is required to be taken up only if it is compatible with the objective of protecting

206

the environment, more so in the context of the location of dolomite mines relatively in close proximity to Kanha National Park. The Tribunal emphasised the duty of the Central Government and the State Government to take steps to protect the environment, which included maintenance of the ecological balance and prevention of damage that may be caused by mining operations.[50]

D. Bombay Marine Ecology[51]

In the Arabian Sea, off the coast of Mumbai, an oil spill occurred due to the sinking of the Panamanian ship 'M.V. RAK', hereinafter referred as the Ship, which was carrying coal for and on behalf of an Indian company at Ahmadabad for its thermal power plant. The Ship was carrying more than 60,054 Mt of coal in its holds. The Ship contained 290 tonnes of fuel oil and 50 tonnes of diesel. Its voyage was from Indonesia to Dahej in Gujarat. On the voyage to its destination, the Ship sank approximately 20 nm from the coast of South Mumbai. As a result of the oil spill, there had been damage to mangroves and the marine ecology of the Bombay coast. The oil spill waste reached the shoreline and coasts and interacted with sediments, such as beach sand and gravel, rocks and boulders, vegetation and terrestrial habitats of both wildlife and humans, causing erosion as well as contamination. It had a definite impact on fish, marine mammals, birds, coastal marshes, mangroves, wetlands, wildlife habitats and their breeding grounds.[52] An application was filed in the Tribunal raising substantial questions relating to the environment, restitution of the environment and compensation commensurate to the damage done to the ecology. The Applicant claimed that on account of damage caused to the aquatic flora and fauna, mangroves and fishermen, and the damage done to the environment including soil, water, land and ecosystem, the Respondents had a joint and several liability to pay compensation claimed in the application[53] on the Polluter Pays Principle.

The Principal Bench, New Delhi, in its 223-page remarkable judgment, observed that on the true and purposive construction of the International Conventions and the statutory provisions, no party from any country in the world has the right or privilege to sail an unseaworthy ship to the Contiguous and Exclusive Economic Zone of India, and in any event to dump the same in such waters, causing marine pollution, damage and degradation thereof.[54]

Having examined various reports, the Tribunal ruled that reports on record clearly show that the documents in favour of the Ship were issued in a biased manner and the Ship was not seaworthy, right from the inception of its voyage. The accident investigation report, the report by NEERI and the report by Annamalai University show that there was serious marine pollution caused by the oil spill.[55] In the Tribunal view, the damage stands established not only to the aquatic life but also to the seawater and shore. There had been degradation and damage to the mangroves, and adverse impacts on human and aquatic life, on the shore and to tourism and activities of the fishermen. The oil spill caused substantial damage; it spread over the water surface and also formed tar balls, affecting the aquatic community. Even the dispersants used for controlling the oil spill had been shown to be harmful for the organisms living in the area. After holding that tremendous damage and loss to the aquatic life and marine environment had been caused, the Tribunal concluded that the Respondents were liable to pay environmental compensation. It referred to the Supreme Court judgment[56] wherein it was held that where the industry had violated the provisions of the Water (Prevention and Control of Pollution) Act, 1974 and had operated without obtaining consent, it was liable to pay damages of Rs. 1 billion for the default period and observed that this judgment had been followed by the Tribunal in a large number of cases.[57] Thus, the Tribunal ordered the damages of Rs. 1 billion to be paid by Respondents no. 5,[58] 7[59] and 11,[60] jointly and severally,[61] to the Ministry of Shipping, Government of India. The Tribunal clarified that above Rs. 1 billion shall be included the expenses incurred by the Coast Guard and other forces for the prevention and control of pollution caused by the oil spill and saving the crew, etc. Out of this amount, a sum of Rs. 69,184,405 shall be adjusted and paid to the respective agencies. The Tribunal also held Respondent no. 6[62] liable to pay Rs. 50 million as environmental compensation for dumping of the cargo in the sea. It also directed that the Ship and its cargo should be removed within a period of six months from the date of submission of the Report of the Committee before the Tribunal.[63]

E. Yamuna biodiversity[64]

Rampant illegal mining of minor minerals like sand, boulders, etc. in Saharanpur and more particularly on the river banks and river bed of river Yamuna – going on in complete violation to the provisions of the Environment (Protection) Act, 1986, the Rules framed therein and Environment Impact Assessment Notification, 2006 – was brought to the notice of the Tribunal. The Applicants alleged that there was a huge demand for sand and other minor minerals which the people were extracting both legally and illegally to earn profits at the cost of disturbing the river banks, weakening the river bed and disturbing the ecology. The Applicants submitted that illegal mining activity has a serious impact on the ecology and biodiversity as well as causes destruction of flora and fauna, including aquatic life, thus causing ecological imbalance and environmental degradation. Accepting the stand of the Applicants, the Principal Bench, New Delhi, ruled that 'such illegal, unauthorised and unscientific mining is bound to have adverse impacts upon the environment, ecology and biodiversity, particularly, of the river. It is even apprehended that because of such an activity, the river may change its course'.[65]

In its 146-page landmark judgment, the Tribunal referred at length to the Central Empowered Committee Report[66] and many cases decided by the Supreme Court of India and High Courts, especially many orders of Allahabad High Court; and examined the content and compliance of ECs granted to mine lease holders and found that the private Respondents and the Noticees failed to comply with conditions attached to the order granting EC.[67] It took serious note of the fact that the MoEF&CC had not filed any inspection report on record to show that the conditions of the order granting EC were strictly adhered to.[68] While quoting the work of leading scholars on the adverse effects of mining on the river bed, the Tribunal emphasised that excessive mining, both in stream and from river banks and floodplains, would deplete the volume of water held in aquifers and the flow from the aquifer to the stream during the lean season, impacting water availability for drinking and agricultural purposes besides affecting aquatic biodiversity.[69] Exploring the ways whereby the groundwater regime of the river is impacted by unregulated and prolonged mining,[70] it held that in-stream roughness elements, including the gravel itself and large woody debris, play a major role in providing structural integrity and complexity to the stream or river ecosystem, and provide habitat critical for several fish and other aquatic organisms. In-stream mining can also result in the loss of fertile streamside land, as well as valuable forest resources and wildlife habitats in the riparian areas, besides the loss of biodiversity and recreational potential.[71] It went on to stress that the operation of heavy equipment like trucks, JCBs and excavators in the channel bed can directly destroy spawning habitat, rearing habitat, the juvenile fish themselves, and macro invertebrates.[72] It emphasised, further, that all species require specific habitat conditions to ensure long-term survival. Native species in streams are uniquely adapted to the habitat conditions that existed before humans began large-scale alterations. These have caused major habitat disruptions

209

that favoured some species over others and caused overall declines in biological diversity and productivity.[73] Expressing its serious concerns for Ghariyals and the river dolphins, the Tribunal concluded that

> on the analysis of the above studies, it is clear that such impacts can be divided into two different categories. First category can relate to general impacts of mining on river ecology and biodiversity which would include physical impacts as well as depletion of water level and recharging or restoration of the minerals. Second category deals with adverse impacts of excessive, more particularly, illegal and unscientific mining on river ecology and bio-diversity.[74]

The Tribunal then ruled that in the present case, the illegal extraction of minor minerals and their transportation had been much in excess to the limits of permissible mining.[75] Once the nexus between the activity, particularly illegal activities, and the consequential damage to the environment and ecology were established, the Tribunal said the liability in terms of Sections 15 and 17 of the NGT Act arises.[76] Thus, a compensation of Rs. 0.5 billion was directed to be paid by each of the private Respondents/Noticees who were carrying on the extraction of minor minerals and Rs. 25 million, respectively, by each of the stone crushers/screening plants which had been running illegally, in an unauthorised manner, without consent of the concerned Pollution Control Board.[77]

III. Some other important judgments on biodiversity

Save Mon Region Federation & Ors v. Union of India[78] (hereinafter referred to as *Naymjang Chhu River*) is a classic example to illustrate that even development of hydro power, which is a renewable source of energy, is sustainable only if it does not cause any irreversible loss to the biodiversity. The Appellants challenged the Environmental Clearance granted for the construction of a hydroelectric dam on the Nyamjang Chhu River, in the Tawang district of Arunachal Pradesh. The proposed location of the dam was one of two wintering sites for the endangered black-necked crane, which is revered by the local Monpa community as the incarnation of the 6th Dalai Lama This material information was not revealed in the scoping process and, accordingly, was not discussed or evaluated in the environmental impact assessment (EIA) or properly considered by the expert appraisal committee (EAC).[79] The Appellants pointed to faults in the EIA procedure and a lack of close scrutiny of the project by the EAC. The EIA also lacked any discussion of the cumulative impacts of other projects being proposed within the river basin. The Principal Bench, New Delhi, held that it is true that the hydro power project provides an eco-friendly renewable source of energy and its development is necessary, however, that such development should be 'Sustainable Development' without there being any irretrievable loss to the environment.[80] The NGT suspended the Environmental Clearance until the Ministry of Environment, Forests and Climate Change undertakes an environmental flow study to determine how to protect the black-necked crane and its habitat.

In *Md Mubeen v. Anees Ahmed Ors.*[81] (hereinafter referred to as *Ghana Jungle*), the Applicant was aggrieved by the action of the Respondents, who indulged in deforestation of private lands which were recorded as Ghana Jungle in the records. The Central Zonal Bench, Bhopal, referring to the case of *T.N. Godavarman Thirumulpad* v. *Union of India*,[82] where the Supreme Court had made it clear that once the area is recorded as forest in the government record, irrespective of its ownership, it attracts the Forest (Conservation) Act, 1980, and observed that for the conservation of forests, the provisions of the Forest Act must apply to all forests, irrespective of the nature of ownership or classification thereof. It held that the word 'forest' covers all statutorily recognised forests, whether designated as reserved, protected or otherwise for the purpose of Section 2(i) of the Forest Conservation Act. The term 'forest land', occurring in Section 2, will not only include 'forest' as understood in the dictionary sense, but also any area recorded as forest in the Government record, irrespective of the ownership.[83]

The Tribunal took note of the photographs produced before it and stated that it was evident that the land under dispute harboured naturally occurring tree growth of almost 0.4 density and qualified to be categorised as dry deciduous forest. It also stated that the term 'forest land' mentioned in Section 2 of the Forest (Conservation) Act, 1980 refers to Reserved Forest, Protected Forest or any area recorded as forest in the Government records. All proposals for diversion of such areas for any non-forest purpose, irrespective of its ownership, would require the prior approval of the Central Government. Therefore, the land in question could not be put to any non-forest activity without the approval of the Central Government as in the revenue record its category was recorded as '*Ghana Jungle*'.[84]

Similarly, in *Shyam Sunder* v. *Union of India*[85] (hereinafter referred to as *Mainpuri Forest*), the Tribunal was approached to protect a region which was notified as forest in earlier records, but there were no trees presently (during the case) growing in that area. The DM, Mainpuri, issued an NOC for the establishment of a petrol pump outlet without taking permission from the Forest Department as prescribed under the Forest (Conservation) Act 1980. The Tribunal noted that the area had already been cleared of trees. However, it held that whether there was any tree standing of forest species or that it was a vacant land does not make any difference. Relying on the case of *T.N. Godaverman* v. *UoI & Ors*[86] where the Apex Court held that if any land is notified as protected forest, then it has to be treated as 'forest' in the eyes of law, the Tribunal held that it was not necessary to examine whether trees of forest species were actually there or not at the site in question,[87] and once the land was declared as protected forest, it was manifestly clear that any use thereof for a 'non-forestry' purpose was impermissible without permission granted by the competent authority[88] and, hence, cancelled the NOC issued by the DM.

It may be mentioned that the Applicant in this case sought directions to the Divisional Forest Officer, Mainpuri, to register a First Information Report (FIR) under the provisions of the Forest (Conservation) Act, 1980 for offenses punishable under it, against District Magistrate Mainpuri along with some other Respondents. Though the Tribunal observed that the 'District Magistrate attempted to assert his authority by circumventing the regular procedure'[89] and arrived at the conclusion that the NOC was 'malafidely issued by the District Magistrate',[90] it failed to pass directions for taking penal action against the DM and some other respondents.

In *Sudiep Shrivastava* v. *State of Chhattisgarh*[91] (hereinafter referred to as *Sarguja Forest*), the Forest Advisory Committee denied the permission of forest clearance of some forest land at Parsa East and Kante-Basan captive coal blocks, situated in the Hasdeo-Arand coal fields in the Hasdeo-Arand forest–South Sarguja Forest Division, Chhattisgarh, in favour of an electricity company; but the MoEF&CC, overriding the FAC's recommendation, granted permission for forest clearance, saying that the project was situated in the fringe area and not within the rich biodiversity area, and that the area was further separated by a well-defined high and hilly ridge.

On appeal, the Principal Bench, New Delhi, held that the geographical situation of an area need not per se define its wealth of biodiversity. Biodiversity can exist or can share more than one watershed.[92] The Tribunal referred to Section 2(b) of the Biological Diversity Act 2002, which defines 'biological diversity' as the variability among living organisms from all sources and the ecological complexes of which they are part and includes diversity within species or between species and of ecosystems; and the Tribunal observed that, going by this definition, the area in question, being fringe area as described by the Ministry, can be regarded as *Ecotonal area*, i.e. area on the edge of the forest.[93] Setting aside the order[94] of the Ministry granting permission, the case was remanded to the MoEF with directions to seek the fresh advice of the FAC on all aspects of the proposal with emphasis on the following aspects viz. type of flora and fauna in terms of biodiversity and forest cover that existed on the date of the proposal; endemic or endangered species of flora and fauna; migratory route/corridor of any wild animal particularly; elephants passing through the area in question; and the significance of conservation value of the area, etc.[95]

In *Devendra Kumar* v. *Union of India* (hereinafter referred to as *Aravali Ridge*), the issue of the protection of the Aravali Ridge Area came before the Tribunal. The Forest Department of the State of Haryana, the Municipal Corporation of Gurgaon and the Haryana State Pollution Control Board had erred in proper implementation of the provisions of the Notification relating to protection of the Aravali Ridge Area.[96] The Tribunal indicted the concerned authorities acting in blatant violation of the Notification. It stated that it was expected from these government functionaries that they should be vigilant and should take appropriate steps for execution of the Notifications issued by the Government of India and without their knowledge, it was practically impossible for such a large number of marble traders to not only establish their shops/go-downs but also enjoy the power back-up facilities and be able to use their machineries.[97]

Applying the Polluter Pays Principle, the Tribunal gave directions which included that nobody shall cut any tree or bush from the area in question. On the contrary, each of the non-Applicants shall plant at least 50 trees in that area. Upon planting such trees, they shall ensure that the trees are looked after till they attain the sustained age and shall also inform the State of Haryana, the Chief Conservator of Forests and the Pollution Control Board, who shall then conduct a joint survey of the area and report to the Tribunal if the terms of the present order have been complied with or not.[98]

The Tribunal highlighted the need for carrying out the compensatory afforestation and re-plantation *pari passu* the project development activities in *Powergrid Corporation of India* v. *Deputy Conservation of Forests*[99] (hereinafter referred to as *Tansa Wildlife Sanctuary*). The Eastern Bench, Pune, allowed the Applicant the construction work of a Transmission Line in the area coming within 10 km from the boundary of Kalsubai Harishchandra and Tansa Wildlife Sanctuary, after it took permission from the Government of India under the Electricity

Act, 2003 and from the MoEF&CC under Section 2 of the Forest (Conservation) Act, 1980, permitting the cutting of 1,696 number of trees in that area, considering that the proposed project aimed at providing efficient transmission of electricity for overall socio-economic development.[100] The Tribunal also accepted the proposal of the Applicant for compensatory afforestation of planting approximately 386,350 saplings, to be planted and maintained by the Forest Department for a period of seven years at the cost of the Applicant, estimated about Rs. 93.5 million. It went a step further to order that in the meantime, the Applicant shall carry out plantation of at least 1,696 trees of native species in this monsoon season (2016) without awaiting the start of the project construction activity through the Forest Department in the affected area. This plantation is over and above as stipulated independently in the Forest Clearance.[101]

In *Hazira Macchimar Smiti & Ors* v. *Union of India & Ors*[102](hereinafter referred to as *Hazira Mangroves*), the Western Bench, Pune, set aside the Environmental Clearance granted by the MoEF&CC for the further development of Port activities at Hajira district, Surat, as such expansion Port activities were to hinder the safe and proper access to seawater for the Appellants, who are traditional fishermen of the village. The Tribunal also took serious note of the fact that Respondent no. 6, namely Adani-Hajaria Port Pvt. Ltd., had already caused massive destruction of mangroves. In view of this, the Tribunal imposed a penalty of Rs. 25 million on Respondent nos. 6 and 7 for compensation and restoration of the environment, to be deposited within four weeks with the Collector, Surat.[103]

The Environmental Clearance granted to the integrated Kashang Hydro-Electric project (243 MW) was challenged before the Tribunal in *Bhagat Singh Kinnar* v. *Union of India & ors*[104] (hereinafter referred to as *Pine & Chilgoza*). One of the issues was related to the over- exploitation of Pine and Chilgoza trees. It was contended that the diverted forest land marked for the execution of Stage II of the project was comprised of Pine and Chilgoza trees. These trees in a cold region have a very slow growth rate, and a Chilgoza tree may take more than one hundred years to bear fruits.[105] Another issue related to the impact of the project on the Lippa Asrang Wildlife Sanctuary and the National Board for Wildlife, as a large number of endangered species of birds and animals were likely to be affected. The Principal Bench, New Delhi, did not quash EC; however, to ensure compliance to the latest standards and ensure proper implementation of mitigative measures, it constituted a Committee to review the terms and conditions of EC, and submit a compliance report to the Tribunal within a period of two months. The Tribunal specifically directed that the option for translocation of Chilgoza trees should be explored and a time-bound action plan should be prepared to stop overexploitation of the Chilgoza trees. It also directed the Committee to examine the issue of proximity, especially of Stage IV of the integrated project, to the Lippa Asrang Wildlife Sanctuary, and make recommendation thereof.[106]

The total land required for the thermal power project was shown to include a diversion of forest land in *Jeet Singh Kanwar* v. *Union of India*[107] (hereinafter referred to as *Korba Forest*). The Environmental Clearance was granted to the project proponent subject to a grant of Forest Clearance. The Tribunal found, after going through the appraisal report, that the project was recommended for EC on the basis of assumption that no diversion of forest land was involved, and thus such an assumption was held to be incorrect by the Tribunal.[108] The Tribunal took note of the fact that the supposed project was to be initiated in a critically polluted area. Therefore, the Tribunal quashed the EC granted to the project proponent by applying the Precautionary Principle and held that the Precautionary Principle requires the authority to examine probability of environmental degradation that may occur and result in damage.

IV. Innovative techniques for progressive environmental jurisprudence

In the cases discussed above and a number of other cases, the Tribunal, assuming wide-ranging powers[109] to adjudicate upon any dispute that involves questions of importance to the environment, has provided a unique direction to the development of environmental jurisprudence in the country. The aggressive and powerful approach to protect the biodiversity and natural resources is evident from the following distinctive methods adopted by the Tribunal.

A. Stakeholders consultative process

In order to achieve a fast and implementable resolution to the serious and challenging environmental issues facing the country, the Principle Bench, New Delhi, has adopted the mechanism of 'Stakeholder Consultative Process in Adjudication'. In the matter of *Indian Council for Enviro-Legal Action & Ors.* v. *National Ganga River Basin Authority & Ors.*,[110] the Tribunal dealt with the issue of cleaning of River Ganga, the most sacred river for the people of India. The Original

Application[111] came before the Tribunal upon a transfer that was originally registered with the Supreme Court of India.[112] The Tribunal decided to deal with the serious issue of cleaning the River Ganga in phases.[113] In the first segment of Phase-1, Gaumukh to Haridwar, the Tribunal dealt with the prevention and control of pollution of River Ganga and restoration of the river and its biodiversity to its pristine form. The Tribunal, in the consultative meetings held with various stakeholders viz. Secretaries from the Government of India, Chief Secretaries of the respective States, concerned Member Secretaries of Pollution Control Boards, Uttarakhand Pey Jal Nigam, Uttar Pradesh Jal Nigam, Urban Development Secretaries from the States, representatives from various Associations of Industries (big or small) and even the persons having the least stakes were required to participate in the consultative meetings, wherein various mechanisms and remedial steps for preventing and controlling pollution of the River Ganga were discussed at length. The purpose of these meetings, as stated by the Tribunal, was primarily to know the intent of the executives and political will of the representative States who were required to take steps in that direction. After having the consultative meetings with various stakeholders, it passed directions of varied dimensions to protect and restore the river ecology to its original status in terms of water quality, river flow, purity and its biodiversity.[114]

B. From 'Seven Legislations' to 'Comprehensive Environment'

Section 14 of the Act providing Original Jurisdiction to the Tribunal prescribes 'The Tribunal shall have jurisdiction over all civil cases where a substantial question relating to environment is involved and such question arises out of the implementation of the enactments specified in Schedule I of the Act'. The Schedule I of the Act contains only seven enactments viz. The Water (Prevention and Control of Pollution) Act, 1974; The Water (Prevention and Control of Pollution) Cess Act, 1977; The Forest (Conservation) Act, 1980; The Air (Prevention and Control of Pollution) Act, 1981; The Environment (Protection) Act, 1986; The Public Liability Insurance Act, 1991; and The Biological Diversity Act, 2002. The Schedule does not even include all modern Environmental Law statutes, for instance Wild Life (Protection) Act, 1972.[115]

When the Central Zonal Bench of Bhopal took *sue moto* cognizance of threats to Tiger corridor in *Kanha National Park*, in Mandla District of Madhya Pradesh, the Tribunal was conscious of the fact that

216

under Schedule-I of the National Green Tribunal Act, 2010, Wildlife (Protection) Act, 1972 is not listed; and therefore, it may be argued that this Tribunal has no jurisdiction to adjudicate the matters related to Wildlife.[116] Giving a detailed description of the term 'Environment' as defined under the Act of 1986, it held that 'Environment' includes water, air and land and the interrelationship which exists among and between water, air and land and human beings, other living creatures, plants, micro-organisms and property. To assume jurisdiction over this matter, it observed that occurrence of wildlife in a particular ecosystem having relation with the environment has to be considered as a part of the environment; and therefore, the matters related to wildlife are liable for adjudication and can definitely be brought before the Tribunal, under the environmental jurisprudence, more so in cases pertaining to ESZs.[117]

C. *All civil cases as well as anticipated actions*

In *Western Ghats-I*, at the very initial hearing, the respondents contended, before the Tribunal, that the dispute raised by the Applicants is not a 'civil case' within the meaning of Section 14 of the NGT Act, and thus the Tribunal does not have the jurisdiction to entertain the application.[118] Rejecting the contention, the Principal Bench, New Delhi, discussed at length the indicators of legislative intent necessary to outline the scope of its jurisdiction. The Tribunal, describing the Preamble to the NGT Act, as the first indicator of the legislative intent, affirmed that a case relating to environmental protection, conservation of forests and other natural resources or even enforcement of legal rights relating to the environment and other matters connected therewith, will fall under its jurisdiction.[119] Further, it referred to the words 'for matters connected therewith or incidental thereto' and observed that these two expressions can only be construed liberally and to provide greater dimension to the mode of access to a person claiming redress of his grievances as well as adjudication by the Tribunal.[120] Referring to the second indicator of legislative intent, the Tribunal observed that the expression 'civil cases' used in Section 14(1) of the NGT Act has to be understood in contradistinction to 'criminal cases' and has to be construed liberally, as a variety of cases of civil nature could arise which would be raising a substantial question of the environment and thus would be triable by the Tribunal.[121]

The Tribunal also relied upon Section 15 of the NGT Act, which specifically vested the Tribunal with the powers of granting reliefs like compensation to the victims of pollution and other environmental damage, for restitution of property damaged and for restitution of the environment for such area or areas; and held that once Section 14 is read with the provisions of Section 15, it can, without doubt, be concluded that the expression 'all civil cases' is an expression of wide magnitude and would take within its ambit cases where a substantial question or prayer relating to environment is raised before the Tribunal.[122]

The Tribunal then referred to the substantial question relating to the environment covered under Section 14(1) providing jurisdiction to the Tribunal. Observing that the disputes must relate to implementation of the enactments specified in Schedule I to the NGT Act, the Tribunal referred to one of the scheduled Acts, i.e. Environment Protection Act (EPA), 1986,[123] and discussed the object and reason for enacting that law; and observed that the objects and reasons of the EPA would have to be read as an integral part of the object, reason and purposes of enacting the NGT Act.[124] Assuming jurisdiction for anticipated action, under sub-sections (1) and (2) of Section 14 of the NGT Act, the Tribunal referred to Section 20 of the NGT Act, which provides that, while deciding cases before it, the Tribunal shall take into consideration, inter alia, the precautionary principle.[125]

D. Foreign jurisdiction

In *Bombay Marine Ecology*, one of the important issues before the Tribunal was as to whether it has jurisdiction over a foreign ship that has caused damaged to ecology in India. Explaining the concept of sovereignty over territorial waters, in detail, referring to extensive case law and conventions such as the International Convention on Civil Liability for Bunker Oil Pollution Damage, 2001; the International Convention for Prevention of Pollution from Ships, 1973 (MARPOL); the Maritime Zones Act, 1976; and Merchant Shipping Act, 1958; the Principal Bench, New Delhi, ruled that it will not be appropriate to state that in relation to the marine environment and pollution being caused in the exclusive economic zone, the concerned State or contracting party is without remedy and has no control whatsoever. The laws of the State would come into play, except where their application is excluded or they are in conflict with the Convention.[126]

Referring to the provisions of the Act of 1958 to indicate the jurisdiction to which the foreign ships will be subject to in the coastal waters of India, it quoted Section 352B that provides that every ship, while it is at a port or place in India or within the territorial waters of India, or any marine area adjacent thereto over which India has or may hereafter have exclusive jurisdiction in regard to control of marine pollution under the Act of 1976, or any other law for the time being in force, civil liability of oil pollution under Chapter XB of the Act, would be applicable.[127] It also mentioned Section 356B of the Act of 1958, in terms of which the 'coastal waters' means any part of the territorial waters of India, or any marine areas adjacent thereto over which India has, or may hereafter have, exclusive jurisdiction in regard to control of marine pollution under the Territorial Waters, Continental Shelf, Exclusive Economic Zone and other Maritime Zone Act, 1976 (80 of 1976) or any other law for the time being in force, would be squarely covered under Chapter XIA for the purpose of prevention and containment of pollution of the sea by oil. The cumulative reading of both of these provisions, according to the Tribunal, clearly shows that a foreign ship will be subject to Indian jurisdiction if the incident has occurred in any of the zones afore-stated.[128]

The Tribunal also rejected the argument that the environmental laws, including the National Green Tribunal Act, would not be applicable to any area beyond the territorial waters of India, as the Central Government has not issued a notification extending the operation of the Act of 2010 in terms of sub-section–7 of Section 7. It clarified that sub-section 7 of Section 7 operates only in relation to the Exclusive Economic Zone and no other zones, i.e. the Continental Shelf and Contiguous Zone of seawaters. The present ship admittedly was 20 nm from the baseline of the coastal area in Mumbai, where she sank. This location falls within the ambit and scope of the Contiguous Zone. In that zone, sovereign rights can be exercised, though for a limited purpose. Even under Sections 6 and 7, the sovereign rights are specifically enforceable inasmuch as India has exclusive jurisdiction to preserve and protect the marine environment and to prevent and control marine pollution.[129]

E. Guesswork compensation

In *Yamuna Biodiversity*, a case involving indiscriminate illegal mining on the river bed of Yamuna resulting in massive destruction to the river ecology and biodiversity, the Tribunal noticed that despite directions of the Court, both the State Governments failed to place on record any report which would define the damage caused due to the wrongful acts by these persons with exactitude and the exact money that would be required for restoration, restitution and revitalization of the environment, ecology and bio-diversity with particular reference to the river Yamuna.[130] The Tribunal was convinced that the respondents caused serious damage and degradation of the environ-ment which they must make good of. In such a situation, with the help of documentary evidence and reports on record, the Principal Bench, New Delhi, applied 'guesswork' while resolving this issue.[131] It took legitimacy from a judgment of the Supreme Court,[132] where it had applied this principle while imposing the compensation of Rs. 1 bil-lion upon the industry which has been operating without the consent of the Pollution Control Board for a long time. After referring to the Supreme Court judgment, it observed that there could be cases where it is not possible to determine such liability with exactitude but that by itself would not be a ground for absolving the defaulting parties from their liability. On a reasonable basis, such defaulters could be called upon to pay the environmental compensation. In the present case, the parties opted not to lead any evidence except the documents and affidavit that they had filed in support of their respective cases.[133] Holding that the respondents have carried on excessive unauthorised mining in a manner that has caused substantial damage and degrada-tion of environment, ecology and biodiversity, a compensation of Rs. 0.5 billion was ordered to be paid by each of the private Respondents/ Noticees who were carrying on the extraction of minor minerals and Rs. 25 million, respectively, by each of the stone crushers/screening plants.[134]

F. Liability for tolerating infringement of law

The most beautiful part of the judgment of *Kaziranga National Park* relates to the serious concerns of the Principal Bench, New Delhi, as to how to deal with the 'tolerance of flagrant violation of law' by those very authorities which are responsible to implement the law. The most interesting part of this discourse is that, here, the situation is not demanding the implementation of law directly created by the legisla-ture for the executive, but of that Notification that has been issued by none else than the executive itself. To deal with such a pathetic and inexcusable situation, the Tribunal rightly fixed the responsibility on such erring authorities and ruled that

> we,—have no hesitation to direct the MoEF and the Govern-ment of Assam to deposit Rs. 1,00,000/- (Rupees one lakh only) each, with the Director, Kaziranga National Park for conservation and restoration of *flora* and *fauna* as well as bio-diversity, eco–sensitive zone, ecology and environment of the

vicinity of Kaziranga National Park in general and within the No Development Zone in particular. The said amount shall be utilised exclusively by the Director, Kaziranga National Park for conservation, protection and restoration as well as for afforestation of suitable trees of the local species in and around the No Development Zone.[135]

Though the Direction passed by the Tribunal, in putting responsibility on the erring authorities for not implementing law, is commendable, it is not clear, the way it is written, if it is a penalty, or just compensation for restoration of environment. In any case, the amount of Rs., one lakh each for Central and State Government, is a very nominal amount. Also, no penal action was recommended by the Tribunal against the industrialists for violating the provisions of law. Nonetheless, the judgment is praiseworthy, despite the above criticisms.

G. Relaxation of locus standi

Having elaborately discussed the case pleaded by the parties and keeping in view the multiple and contradictory pleas raised by the parties for consideration, the Principal Bench, New Delhi, in *Bombay Marine Ecology*, formulated and discussed some issues of national and international importance pertaining to environmental jurisprudence.[136] One of such issues related to the *locus standi* of the Applicants to institute the application. It was contended that a person filing an application before the Tribunal must have suffered personal injury, and the Applicant in the present case is not a person as contemplated under Section 18(2) of the Act of 2010; and *locus standi* under the provisions of the Act of 2010 cannot be expanded like in a public interest litigation before the higher courts. Rejecting this contention, however, the Tribunal observed that the expression 'any person aggrieved' appearing in Section 18(2)(e) of the Act of 2010 is to be given a wider meaning. The provision of this sub-section has to be interpreted in light of the objects and reasons of the statute and the definition of 'environment' and scope and ambit of Section 14 of the Act of 2010. This expression is not to be restricted only to a person who has suffered personal injury. The Tribunal emphasised that it is to be interpreted and understood in contradistinction to Section 18(2) (a) where the person who has suffered injury personally can invoke the jurisdiction of the Tribunal under Sections 14, 15 and 16 of the Act of 2010.[137]

H. Merit and judicial review

In *Yamuna Biodiversity*, the Principal Bench, New Delhi, after referring to one of its previous orders,[138] reiterated its power of merit and judicial review. In its own words,

> Tribunal exercising its appellate power and Original jurisdiction in terms of Section 14 and 16 of the Act of 2010, has the powers of merit and judicial review and is competent to issue such directions as it may deem necessary in terms of the said provisions including Section 18 of the NGT Act, 2010.

I. Suo moto power

The Central Zonal Bench, Bhopal, exercising *suo moto* power, in *Kanha National Park*, took notice of a news item regarding a threat to Tiger corridor due to dolomite mining in the Mandla district of Madhya Pradesh and passed necessary directions for the protection of the environment. The NGT Act does not expressly confer this power on the Tribunal. Prior to *Kanha National Park*, and after this judgment, in a number of cases the Tribunal exercised similar jurisdiction.[139]

J. Circuit benches

The NGT has its headquarters in Delhi and four Regional Benches at Bhopal, Calcutta, Pune and Chennai. It is structurally inadequate to satisfy people's needs in a vast country[140] like India. Due to lack of access, thousands of grievances related to environmental issues do not reach the adjudicatory mechanism at all. This is also antithetical to the concept of 'just, quick and cheap' judicial remedy. It has started holding Circuit Benches at various places like Shimla, Jodhpur, Shillong, etc.[141] It needs to have more Regional Benches as well as to hold more Circuit Benches for better accessibility to justice for the masses.

V. Growing tension for powerful stakeholders

The above innovative techniques, adopted by the Tribunal, are making its functioning extremely successful to protect biodiversity and a wholesome environment. Interestingly, the very success of the NGT has been a cause of anxiety for powerful interests, such as the MoEF&CC, State

Governments, and also the concerns of some of the High Courts.[142] For instance, the Chief Minister of Meghalaya urged the Prime Minister to allow the state government to regulate the largely unregulated mining activities in view of the peculiar ground conditions in the Hill state. This intervention was the result of a wave of protests against the NGT's ban on unscientific rat-hole mining in Meghalaya. Coal mining is a major source of livelihood for people of the State and of revenue for the Government. However, the NGT decided that the unscientific, unlicensed and illegal coal mining affected water sources and the landscape, and overall, it degraded the ecology.[143]

In January 2014, the Madras High Court held that the NGT had no statutory jurisdiction to take up cases *suo motu*, thereby restraining the Tribunal from initiating litigation on its own on the basis of media reports and other sources. A month thereafter, it held that High Courts did have jurisdiction to entertain appeals against the orders of the NGT.[144] It may be noted that under Section 22 of the NGT Act, appeals from the NGT go to the Supreme Court. But the Madras High Court stressed that appeal from the NGT had to go to the High Court first before going to the Apex Court. The growing conflicts between the NGT and the High Courts became evident from these kinds of judgments from the High Court.

With a view to clip its wings, the MEF&CC, on August 29, 2014, constituted a High Level Committee (HLC) to review Environmental Laws under the chairmanship of former Cabinet Secretary T. S. R. Subramanian. The Committee submitted its report to the Central Government in less than three months,[145] suggesting the Parliament enact an umbrella legislation known as the Environmental Laws (Management) Act (ELMA) that would constitute the 'National Environment Management Authority' (NEMA) at the Centre and 'State Environment Management Authority' (SEMA)[146] in States to deal with applications for clearances and permissions under environment-related laws.[147] To hear appeals against the decision of the NEMA or SEMA, it proposed the creation of an Appellate Tribunal, to be constituted by the Government of India and presided over by a retired judge of the High Court with two senior officers of the rank of Additional Secretary to the Government of India.[148] The role of the NGT was proposed to be limited only to the judicial review of the decisions of the Appellate Boards.[149] The MEF&CC would have the powers to issue directions to the NEMA and SEMA in all matters.[150] It proposed to ban the jurisdiction of the NGT, stating 'The decisions of the Government, NEMA or SEMA shall not be questioned before nor enquired in to by any court or tribunal either *suo moto* or at any ones behest on any ground what so ever'.[151]

The HLC Report aiming to defang the National Green Tribunal has been heavily criticised[152] by various stakeholders. The Parliamentary Standing Committee (PSC) on Science & Technology,[153] fortunately, after reviewing the HLC report, has rejected it. The PSC has instead recommended that the Government should constitute a new committee to review the laws, noting that the 'three-month period given to the HLC for reviewing six environmental Acts was too short' and that 'there was no cogent reason for hurrying through with the report without comprehensive, meaningful and wider consultations with all the stakeholders'.[154]

VI. Conclusion

The above analysis of some of the judgments of the NGT's Principal and Zonal Benches on biodiversity reveals that India's NGT is coming out as a true guardian for the preservation of biodiversity, be it the Kaziranga National Park, Kanha National Park, Ghana Jungle, Sarguja Forests, Aravali Ridge Area, Yamuna Biodiversity, Ganga Biodiversity, Bombay Marine Ecology, etc. It is, jealously, justifying its existence, in disposing of cases relating to environmental protection and conservation of forests and other natural resources, in an effective and expeditious manner. It's a serious concern, for the enhancement of biodiversity and a wholesome environment is evident in its flexible approach, liberal interpretation and innovative techniques to address environmental issues, giving new dimensions to environmental jurisprudence in India. It is seen as a powerful adjudicatory body, fearlessly applying the law for Union Government, State Governments, Pollution Control Boards, big Corporate Bodies, big Spiritual Leaders, etc. Its unswerving and belligerent attitude in strict implementation of environmental laws and innovative methods for expanding its jurisdiction is winning the hearts of environmentalists but sinking the hearts of powerful stakeholders. In this scenario, some scholars have expressed apprehension, predicting the future of the NGT as uncertain.[155] However, looking to the latest developments, wherein the Parliamentary Standing Committee has rejected the report of the HLC that targeted to make the NGT ineffective and powerless, this uncertainty has leaned in favour of strengthening the NGT. Within a short span of five years, India's NGT has emerged as an extremely effective forum, setting high standards for environmental adjudication, especially under the dynamic leadership of its current Chairperson, Justice Swatanter Kumar. The kind of strong foundation that is being laid down during his leadership will make the NGT achieve tall heights in preserving and enhancing the biodiversity and natural resources of the country.

It goes without saying that the level of credibility that the NGT has acquired in protecting the environment makes a strong argument for strengthening it against powerful interests such as the MoEF&CC, which instead of adopting an adversarial approach, should follow a cooperative approach towards the NGT.

Notes

* I acknowledge the research assistance rendered by Aditya N. Prasad, Akash Anand and Varun Bansal.

1 According to the 2012 Revision of the official United Nations population estimates and projections, the world population is projected to increase by almost 1 billion people within the next 12 years, reaching 8.1 billion in 2025, and to further increase to 9.6 billion in 2050 and 10.9 billion by 2100. See *World Population Projects: The 2012 Revision, Key Findings and advance Tables*, United Nations, Department of Economic and Social Affairs/Population Division (2012) 1.

2 L. Fahrig, 'How Much Habitat Is Enough?' *Biological Conservation*, 2001, 100: 65–74, 65.

3 U. Tandon, 'Population Growth, Climate Change, and the Law With Special Reference to India', in U. Tandon (ed.), *Climate Change: Law, Policy and Governance*, Lucknow: Eastern Book Company, 2016, pp. 294–324.

4 See *Impact of urbanisation on biodiversity–Case studies from India*, Report, WWF India, 2011.

5 Due to the complex techno-science issues, various countries the world over have established either Environmental Courts or Green Benches in the ordinary Courts. See G. Pring and C. Pring, *Greening Justice: Creating and Improving Environmental Courts and Tribunals* (The Access Initiative 2009), www.accessinitiative.org/resource/greening–justice (accessed on 30 July 2016). Scholars like Lord Woolf and Alison strongly support this specialised environmental forum. See Lord Woolf, 'Are the Judiciary Environmentally Myopic', *Journal of Environmental Law*, 1992, 4: 1; A. C. Flournay, 'Scientific Uncertainties in Protection of Environmental Decision Making', *Harvard Environmental Law Review*, 1991, 15: 333.

6 The National Green Tribunal Act of 2010, Preamble and Section 14 (1).

7 Act No. 19 of 2010 (hereinafter referred to as the NGT Act).

8 The concerns for establishing Environmental Courts in India originated from a proactive judiciary, when in 1986 in *M.C. Mehta v. Union of India*, the Supreme Court of India expressed difficulty in solving techno-science disputes handling environmental litigation. In 2003, the Law Commission of India also recommended for the constitution of State-level Environmental Courts.

9 For a comparative account, see B. H. Desai and B. Sidhu, 'On the Quest of Green Courts in India', *Journal of Court Innovation*, 2010, 3(1): 79, https://law.pace.edu/sites/default/files/IJIEA/jciDesai_India_3–17_cropped.pdf (accessed on 20 September 2016).

10 Ministry of Environment, Forest and Climate Change (MoEF&CC). Before 2014, it was known as the Ministry of Environment and Forest (MoEF).

11 S. Ghosh, 'Environmental Litigation in India', 2012, www.thehindubusi nessline.com/opinion/environmental–litigation–in–india/article2848051. ece (accessed on 20 September 2016).
12 Ministry of Environment and Forests (MoEF), Government of India, Notification, 17 August 2011, SO 1908 E. (It also holds Circuit Benches at various places like Shimla, Jodhpur, Shillong, etc.).
13 NGT Act, Preamble.
14 NGT Act, Section 15(1).
15 NGT Act, Section 20.
16 Justice Lokeshwar Singh Panta was its founding Chairperson. Justice Swatanter Kumar has held this position since December 2012.
17 NGT Act, Section 4.
18 NGT Act, Section 5(2 a).
19 NGT Act, Section 5(2 b).
20 NGT Act, Section 6(1).
21 NGT Act, Section 21.
22 NGT Act, Section 21 first Proviso.
23 NGT Act, Section 21 second Proviso.
24 G. Sahu, 'Impact of the National Green Tribunal on Environmental Governance in India: An Analysis of Methods and Perspectives', *Journal on Environmental Law, Policy and Development*, 2016, 3(28): 33.
25 *Goa Foundation & Ors.* v. *Union of India & Ors.*, Application no. 26 of 2012 dated July 18, 2013 (hereinafter referred to as *Western Ghats-I*).
26 The Western Ghats cover States like Gujarat, Maharashtra, Goa, Karnataka, Kerala and Tamil Nadu. Nearly 44 districts fall under the eco- sensitive area of the Western Ghats.
27 *Goa Foundation & Ors.* v. *Union of India* & Ors., Application no. 26 of 2012 dated September 25, 2014 (hereinafter referred to as *Western Ghats-II*) para 13, p. 12.
28 *Western Ghats-I*, para 1, p. 4.
29 Respondents included Union of India, States of Maharashtra, Karnataka, Goa, Kerala, Gujarat, Tamil Nadu and State Pollution Control Boards as well as State-level Environment Impact Authorities of these six States.
30 *Western Ghats-II*, para 2, p. 6.
31 *Western Ghats-II*, para 10, p. 11.
32 *Western Ghats-II*, para 14, p. 13.
33 *Western Ghats-I*, para 9, p. 9.
34 *Rohit Choudhury* v: *Union of India & Ors.*, Application no. 38/2011 dated September 7, 2012 (hereinafter referred to as *Kaziranga National Park*).
35 *Kaziranga National Park*, para 2, p. 5.
36 *Kaziranga National Park*, para 2, p. 6.
37 Dated July 5, 1996. The said Notification was issued in exercise of the powers conferred under Section 5 of the Environment (Protection) Rules, 1986.
38 1993(3) SCALE 579.
39 *Kaziranga National Park,* para 1, p. 5.
40 *Kaziranga National Park,* para 22, p. 23.
41 *Kaziranga National Park,* paras 23 and 24, p. 24.
42 *Kaziranga National Park,* para 19, p. 17.

43 *Kaziranga National Park*, para 25, pp. 24–25.
44 *Kaziranga National Park*, para 28, p. 28.
45 Such as *Indian Council for Enviro-Legal Action* v. *Union of India and Others*, M.C. Mehta v. *Union of India* (2006) 3 SCC 399, M.C. Mehta v. *Union of India* (2006) 3 SCC 399; see *Kaziranga National Park*, paras 29–31, pp. 28–30.
46 *Kaziranga National Park*, para 32, p. 31.
47 *Kaziranga National Park*, para 33, pp. 32–36.
48 *Tribunal at Its Own Motion* v. *The Secretary Ministry of Environment & Forests, Govt. of India, New Delhi,* Original Application no. 16/2013 (CZ) dated April 4, 2014, (hereinafter referred to as *Kanha National Park*).
49 *Centre for Environment Law, WWF-I* v. *Union of India & Others (I.A. No. 100 in Writ Petition (Civil) No. 337 of 1995)*, *Kanha National Park*, para 31, p. 22.
50 *Kanha National Park*, para 30, p. 21.
51 *Samir Mehta* v. *Union of India & Ors.,* Original Application no. 24 of 2011 and M.A. no. 129 of 2012, M.A. nos. 557 & 737 of 2016 dated August 23, 2016 (hereinafter referred to as *Bombay Marine Ecology*).
52 *Bombay Marine Ecology*, para 4, p. 6.
53 *Bombay Marine Ecology*, para 6, p. 8.
54 *Bombay Marine Ecology*, para 147, p. 220.
55 *Bombay Marie Ecology*, para 145, p. 219.
56 *Sterlite Industries India Ltd.* v. *Union of India* 2013 (4) SCC 575.
57 *Bombay Marie Ecology*, see para 146, p. 220.
58 Delta Shipping Marine Services SA, Panama.
59 Delta Navigation W.L.L. Qatar.
60 Delta Group International, Qatar.
61 *Bombay Marie Ecology*, para 146, p. 220.
62 Adani Enterprises Limited, Gujarat.
63 *Bombay Marie Ecology*, para 147, p. 221.
64 *Gurpreet Singh Bagga* v. *MoEFCC & Ors.,* Original Application no. 184 of 2013 and *Jai Singh & Ors.* v. *UOI,* Original Application no. 304 of 2015 dated February 18, 2016 (hereinafter referred to as *Yamuna Biodiversity*).
65 *Yamuna Biodiversity*, para 64, p. 107.
66 CEC had submitted its report dated January 4, 2012, in furtherance of the order of the Hon'ble Supreme Court of India dated November 25, 2011, with Deepak Kumar dealing comprehensively with the issue of illegal mining in district Saharanpur, Uttar Pradesh; district Alwar, Rajasthan; and the areas identified for mining in Haryana.
67 *Yamuna Biodiversity*, para 70, p. 116.
68 *Yamuna Biodiversity*, para 71, p. 117.
69 *Yamuna Biodiversity*, para 79, p. 124.
70 *Ibid.,* pp. 124–125.
71 *Yamuna Biodiversity*, para 85, pp. 127–128.
72 *Yamuna Biodiversity*, para 86, p. 128.
73 *Yamuna Biodiversity*, para 87, p. 129.
74 *Yamuna Biodiversity*, para 90, p. 130.
75 *Yamuna Biodiversity*, para 91, p. 131.

76 *Yamuna Biodiversity*, para 92, p. 135.
77 *Ibid.*, p. 136.
78 Hereinafter referred to as *Naymjang Chhu River*, Appeal No. 39 of 2012, dated April 7, 2016.
79 *Nyamjang Chhu River*, see paras 15–17, pp. 18–21.
80 *Nyamjang Chhu River*, para 25.
81 Original Application No. 33/2013 (CZ) dated December 6, 2013 (hereinafter referred to as *Ghana Jungle*).
82 (1997) 2 SCC 267 and (2006) 1 SCC 1.
83 *Ghana Jungle*, para 9, p. 9.
84 *Ghana Jungle*, paras 10 and 11, p. 10.
85 Application no. 7/2013 (THC) dated July 18, 2013 (hereinafter referred to as *Mainpuri Forest*).
86 (1997) 2 SCC 267.
87 *Mainpuri Forest*, p. 13.
88 *Ibid.*
89 *Mainpuri Forest*, para 24, p. 17.
90 *Mainpuri Forest*, para 26, p. 18.
91 Appeal no. 73/2012 dated March 24, 2014 (hereinafter referred to as *Sarguja Forest*).
92 *Sarguja Forest*, para 39, p. 39.
93 *Ibid.*
94 *Sarguja Forest*, para 49, p. 47.
95 *Ibid.*, p. 48.
96 Application no. 91 of 2012 dated March 16, 2016 (hereinafter referred to as *Aravali Ridge*).
97 *Aravali Ridge*, p. 12.
98 *Ibid.*, p. 13.
99 Execution Application no. 16/2016 In Application no. 135/2013 (THC) (disposed of on January 13, 2014) dated July 8, 2016 (hereinafter referred to as *Tansa Wildlife Sanctuary*).
100 *Tansa Wildlife Sanctuary*, para 10, p. 8.
101 *Tansa Wildlife Sanctuary*, para 12, p. 9.
102 Appeal no. 79 of 2013 dated January 8, 2016 (hereinafter referred to as *Hajira Mangroves*).
103 *Hajira Mangroves*, para 13, p. 20.
104 Appeal no. 14/2011 (T) dated January 28, 2016 (hereinafter referred to as *Pine & Chilgoza*).
105 *Pine & Chilgoza*, para 10, p. 14.
106 *Pine & Chilgoza*, para 33, p. 41.
107 Appeal no. 10/2011 (T) dated April 16, 2013 (hereinafter referred to as *Korba Forest*).
108 *Korba Forest*, para 21, p. 18.
109 The Tribunal locates it, *inter alia*, in Section 19 of the NGT Act, conferring on it the power to regulate its own procedure, as well as invokes the inherent power which, it believes, is vital for effective functioning of the Tribunal.
110 Original Application no. 10 of 2015, dated December 10/18, 2015 (hereinafter referred to as *Ganga Biodiversity*).
111 No. 200 of 2014.

112 Vide Civil Writ Petition no. 3727 of 1985. See para 2, p. 6.
113 *Ganga Biodiversity,* para 5, p. 9.
114 *Ganga Biodiversity,* see pp. 129–155.
115 Wildlife (Protection) Act, 1972.
116 *Kanha National Park,* p. 15.
117 *Kanha National Park,* para 16, p. 22.
118 *Western Ghats-I,* para 14, p. 15.
119 *Western Ghats-I,* para 17, p. 16.
120 *Ibid.,* p. 18.
121 *Western Ghats-I,* para 20, p. 20.
122 *Western Ghats-I,* para 21, p 20.
123 *Western Ghats-I,* para 24, pp. 23–24.
124 *Ibid.,* p. 24.
125 *Western Ghats-I,* para 42, pp. 37–38.
126 *Western Ghats-I,* para 79, p. 136.
127 Para 80, p. 139.
128 Para 81, p. 139.
129 *Western Ghats-I,* para 82, p. 140.
130 *Yamuna Biodiversity,* para 91, p. 133.
131 *Ibid.*
132 *Sterlite Industries (India) Ltd. v. Union of India (UOI) and Ors.,* (2013) 4 SCC 575.
133 *Yamuna Biodiversity,* para 92, p. 135.
134 *Ibid.,* p. 136. Yamuna Biodiversity is not the first case where the Tribunal has applied the principle of 'guesswork' to determine compensation. Before this case, this principle was applied in *Krishan Kant v. NGRBA* (2014) ALL (I) NGT Reporter (3) (Delhi) 1; and *S.P. Muthuraman* v. *Union of India & Ors.* (2015) ALL (I) NGT Reporter (2) (Delhi) 170.
135 *Yamuna Biodiversity,* para 35, pp. 38–39.
136 *Yamuna Biodiversity,* para 29, p. 59.
137 *Yamuna Biodiversity,* para 31, p. 63. To substantiate this decision, the Tribunal referred to a number of cases wherein the issue of *locus standi* had already been settled by the Tribunal viz. *Kishan Lal Gera* v. *State of Haryana & Ors.,* 2015 ALL (I) NGT Reporter (2) (Delhi) 286; *Goa Foundation* v. *Union of India* (2013) ALL (I) NGT Reporter (Delhi) 234; *Wilfred J.* v. *Ministry of Environment & Forests* (2014) ALL (I) NGT Reporter (2) (Delhi) etc.
138 *S.P. Muthuraman* v. *Union of India & Ors.* (2015) ALL (I) NGT Reporter (2) (Delhi) 170.
139 See, for instance, *Tribunal on Its Own Motion* v. *State of Himachal Pradesh,* NGT Judgment, February 6, 2014; *Tribunal on Its Own Motion* v. *Government of NCT, Delhi,* NGT Order, June 19, 2015.
140 K. Raj, 'Decentralizing Environmental Justice', *Economic & Political Weekly,* ISSN (online)–2349–8846, www.epw.in/journal/2014/48/web–exclusives/decentralising–environmental–justice.html (accessed on 25 September 2016).
141 The power is vested under clause (d) of sub-section 4 of Section 4 of the National Green Tribunal Act 2010 read with Rules 3, 4 and 6 of the National Green Tribunal (Practices and Procedure Rules) 2011, wherein the Chairperson may direct to have sitting of the NGT at any place other

than the place it ordinarily sits and may prescribe circuit procedure in that regard.

142 See G. N. Gill, 'Environmental Justice in India: The National Green Tribunal and Expert Members', *Transnational Environmental Law*, 2015: 29, available on CJO 2015 doi:10.1017/S2047102515000278 (hereinafter referred to as Gill, *Environmental Justice*).

143 *Impulse NGO Network* v. *State of Meghalaya,* Original Application no. 13 of 2014 and Ors. Before the National Green Tribunal, Circuit Bench at High Court of Meghalaya, Shillong, June 9, 2014, cited in Gill, *Environmental Justice*).

144 A. Subramani, 'High Court Can Hear Appeal Against National Green Tribunal Orders', *The Times of India*, February 3, 2014, http://timesof india.indiatimes.com/city/chennai/High-court-can-hear-appeal-against-National-Green-Tribunal-orders/articleshow/29876486.cms. (accessed on 11 August 2016).

145 *Report of High Level Committee to review various Act Administered by MEF&CC*, November 2014, Ministry of Environment, Forest and Climate Change (MOEF&CC), Government of India (hereinafter referred to as *HLC*).

146 *HLC*, para 8.2, p. 62.

147 The NEMA and SEMA, when established under ELMA, shall replace the Central Pollution Board and State Pollution Boards. *HLC*, para 8.2 (v), p. 63.

148 *HLC*, para 8.7, p. 64.

149 *HLC*, para 8.8, p. 65.

150 *HLC*, Sec 6.2 of Proposed ELMA, p. 71.

151 *HLC*, Section 15 of proposed ELMA, p. 75.

152 R. Dutta, M. Misra and H. Thakkar 'The High Level Committee Report on Environmental Law a Recipe for Climate Disaster and Silencing People's Voice', 7, www.ercindia.org/files/erc_desk/HLC%20Report.pdf. (accessed on 12 July 2016).

153 The Report was tabled before the Rajya Sabha in the current monsoon session of the Parliament. It went to the Rajya Sabha on July 21 and to the Lok Sabha on July 22. See www.downtoearth.org.in/news/par liamentary-standing-committee-rejects-tsr-subramanian-report-on-environmental-laws-50577 (accessed on 30 September 2016).

154 S. Banerjee, 'Parliamentary Standing Committee Rejects TSR Subramanian Report on Environmental Laws', July 24, 2015', www. downtoearth.org.in/news/parliamentary-standing-committee-rejects-tsr-subramanian-report-on-environmental-laws-50577 (accessed on 30 September 2016).

155 Gill, *Environmental Justice*.

10

ROLE OF THE JUDICIARY IN BIODIVERSITY ENFORCEMENT AND COMPLIANCE

Comparison between the International
Court of Justice and the Indian Judiciary

Stellina Jolly

The world's biodiversity, which sustains the life on the planet, is facing a serious threat. Major reasons which could be attributed for this inevitable decline include exponential growth of population, growing urbanization, mega developmental projects, destruction of habitats, environmental destruction, degradation and recently emerged climate change.[1] Multilateral efforts have been made at the international level to adopt regulatory and legislative measures aimed at countering the multitude of biodiversity hazards and restoring the lost biodiversity.[2] The results of these initiatives have been a host of robust and dynamic legislative and institutional mechanisms at the international and national levels.[3] However, even after decades of legislative and institutional initiatives aimed at biodiversity protection, studies and reports paint a grim picture of massive and unrelenting environmental degradation, which points to major critical gaps in biodiversity governance.[4] The focus of environmental and biodiversity governance has slowly shifted from preventive mechanisms to strong and effective compliance and implementation mechanisms.[5] Among the implementing institutions, the judiciary is strongly positioned to play a progressive role in creating and shaping a strong compliance mechanism.[6] The Supreme Court of India rendered a momentous verdict in the case of *Vedanta*[7] that emphasized the local communities' participation and access to information before a mining decision is made.

Judicial insistence on consultation is not merely an extension of the administrative principle of natural justice, but indicates its solemn

231

commitment to advance the declared principles of international environmental law adopted at multilateral conventions, including the Convention on Biodiversity (CBD)[8] and the Rio Declaration on Environment and Development (Rio Declaration).[9] Therefore, this chapter attempts to explore the central and pivotal role that the judiciary plays in enforcing biodiversity laws while exploring the role played by the International Court of Justice for biodiversity protection through analyzing the case of Pulp Mills[10] and Gabcikovo-Nagymaros.[11] Part I of the chapter traces the brief development of international environmental law and principles. Part II discusses the role played by the International Court of Justice in furthering biodiversity and environmental protection. Further, this chapter, through a catena of cases, dwells into the role played by the Indian judiciary in protecting and bringing compliance to India's international commitment at biodiversity protection and attempts to see whether it can be replicated at the international level.

I. Modern environmental law: an overview

Environmental issues do not follow the territoriality principle and often have trans-boundary impacts. The interdependent nature of environmental issues has led to gradual proliferation of multilateral legal instruments and a multitude of actors ranging from states and corporations to civil societies engaged in environmental conservation and protection.[12] There are more than 250 international treaties and other agreements related to the environment operating at the multilateral level.[13]

The initial phase of international environmental law was characterized by the concern for the conservation of nature, which spurred the first concerted international environmental regulations in the early 1900s.[14] Many consider the first phase of environmental law development as a conservation phase.[15] Gradually the focus of environmental protection got widened to include pollution aspects and other concerns.[16] However, these developments were not truly comprehensive and global; they were mostly measures introduced in the nature of an ad hoc and piecemeal approach.[17]

The real beginning of modern environmental law is traced to the historic United Nations (UN) Conference on the Human Environment, which produced a fundamental declaration of non-binding principles.[18] The Stockholm Declaration[19] for the first time saw the international community focusing its concerted attention on environmental problems.[20] However, developing countries were sceptical about the whole process and participated at the forum reluctantly. They questioned the need for such a conference and viewed it not just as a distraction, but also as a threat to their development pursuits.[21] The developing countries' main concern was on addressing questions of legitimacy in environmental governance rather than on the effectiveness of environment protection.[22] In order to bring developing countries to participation, the Stockholm Declaration linked the environment with a human right.[23] From a legal point of view, two developments are important: the Stockholm conference led to the creation of a new institutional structure like the United Nations Environment Programme to coordinate and assist in the progressive development of international law;[24] and the principle of no trans-boundary harm, enunciated in Stockholm, has emerged as a core legal norm of international environmental law through subsequent multilateral instruments.[25]

Stockholm was taken to its logical development through the 1992 United Nations Conference on Environment and Development (Rio Declaration). The Rio Declaration emphatically acknowledged an unqualified 'Right to Development';[26] however, it fell short of declaring a right to the environment as a human right.[27] The Rio Declaration saw the formulation, enunciation, refinement and emergence of the core principles of international environmental law.[28] Some of the major declared principles include sustainable development, intergenerational equity, common but differentiated responsibility, precautionary principle, polluter pays principle, environmental impact assessment and public participation, and access to information.[29] Though the Rio Declaration and accompanied documents are framed with flexibility and are in the form of soft law mechanisms, it is supplemented with adequate strong legal terms and intentions.[30]

The concept of sustainable development, which had its origin in the report of the World Commission on Environment and Development (WCED 1986),[31] is a core principle of the Rio Declaration. The WCED defined sustainable development as 'development, which meets the needs of the present generation without compromising the ability of future generations to meet their own needs.' The primary focus of the concept is on equity consisting of intragenerational as well as intergenerational equity.[32] The concept has its origin in economics and attempted to integrate economic factors with environmental and social factors.[33] Writers have struggled to bring the core components of sustainable development.[34] The term currently incorporates both substantive and procedural aspects. The substantive aspects include intergenerational equity, the precautionary principle, the polluter pays principle, the right to development and the principles of integration.[35] Procedural aspects cover public participation and environmental impact assessment.[36] The Rio Declaration incorporates both substantive and procedural aspects of sustainable development.[37]

Apart from being a mere component of sustainable development, the above-noted principles are self-standing and have been incorporated under the Rio Declaration. Jacqueline Peel writes that the Rio Declaration's endorsement of the precautionary principle has changed the discourse and reactions of environmental law regarding considerations of environmental risk.[38] Principle 15 of the Rio Declaration elaborates on the context and application of the precautionary principle. It states 'when serious threats and irreversible damage to environment is foreseen in situations, the absence of a full and comprehensive scientific certainty cannot be used as an excuse to postpone or avoid taking cost-effective preventive mechanisms.'[39] The polluter pays principle developed by the Council of the Organization for Economic Cooperation and Development (OECD) in 1972 was strengthened by the Rio Declaration.[40] An environmental impact assessment (EIA) is mandated for proposed activities which are likely to have a significant impact on the environment and society.[41] Principle 10 of the Rio Declaration sets out the fundamental elements for good environmental governance, focusing on access to information, public participation and access to justice.[42] These principles embody the universal consent achieved in international environmental law and are incorporated in multilateral legal instruments.[43]

The Rio Declaration also produced important environmental instruments, including the UN Framework Convention on Climate Change,[44] the CBD[45] and Agenda 21.[46] The Declaration was a compromise between developed and developing nations and tilted majorly in favour of developing nations due to their sheer numerical majority in the negotiations.[47] Developed nations came to the negotiation with a view to produce a legally binding document elaborating on the rights and duties of states. However, for developing nations, developmental concerns were the priority.[48] As a compromise, the Rio Declaration acknowledged the unqualified right to development of states, and developing nations agreed to pursue a development path based on sustainable development.[49]

In relation to biodiversity protection, early attempts to protect biodiversity were fragmented and were scattered among international rules that addressed the problem.[50] The Stockholm Declaration marked a fundamental change of attitude towards a more concrete form of biodiversity protection. Wildlife protection was the initial area to get prominence. Principle 4 of the 1972 Stockholm Declaration states

> Man has a special responsibility to safeguard and wisely manage the heritage of wildlife and its habitat, which are now gravely imperilled, by a combination of adverse factors. Nature conservation, including wildlife, must therefore receive importance in planning for economic development.[51]

The resolution to build up a comprehensive approach to protect the earth's biological diversity and to prevent its continuing degradation led the way for a global cooperative mechanism and culminated in the UN Convention on Biological Diversity.[52] When the CBD was negotiated and adopted, the key principles of environment protection were still struggling to find their foothold in a significant number of multilateral environmental instruments, which were advocating for a comprehensive approach.[53] The Rio Declaration was not only comprehensive in its approach, but also acted as an adhesive to environmental principles. The principle of sustainable development runs through the breadth and length of the convention. The declared objectives of the convention are a reflection of sustainable development, and include conservation of biological diversity, the sustainable use of its components and the fair and equitable sharing of the benefits arising out of the utilization of genetic resources.[54] It elaborates on the precautionary principle in the objective part of the text, where it states 'where there is a threat of significant reduction or loss of biological diversity, lack of full scientific certainty should not be used as a reason for postponing measures to avoid or minimize such a threat.'[55] The precautionary principle assumed critical importance in biodiversity protection and has been extensively made part of varied components, issues, approaches and types of biodiversity, including biosafety, marine and coastal biodiversity and the ecosystem approach.[56] The EIA has been conceived as a procedural mechanism whereby the developmental requirements and biodiversity concerns could be harmonized.[57] The CBD advances the principle of equity and implements the common but differential responsibility principle concentrating on technology transfer and a benefit-sharing mechanism founded on fair and equitable benefit sharing.[58] In addition to the recognized general principles of environmental law, certain new principles found their foothold in the CBD in the form of the access and benefit-sharing mechanism.[59] The trend of the Rio Declaration in enunciating environmental principles was carried further by the subsequent multilateral biodiversity agreements,[60] which devote whole articles to principles.[61] Currently, international environmental law is best with principles, some of them unique and exclusive to environment protection while others are borrowed from general international law for providing a framework to negotiate and implement conventions and treaties.[62] It reveals the trend of legal evolution and provides guidance for the future course, indicating the essential characteristics of law and institutions.[63] In spite of the vast and pervasive presence of these principles, a general consensus is lacking on the form, content and application to environmental issues.[64]

II. International environmental principles: judicial developments

Numerous multilateral conventions and conferences have produced a host of environmental principles forming the corpus of international environmental law.[65] In the words of Owada,

> it should be axiomatic that the task for ensuring observance of international environmental law should be looked at as much from the angle of the mechanism for preventing and monitoring of possible infractions of norms of international environmental law, as from the angle of the mechanism for dispute settlement and enforcement against such infractions ex post facto.[66]

Jorge Vinuales identified two main periods of cases in the International Court of Justice (ICJ) jurisprudence relating to international environmental law.[67] The first wave covers the Corfu Channel case[68] and an important obiter dictum made in the Barcelona Traction case.[69] In the Corfu Channel case, general principles of state liability were articulated. The court pointed out the state's obligation to not knowingly allow its territory to be used for acts contrary to the rights of other states.[70] However, the exact scope, standard of care and manner of application of the principle have been unclear.[71] In the Barcelona Traction case, the *erga omnes* nature of international obligations was pointed out. The first wave of cases did not directly concern the environment, but the ICJ attempted to put the existing principles of international law to environmental concerns, and no specific principles of environment were developed by the judiciary.[72] The ICJ developed a founding principle of trans-boundary liability based on due care and obligation. Even though this principle was not pronounced in the context of the environment, it had a strong bearing on the future development of international environmental law. The duty of no trans-boundary harm has been recognized under the later multilateral environmental agreements and subsequent judicial decisions.[73]

In the case of the Legality of Nuclear Weapons,[74] specific environmental concerns were raised and the court emphasized the protection of environment.[75] It observed that 'the environment is under daily threat and that the use of nuclear weapons could constitute a catastrophe for the environment.'[76] The court further observed, 'the environment is not an abstraction but represents the living space, the quality of life and the very health of human beings, including generations unborn.'[77] Kiss and Shelton point to the social and ethical dimensions of the environment, highlighted in the judgment as a significant contribution.[78] The court reiterated the general obligation of the states to prevent activities causing harm to other states and further extended the obligation to areas beyond national jurisdiction as part of the corpus of international law relating to the environment. The judgment enunciated a principle of liability in the nature of an *erga omnes* obligation.[79] The majority opinion, however, did not look into the legal status of various developed principles of international environmental law. From a legal perspective, this was a narrow interpretation, as it did not clarify the legal status of this corpus of international law, even when presented with an issue having substantial consequences to the environment.[80] Justice Weeramnatry, in his dissenting opinion, did not agree with this and noted that

> Environmental law incorporates a number of principles which are violated by nuclear weapons. The principle of intergenerational equity, the common heritage principle, the precautionary principle, the principle of trusteeship of earth resources, polluter pays principle, these principles of environmental law thus do not depend for their validity on treaty provisions. They are part of customary international law. They are part of the sine qua non for human survival.[81]

Justice Weeramantry elevated environmental law principles to the status of customary law. Specific attention was paid to the doctrine of the precautionary principle; the prohibition of nuclear weapons was advocated as in consonance with the precautionary principle. He further stated that the burden of proof of proving safety is on the party who has created the risk.[82] However, this reflects the view of Judge Weeramantry alone, who has often taken very progressive stances with respect to the environment.[83] Many consider it as a lost opportunity for the ICJ to apply the environmental principles and clarify their normative content. In this context, Koskenniemi's analysis may be significant; he believes that the focus of ICJ decisions is not adversarial and the ICJ generally searches for a compromise solution. In the national sphere, the purpose of litigation is to demarcate the liability, responsibility and breach of law explicitly, even if parties oppose it.[84] He links these phenomena to proceduralization, which is the characteristic feature of multilateral negotiations. Lack of clarity is one of the hallmarks of the environmental regimes. He believes that most of the regime building is a mere matter of political consensus rather than aiming for the most effective solution to the issue at hand.[85] In this process, the countries, parties to the negotiation generally agree for a framework and defer the difficult questions to be resolved thorough resolution by diplomatic efforts or legal dispute settlement.[86] The pointed context refers to a situation where all spheres of institutional structure of international governance operate on the basis of consensus.

The *Gabcikovo–Nagymaros Project*[87] was the first case before the International Court of Justice which directly pertained to questions of international environmental law and biodiversity protection.[88] The facts of the case related to a dispute which arose between Hungary and Czechoslovakia on the interpretation of a water management treaty signed between both nations in 1997.[89] Following the collapse of the communist government in 1980, Hungary witnessed a public outcry against ecological destruction.[90] As a result, the government halted all construction and work on Nagymaros in 1989.[91] The unrest and threat of ecological concern finally forced Hungary to unilaterally terminate the treaty.[92] The main argument of Hungary was based on ecological necessity and environmental deterioration, which they argued had clear bearing on biodiversity protection.[93]

The question before the court was on the legality of the Hungarian action of the abandonment of the project and ultimate termination of the treaty, and the ability of Czechoslovakia to proceed with Variant C unilaterally.[94]

The ICJ held that Hungary was not legally entitled to breach the treaty by abandoning works on the project. The court declared the termination of the treaty by Hungary to be invalid. During the course of arguments, many of the environmental principles were argued and examined. The concept of sustainable development was acknowledged in the opinions of both the majority and in the minority opinion of Justice Weeramantry.

The court observed that

> throughout the ages, mankind has, for economic and other reasons, constantly interfered with nature. Owing to new scientific insights and to a growing awareness of the risks for mankind for present and future generations of pursuit of such interventions at an unconsidered and unabated pace, new norms and standards have been developed, set forth in a great number of instruments during the last two decades. Such new norms have to be taken into consideration, and such new standards given proper weight, not only when States contemplate new activities but also when continuing with activities begun in the past. This need to reconcile economic development with protection of the environment is aptly expressed in the concept of sustainable development.[95]

The majority opinion looked at sustainable development as a mere concept. The implication which flows out of it is that though sustainable development has certain normative values under international law, it cannot claim to have achieved a legally binding position in the sense of custom or general principles of law.[96] But Justice Weeramantry, who expressed a dissenting opinion, considered it emphatically as forming a customary international law.[97] The utility of sustainable development lies in its reconciliation of economic considerations with the need to protect the environment.[98] However, it will be difficult to conclude whether the court was able to apply the principle in the context of the case and decide on its violation or non-violation. The reason could be, as Klabbers pointed out, 'judges instead of looking at the gravity of the provision breached was focusing on the gravity of breach itself.'[99]

The major line of the Hungarian arguments was based on ecological necessity. The court accepted the point that ecological concerns could constitute a state of necessity; however, the argument was rejected on the facts of the case.[100] The standard of proof required for the invocation of the doctrine of state necessity has to be an imminent, actual and certain threat.[101] A mere perception of threat is not satisfactory. But on the purpose and application, the court had different interpretations. The court observed that the existence of a state of necessity is not a ground for the termination of a treaty.[102] The court thus seemed to implicitly reject the precautionary principle's notion that uncertain harm was sufficient to prevent an action. Further, Hungary's claim was that the principle of duty to cause no harm has 'evolved into an erga omnes obligation of prevention of damage pursuant to the 'precautionary principle.' Slovakia, although it did not object to the importance of the precautionary principle itself, denied that it had 'risen to norms of *jus cogens* that would override the Treaty'.[103] The court did not refer directly to the precautionary principle and only recognized that the new norms and standards, which have been developed through a number of instruments, have to be taken into consideration not only when states plan new activities, but also when they continue with activities.[104] The opinion and observations of the court on the precautionary principle cannot be considered to be positive, as it did not invoke and apply the principle in the instant case.

The Pulp Mills case[105] between Argentina and Uruguay raised pertinent questions of international environmental law and directly addressed issues of biodiversity protection. Argentina alleged that Uruguay had breached its obligations under the *Statue of the River Uruguay*, a 1975 bilateral treaty authorizing construction of pulp mills on the River Uruguay and creating environmental hazards.[106] The treaty envisaged for the establishment of an Administrative Commission of the River Uruguay (CARU) to oversee the use and protection of the river. Argentina raised serious environmental violations. It contended that authorization and construction of the pulp mills by Uruguay had resulted in procedural violations, including failure to provide for the optimum utilization of water, the obligation to provide for prior notices, and the obligation to protect the aquatic environment and biodiversity. Argentina also pointed out that enough measures were not undertaken by Uruguay to control pollution, to cooperate and to undertake a comprehensive environmental impact assessment.[107]

The ICJ distinguished Argentina's claims on the basis of procedural claims and substantive claims.[108] It ruled that Uruguay's procedural obligation to notify and consult with Argentina prior to the authorization and construction of the mills had been breached.[109] However, the court negated the contention that procedural violations of obligations will automatically result in substantial violations of international obligation.[110] The request of Argentina for provisional measures was not considered even on the basis of sustainable development.[111] For the first time, it explicitly held that a trans-boundary environmental impact assessment is a requirement of customary international law.[112] The court noted that it is essential and mandatory to conduct an environmental impact assessment.[113] Through a progressive and ecological-centric approach, the court observed that while interpreting rules in treaties must be dynamic, new norms of environmental principles like the EIA need to be read into them.

But the court did not go into the detailed requirements of the EIA and its parameters. The court stated that it should be left to the discretion of states to determine the modalities and parameters through domestic legislations and policies. However, the court explicitly stated that the performance of an environmental impact assessment is a prior requirement which cannot be waived off and completed before the implementation of the project.[114] A negative outcome of the ruling was on public consultation of the affected populations. The court expressed the view that the parties had no international legal obligation to consult with the populations affected by the approval and construction of the mills.[115]

The court also approved the possibility for the customary status of the precautionary principle.[116] The court accepted that 'a precautionary approach may be relevant in the interpretation and application of the provisions of the Statute'.[117] The court argued for the 'ecosystems approach' to environmental protection, a largely precautionary device.[118] However, the court did not, in clear terms, declare the normative status of the precautionary principle. In the instant case, while declaring EIA to be a customary international norm, the ICJ indirectly rejected the normative value of the precautionary principle. This is surprising given the general understanding that the EIA is essentially a mechanism of precaution. The court in the instant case specifically lowered the standards of proof in international litigation and mandated that individual countries take a precautionary approach via environmental impact assessments when there is a serious environmental risk. One of the plausible arguments could be that lowering the standard would encourage states to bring environmental disputes against other states and the court could then play a proactive role.[119]

The case of *Ecuador v Colombia* (Aerial Herbicide Spraying Case) was an eagerly awaited decision which would have clarified the environmental and human rights aspects in a trans-boundary cases. However, before the hearing of the case on merits, the case was removed following a settlement agreement between both states.[120] Overall, the judgments affirm the importance of environmental considerations in addressing the rights and obligations of states.[121] However, it has fallen disappointingly short in declaring the normative values of environmental principles.[122]

III. International environmental law principles in India

The beginning of environmental law consciousness in India, as understood in the current legal lexicon of modern international environmental law, is attributed to the first United Nations Conference on the Human Environment, 1972.[123] The Stockholm Declaration was immediately followed by the enactment of a catena of environmental legislations.[124] The Forty-Second Amendment to the Indian Constitution in 1976 introduced principles of environmental protection in an explicit manner through Articles 48A and 51A(g) as Directive Principles of State Policy and Fundamental Duties, respectively.[125] Since these provisions form part of the Directive Principles of State Policy and Fundamental Duties, they are not legally and directly enforceable for individuals and groups.

With regard to specific matters of biodiversity protection, several legislative initiatives have been undertaken at the national level.[126] These statutes clearly acknowledge the role of international law in their enactment and implementation.[127] The Supreme Court of India has also pointed out, time and again, that the UN Conference on the Human Environment created awareness for environmental protection.[128] Unfortunately, the exponential growth of the legislative and institutional framework has been ineffective in protecting and preventing the downgrading of environmental standards in the country.[129] In this context, the judiciary stepped into the mantle and a judicial revolution was witnessed by the country beginning in 1980, which substantially altered the corpus of environment jurisprudence.[130] The initial phase of judicial movement was characterized by a positivist approach of restrictive legal interpretation.[131] An emergency saw the absolute negative verdict on fundamental rights.[132] The Bhopal environmental disaster threw open a host of challenges for the Indian judiciary, and also produced a flurry of legislative measures.[133] In this background, the judiciary was eager to project a positive image and not merely remain as a passive spectator.[134] The journey of Indian judicial activism in the field of the environment has been spectacular and has received wide attention and accolades.[135]

The Supreme Court adopted the care for the environment approach through a host of procedural and substantive modifications.[136] Procedurally, the locus standi rule was modified and the concept of public interest litigation was brought into operation; and substantially, an expansionist interpretation was given to Article 21, which incorporates the right to life.[137] Through the expanded notion of Article 21, the right to life was interpreted to include a healthy and clean environment.[138] The Supreme Court also utilized this expansion to adopt and apply the recognized international environmental principles to ensure environmental protection and protection of human rights.[139]

In *Narmada Bachao Andolan* v. *Union of India*, the court debated on the contents of sustainable development and observed that

'Sustainable Development means what type or extent of development can take place which can be sustained by nature or ecology with or without mitigation.'[140] In *Vellore Citizens' Welfare Forum* v. *Union of India*,[141] public interest litigation under Article 32 of the Constitution of India was filed against the massive pollution caused by the enormous discharge of untreated effluent by the tanneries and other industries in the State of Tamil Nadu. The court elaborately discussed the concept of 'sustainable development' and observed, 'The traditional concept that development and ecology are opposed to each of her, is no longer acceptable.'[142] The court held that 'sustainable development' as a balancing concept between ecology and development has been accepted as a part of the customary international law.[143] While pronouncing the customary nature of sustainable development, the court also invoked the common but differential responsibility by emphasizing the financial assistance to developing nations.[144]

The polluter pays principle has been held to be part of environmental jurisprudence in *Indian Council for Enviro-Legal Action* v. *Union of India*,[145] and in *Vellore Citizens Welfare Forum* v. *Union of India*, the court observed,

> Once the activity carried on is hazardous or inherently dangerous, the person carrying on such activity is liable to make good the loss caused to any other person by his activity irrespective of the fact whether he took reasonable care while carrying on his activity.[146]

In *A.P. Pollution Control Board* v. *Prof. M.V. Nayudu*,[147] the Supreme Court discussed the development of the precautionary principle and declared it to be part of the law of the land. In *Narmada Bachao Andolan* v. *Union of India*, the court explained that:

> When there is a state of uncertainty due to lack of data or material about the extent of damage or pollution likely to be caused then, in order to maintain the ecology balance, the burden of proof that the said balance will be maintained must necessarily be on the industry or the unit which is likely to cause pollution.[148]

The precautionary principle was followed in the case of *Research Foundation for Science Technology National Resource Policy* v. *Union of India and Anr.*[149] In *M.C. Mehta* v. *Kamal Nath*, applying the public trust doctrine, the court held that common properties such as rivers, forests, seashores and the air are to be held for the benefit of the public.[150] In the case of *State of Himachal Pradesh and Ors v. Ganesh Wood Products and Ors.*,[151] the court recognized the significance of intergenerational equity and held a government approval to establish that forest-based industry is contrary to the public interest involved in preserving forest wealth, maintenance of the environment and ecology, and considerations of sustainable growth and intergenerational equity.

With regard to biodiversity protection, in *T.N. Godavarmanv Thirumulpad v. Union of India*,[152] a writ petition was filed to protect a part of the Nilgiris forest from deforestation by illegal timber felling.[153] The court suo moto expanded the mandate and considered the issue to be of national importance.[154] The Supreme Court's vast assumption of powers concerning environmental issues had no precedence.[155] The decision touched upon vital concerns of forest management ranging from definition of forest, working plans, 'encroachment,' ecological aspects of forests, the economic role of forests, salaries of staff, and the role and constitution of the Forest Advisory Committee.[156] The Supreme Court defined a 'forest' in the absence of a definition in the Forest (Conservation) Act, 1980, and expanded the definition to cover all categories of forests, including privately held ones.[157] The order required state governments to constitute expert committees to measure the sustainability of forests.[158] Holding the adherence to the principle of sustainable development as a constitutional requirement, the court invoked 'continuing mandamus' to pass orders and directions.[159]

However, this case raises certain fundamental questions of constitutionalism in India regarding the extent of judicial power and the separation of powers doctrine.[160] Armin Rosencranz, in his powerful criticism, highlighted the negative fallout of an over-active judiciary. He points out that the court extended its assumption of powers beyond any reasonable time frame.[161] In its over-enthusiasm to protect the constitutional mandate and rights, the Supreme Court, in the instant case, travelled beyond its assigned role of interpreter of the constitution and transgressed into the field of the executive.[162]

In 2003, an environmental clearance was sought from the Ministry of Environment and Forests (MoEF) India by Vedanta Alumina Limited for the construction of an alumina refinery project near the Niyamgiri Hills in the Eastern Ghats.[163] The Dongria Kondh, an Adivasi community that for generations has depended entirely on the area for its economic, physical and cultural existence, occupies the Niyamgiri Hills.[164]

The Central Empowered Committee (CEC), part of the MoEF, received petitions opposing the construction of Vedanta's alumina refinery.[165] The MoEF-appointed panel reported that the company should not be given permission to mine bauxite in Niyamgiri.[166] In a landmark judgment, the Supreme Court of India held that mining could proceed only with consent of the Dongria Kondh communities.[167] In August 2013, all 12 tribal villages voted against Vedanta's project in the Niyamgiri Hills.

The Supreme Court decision legitimized the role of the gram sabha as a democratic decision-making forum.[168] The decision has a far-reaching significance and depicted the importance attached by the Supreme Court to the recognized human right to information and participation of affected stakeholders recognized at the international level, a principle of significance proclaimed at the Rio Declaration.[169] Principle 10 of the Rio Declaration emphasizes access to information, public participation and access to justice. It requires that each individual shall have appropriate access to information concerning the environment that is held by public authorities, and the opportunity to participate in decision-making processes.[170] States shall facilitate and encourage public awareness and participation by making information widely available. Access to justice to vindicate environmental legal rights has become a customary norm.[171] The Supreme Court order is a good precedent for arriving at a decision based on democratic practices and a good illustration to see the judicial innovativeness that has balanced environmental and social impacts of developmental projects having an effect on biodiversity.

In *Centre for Environmental Law WWF-1* v. *Union of India and Others*,[172] the court has called for revamping the standards for endangered species conservation. The court instructed the ministry to formulate plans to recover and preserve some of the threatened species for their intrinsic worth. It also stressed revamping the legislative structure and stressed the necessity of having an exclusive legislation for the preservation and protection of endangered species.

The journey of judicial activism in India has been anything but phenomenal. Decisions have not only provided much-needed clarity to the statutory provisions, it has created new norms and provided directions to Indian environmental jurisprudence based on social, economic and environmental justice. For the Indian Judiciary, international law has been a catalyst in their judicial pursuit and they have liberally borrowed from the international principles to steer forward countries environmental governance and jurisprudence. Even when the status of environmental principles at the international level is not made certain, the judiciary in India has declared the customary nature of those principles. The influence of international law can be clearly seen in cases like *Research Foundation for Science Technology and Natural Resources Policy* v. *Union of India* where the court directed the government to take precautionary measures on the basis of the Basel Convention on Hazardous Wastes and asked for clarification on what steps are to be taken by the government in implementing provisions of the Basel Convention.

IV. Conclusion

The concerns for global environmental problems have resulted in a growing number of legal instruments in the form of multilateral environmental agreements. In comparison to environment disputes, the ICJ has contributed to the substantive developments in other fields and new rules have emerged from the pronouncements of the ICJ, or existing rules have been moulded or given new interpretation by it.[173] The case of Reparation Injuries advisory opinion[174] has provided much clarity and normative framework to international law. By pronouncing the legal personality of the United Nations to bring a claim against a state on behalf of its personnel, who had been injured by that state, it has created a new norm. In the context of the environment, even when presented with an opportunity, the ICJ did not clarify the legal position of liability in the nuclear testing case. Justice Weeramantry has attempted to link the right to environment as a human right; however, it was a minority opinion reflective of his individual viewpoint. With respect to sustainable development, in spite of the multitude of convention provisions eloquently speaking the language of sustainable development, the ICJ failed to recognize the clear normative content of the principle.

The structural and functional elements of the ICJ could be responsible for this discrepancy; ICJ decisions do not operate as precedent and it cannot enforce its own judgments.[175] It relies on the good will of states to enforce the decisions. The field of environmental law is a new area, still undergoing major transformation. In this scenario, establishing aggressive and strict decisions will only weaken the enforcements and countries may be discouraged from approaching the ICJ. This point may become pertinent given the scenario where environmental principles are couched in the language of soft law. This situation could change with further development and strengthened environmental law.

The very fact that Justice Weeramantry, through his minority opinion, was able to pronounce on the core principles of international law, is evidence that it could be replicated at the international level also. The pronouncements of the ICJ can undoubtedly alter the legal status and behaviour of states regarding international environmental principles. In the opinion of Roslyn Higgins, for the progressive interpretation of rules and principles, the 'International Court to is to continue doing what it does, namely meticulously apply international law in an impartial manner to the disputes before it.'[176]

At the national level, constitution and statutes define the role and powers of judiciary. The judiciary has been entrusted to apply and interpret the laws made by the legislature. The power of judicial review has made Indian judiciary one of the most powerful institutions in the world. Through the application of judicial review, judiciary has been creating norms and positively interpreting the contours of law. Unlike national legal systems where certainty is the norm, international law operates in a space plagued by uncertainty and power of judicial review has not been conferred on the ICJ.

At the international level, very few cases have been addressed relating to environmental concerns. States generally base their arguments on other treaty mechanisms rather than on environmental law provisions. One of the factors which contribute to the reluctance of the states could be attributed to the fact that most of the environmental law provisions are couched in the form of soft laws, lacking the binding force, and only a few customary principles have been developed in the context of environmental protection. However, the lack of an established and clear law should not be a ground for rejecting the decision in a given case.[177]

The perusal of cases decided by the ICJ and Indian judicial bodies reveals that, at the domestic level, the judiciary has been more successful in terms of articulating, applying and ensuring compliance of those environmental principles. The saga of judicial activism in India is a testimony to the conviction people had in the efficacy and effectiveness of the judiciary.

Notes

1 'Global Bio Diversity Outlook 3, Convention on Bio Diversity', 2011, 15, www.cbd.int/doc/publications/gbo/gbo3–final–en.pdf (accessed on 3 July 2016); see generally, A. Gillespie, *Conservation, Biodiversity and International Law: New Horizons in Environmental and Energy Law Series*, London: Edward Elgar Publishing Ltd, 2013.
2 See generally, P. Sands et al., *Principles of International Environmental Law*, Cambridge: Cambridge University Press 2012; P. Sands, *Greening International Law*, London: Earthscan Publications, 2011.
3 M. Bowman, 'The Nature, Development and Philosophical Foundations of the Bio Diversity Concept in International Law', in M. Bowman and

C. Redgwell (eds.), *International Law and the Conservation of Biological Diversity*, Hague: Kluwer Law International, 1996, pp. 5-32.

4 Global Bio Diversity Outlook 15; see K. J. Markowitz and Jo J. A. Gerardu, 'The Importance of the Judiciary in Environmental Compliance and Enforcement', *Pace Environmental Law Review*, 2012, 29(2): 538-554, at 539, http://digitalcommons.pace.edu/pelr/vol29/iss2/5 (accessed on 3 June 2016).

5 D. Bodansky, *The Art and Craft of International Environmental Law*, Cambridge: Harvard University Press, 2010, p. 205.

6 Markowitz and Gerardu, 'Environmental Compliance and Enforcement', p. 540.

7 *Orissa Mining Corporation* v. *Ministry of Environment and Forests & Others*, 2013, 6 S.C.R. 881.

8 Convention on Biological Diversity, June 5, 1992, United Nations Treaty Series, Vol. 1760, p. 79 (No. 30619).

9 Rio Declaration on Environment and Development, adopted June 14, 1992, UN Doc. AICONF. 1515/Rev. 1 (1992) (hereinafter Rio Declaration).

10 Case Concerning Pulp Mills on the River Uruguay (*Argentina* v. *Uruguay*) (judgment) [2010] ICJ Rep. 14, General List No.135.

11 *Gabcikovo–Nagymaros Project (Hungary* v. *Slovakia)* [1997] ICJ Rep. 7.

12 Sands et al., *Principles of International Environmental Law*, p. 3.

13 D. F. C. Joseph, *The Global Environment and International Law*, Austin: University of Texas Press, 2003.

14 D. Bodansky, 'Craft of International Environmental Law', in A. C. Kiss and D. Shelton (eds.), *International Environmental Law*, Volume 4, New-York: Transnational Law Publishers, 2004, p. 40.

15 Kiss and Shelton, *International Environmental Law*, 2004; Sands, *International Environmental Law*, 2003.

16 *Ibid.*

17 Bodansky, 'Craft of International Environmental Law'.

18 See generally, P. Birnie, A. Boyle and C. Redgwell, *International Law and the Environment*, Oxford: Oxford University Press, 2009.

19 Declaration of the United Nations Conference on the Human Environment, adopted June 16, 1972, UN Doc. A/CONF. 48/14, reprinted in 11 I.L.M. 1416 (1972) (hereinafter Stockholm Declaration).

20 E. B. Weiss, 'The Evolution of International Environmental Law', *Japanese Year Book International Law*, 2011, 5(4): 1–27 at 5; see generally Birnie, Boyle and Redgwell, *International Law and the Environment*, 2009.

21 A. Najam, 'Developing Countries and Global Environmental Governance: From Contestation to Participation to Engagement', *International Environmental Agreements*, Springer, 2005(5): 303–321 at 307.

22 *Ibid.*, p. 307.

23 Weiss, 'The Evolution of International Environmental Law', p. 16; M. Anita Halvosren, 'The Origin and Development of International Environmental Law', in S. Alam et al. (eds.), *Routledge Hand Book of International Environmental Law*, London: Routledge, 2013, pp. 25–42.

24 C. A. Petsonk, 'The Role of the United Nations Environment Programme (UNEP) in the Development of International Environmental Law', *American University International Law Review*, 1991, 5(2): 351–352.

25 M. Pallemarts, 'International Environmental Law from Stockholm to RIO: Back to the Future', in P. Sands (ed.), *Greening International Law*, London: Earthscan Publications, 1993, pp. 1–19.

26 Rio Declaration, Principle 3.

27 D. Shelton, 'Human Rights and the Environment: Substantive Rights', in M. Fitzmaurice, D. M. Ong and P. Merkouris (eds.), *Research Handbook on International Environmental Law Series*, London: Edwin Elgar Publishers, 2011, pp. 265–284.

28 S. Atapattu, *Emerging Principles of International Environmental Law*, London: Brill, 2007, pp. 24 and 79.

29 S. Tareq et al., 'Principles of International Environmental Law', in S. Alam et al. (eds.), *Routledge Handbook of International Environmental Law*, London: Routledge, pp. 42–60.

30 D. A. Wirth, 'The Rio Declaration on Environment and Development: Two Steps Forward and One Back, or Vice Versa', *Georgia Law Review*, 1995, 29: 599–653.

31 World Commission on Environment and Development (WCED), *Our Common Future*, Oxford: Oxford University Press, 1987, p. 8.

32 E. B. Weiss, 'In Fairness to Future Generations and Sustainable Development', *American University International Law Review*, 1992, 8(1): 19–26 at 19.

33 D. McGoldrick, 'Sustainable Development and Human Rights: An Integrated Conception', *International and Comparative Law Quarterly*, 1996, 45: 796–818.

34 Boyle, Birnie and Redgwell, *International Law and the Environment*, 2009, p. 125.

35 M. Fitzmaurice, *Contemporary Issues in International Environmental Law*, London: Edwin Elgar Press, 2009.

36 *Ibid.*, p. 68.

37 *Ibid.*

38 J. Peel, 'Changing Conceptions of Environmental Risk', in J. E. Viñuales (ed.), *The Rio Declaration on Environment and Development: A Commentary*, Oxford: Oxford University Press, 2015, 75–85, p. 76.

39 Rio Declaration, Principle 15.

40 S. Atapattu, 'Sustainable Development, Myth or Reality? A Survey of Sustainable Development Under International Law and Sri Lankan Law', *Georgetown International Environmental Law Review*, 2001(14): 265–300 at 268.

41 A. Rieu–Clarke, *International Law and Sustainable Development*, London: IWA Publishing, 2005, p. 89; see Rio Declaration, Principle 17.

42 D. Banisar et al., 'Moving From Principles to Rights: Rio 2012 and Access to Information, Public Participation, and Justice', *Sustainable Development Law & Policy*, 2012, 12(3): 8–14.

43 S. Bhutani and A. Kothari, 'The Biodiversity Rights of Developing Nations: A Perspective From India', *Golden University Law Review*, 2002, 32(4): 587–627 at 590.

44 'United Nations Framework Convention Climate Change (UNFCC)', May 21, 1992, United Nations Treaty Series, Vol. 1771, p. 107 (No. 30822).

45 Convention on Biodiversity.
46 United Nations Conference on Environment and Development, Agenda 21, UN Doc. A/ CONF.151/26/Rev.1 (1992).
47 Najam, *Global Environmental Governance*, p. 311.
48 *Ibid.*, p. 310.
49 Rio Declaration principle 3; Attapattu, 'Sustainable Development, Myth or Reality', p. 270.
50 Sands and Peel, *Principles of International Environmental Law*, p. 450; See generally J. Scanlon, and F. Burhenne-Guilmin, *International Environmental Governance: An International Regime for Protected Areas*, Gland, Switzerland and Cambridge: IUCN, 2004, p. 31.
51 Stockholm Declaration, Principle 4.
52 Convention on Biodiversity.
53 C. Oguamanam, 'Biological Diversity', in S. Alam et al. (eds.), *Routledge Handbook of International Environmental Law*, pp. 209–227 at 220.
54 Convention on Biodiversity, Article 1.
55 R. Cooney, 'The Precautionary Principle in Biodiversity Conservation and Natural Resource Management: An Issues Paper for Policy-Makers, Researchers and Practitioners', Gland, Switzerland and Cambridge: IUCN 2004. 24; Convention on Biodiversity, Article 17.
56 Cooney, *The Precautionary Principle*, p. 24.
57 Convention on Biodiversity, Article 14.
58 Oguamanam, *Biological Diversity*, p. 221.
59 Oguamanam, *Biological Diversity*, p. 222.
60 Scanlon, Guilmin and Burhenne, *International Environmental Governance*, p. 31. The Liaison Group of the Biodiversity-Related Conventions formed in June 2004. It identifies the following main biodiversity-related conventions: Convention on Biological Diversity, 1992 (CBD); Convention on International Trade in Endangered Species of Wild Fauna and Flora, 1973 (CITES), 12 I.L.M. 1055 (1973); Convention on Migratory Species of Wild Animals, 1979 (CMS), 19 I.L.M. 15 (1980); Ramsar Convention on Wetlands of International Importance especially as Waterfowl Habitat (Ramsar), 1971, 11 I.L.M. 963 (1972); Convention Concerning the Protection of World Cultural and Natural Heritage (WHC), 1972.
61 S. Atapattu, 'The Significance of International Environmental Law Principles in Reinforcing or Dismantling the North–South Divide', in S. Alam et al. (eds.), *International Environmental Law and the Global South*, New York: Cambridge University Press, 2015, pp. 74–108 at 75.
62 Atapattu, 'Reinforcing or Dismantling the North–South Divide', p. 75.
63 L. Kurukulasuriya and A. Nicholas Robinson, *Training Manual on International Environmental Law*, Nairobi: UNEP, 2006, p. 22.
64 Atapattu, 'Reinforcing or Dismantling the North–South Divide', p. 75.
65 Weiss, 'The Evolution of International Environmental Law', p. 54.
66 H. Owada, 'International Environmental Law and the International Court of Justice', Inaugural Lecture at the Fellowship Programme on International and Comparative Environmental Law, Washington, DC, 2006, p. 14. http://ias.jak.ppke.hu/hir/ias/200634sz/owada.pdf (accessed on 23 August 2016).
67 E. J. Viñuales, 'The Contribution of the International Court of Justice to the Development of International Environmental Law: A Contemporary

Assessment', *Fordham International Law Journal*, 2008, 32(1): 232–258 at 236.
68 [1949] ICJ Rep. 4.
69 Barcelona Traction (*Belg.* v. *Spain*), 1970 I.C.J. 3, J. Vinuales, 'The Contribution of the International Court of Justice', p. 236.
70 Corfu Channel, Merits, Judgment, I.C.J. Reports 1949, 22.
71 Fitzmaurise, 2007: 300.
72 Vinuales, 'The Contribution of the International Court of Justice', p. 240.
73 Ibid., p. 239. Principle 21 of the Stockholm Declaration and Principle 2 of the Rio Declaration have incorporated the no harm principle.
74 Legality of the Threat or Use of Nuclear Weapons, Advisory Opinion, 1996 I.C.J. 266.
75 E. Louka, *International Environmental Law: Fairness, Effectiveness, and World Order*, Cambridge: Cambridge University Press, 2006, p. 45.
76 Legality of Nuclear Weapons, para 29.
77 *Ibid.*
78 P. P. Rogers, K. F. Jalal and J. A. Boyd, *An Introduction to Sustainable Development*, Cambridge: Harvard University Press, 2005.
79 Louka, *Fairness, Effectiveness*, p. 241.
80 *Ibid.*
81 Legality of Nuclear Weapons, pp. 502–504.
82 *Ibid.*
83 Vinuales, 'The Contribution of the International Court of Justice', p. 247.
84 T. Stephens, *International Courts and Environmental Protection*, Cambridge: Cambridge University Press, 2009.
85 *Ibid.*, p. 98.
86 *Ibid.*, p. 99.
87 *Gabcikovo–Nagymaros Project (Hungary* v. *Slovakia)* [1997] ICJ Rep. 7.
88 A. A. Khavari and D. R. Rothwell, 'The ICJ and the Danube Dam Case: A Missed Opportunity for International Environmental Law?' *Melbourne University Law Review*, 1998, 22: 505–537 at 507; E. A. Boyle, 'The Gabcikovo–Nagymaros Case: New Law in Old Bottles', *Yearbook of International Environmental Law*, 1997(8): 1–20 at 3.
89 A. Schwabach, 'Diverting the Danube: The Gabcikovo–Nagymaros Dispute and International Freshwater Law', *Berkeley Journal of International Law*, 1996(14): 291–343 at 292, http://scholarship.law.berkeley.edu/bjil/vol14/iss2/2 (accessed on 12 August 2016).
90 *Ibid.*, p. 292.
91 H. Denise, 'Danube "Blues" Alter Politics for Hungarians', *Los Angeles Times*, August 13, 1989, p. 13, Part 1.
92 *Gabcikovo–Nagymarospara*, pp. 25 & 27.
93 C. Katona, 'Case Concerning the Gabčíkovo–Nagymaros Project: Hungary Versus Slovakia', *Glendon Journal of International Studies*, 1997(7): 1–15 at 5.file:///Users/sau/Downloads/37120–45826–1–PB.pdf (accessed on 2 May 2016).
94 *Gabcikovo–Nagymaros.*
95 *Gabcikovo–Nagymaros*, para 140, p. 78.
96 J. M. Gillroy, 'Adjudication Norms, Dispute Settlement Regimes and International Tribunals: The Statue of "Environmental Sustainability" in International Jurisprudence', *Stanford Journal of International Law*, 2006,

42(1): 27–30; see P. Taylor, 'Case Concerning the Gabcikovo–Nagymaros Project: A Message from the Hague on Sustainable Development', *New Zealand Journal of Environmental Law*, 1999(3): 109, 114; see T. Lauren, 'The International Court of Justice's Treatment of "Sustainable Development" and Implications for Argentina v. Uruguay', *Sustainable Development Law & Policy*, 2009, 10(1): 40, 85.

97 *Gabcikovo–Nagymaros.*

98 V. Barral, 'Sustainable Development in International Law: Nature and Operation of an Evolutive Legal Norm', *The European Journal of International Law*, 2012, 23(2): 377–400 at 387.

99 J. Klabbers, 'The Substance of Form: The Case Concerning the Gabcikovo–Nagymaros Project, Environmental Law, and the Law of Treaties', *Yearbook of International Environmental Law*, 1997, 8: 32–40, 34.

100 J. Howley, 'The Gabcikovo–Nagymaros Case: The Influence of the International Court of Justice on the Law of Sustainable Development', *Queensland Law Student Review*, 2009, 2(1): 1–19 at 8.

101 Gabcikovo–Nagymaros, at para 94, p. 60; Thomson Lee, 'The ICJ and the Case Concerning the Gabcikovo–Nagymaros Project: The Implications for International Watercourses Law and International Environmental Law', *CEPMLP Annual Review*, 2009, www.dundee.ac.uk/cepmlp/gateway/?news=27951 (accessed on 4 April 2016).

102 *Gabcikovo–Nagymaros*, para 95, p. 60; Thomson, Lee, *Case Concerning the Gabcikovo–Nagymaros*, p. 42.

103 *Ibid.*, para 97, p. 62.

104 *Ibid.*, para 140, p. 78.

105 Case Concerning Pulp Mills.

106 Case Concerning Pulp Mills, para 25.

107 Case Concerning Pulp Mills, para 22.

108 D. K. Anton, 'Case Concerning Pulp Mills on the River Uruguay' (*Argentina v. Uruguay*) (judgment) [2010] ICJ Rep. (20 April 2010), *Australian International Law Journal*, 2013(13): 213–223.

109 Anton, 'Case Concerning Pulp Mills', p. 214, Case Concerning Pulp Mills, at paras 67–168.

110 Case Concerning Pulp Mills, paras 169–266.

111 Case Concerning Pulp Mills, para 8.

112 Case Concerning Pulp Mills, paras 203–219; M. Panos, 'Case Concerning Pulp Mills on the River Uruguay (*Argentina v. Uruguay*): Of Environmental Impact Assessments and "Phantom Experts" Case Concerning Pulp Mills on the River Uruguay (*Argentina v. Uruguay*)', Judgment of April 20, http://haguejusticeportal.net/Docs/Commentaries%20PDF/Merkouris_Pulp%20Mills_EN.pdf (accessed on 6 May 2016).

113 C. Payne, 'Pulp Mills on the River Uruguay: The International Court of Justice Recognizes Environmental Impact Assessment as a Duty Under International Law', *American Society of International Law Insights*, 2010(14): 9, www.asil.org/insights100422.cfm (accessed on 6 May 2016).

114 Payne, 'Pulp Mills on the River Uruguay', p. 10.

115 Anton, 'Case Concerning Pulp Mills', p. 221.

116 A. Boyle, 'Pulp Mills Case: A Commentary', www.biicl.org/files/5167_pulp_mills_case.pdf (accessed on 3 March 2016).

117 Case Concerning Pulp Mills, para 164.

118 Boyle, 'Pulp Mills Case: A Commentary'.

119 D. Kazhdan, 'Precautionary Pulp: Pulp Mills and the Evolving Dispute Between International Tribunals Over the Reach of the Precautionary Principle', *Ecology Law Quarterly*, 2011, 38: 528–552 at 547. http://scholarship.law.berkeley.edu/elq/vol38/iss2/11 (accessed on 3 May 2016).

120 *Aerial Herbicide Spraying Case (Ecuador* v. *Colombia)* (Application Instituting Proceedings) Pleading, 2008 ICJ, March 31, 2008, www.icj-cij.org/docket/files/138/14474.pdf (accessed on 9 August 2016).

121 P. Sands, International Courts and the Concept of 'Sustainable Development', 1994, 3 Max Plank Yearbook of United Nations Law, 89–405 at 477.

122 Taylor, *Case Concerning the Gabcikovo–Nagymaros Project*, p. 114.

123 S. Jolly, 'Application of Solar Energy in South Asia: Promoting Intergenerational Equity in Climate Law and Policy,' *International Journal of Private Law*, 2014, 7(1): 20–39 at 23; C. M. Abraham, *Environmental Jurisprudence in India*, Hague: Kluwer Law International, 1999, p. 2.

124 The Water (Prevention and Control of Pollution) Act of 1974, The Forest Conservation Act (1981), The Air Prevention and Control of Pollution Act (1986), The Environmental Protection Act (1986).

125 Jolly, 'Application of Solar Energy in South Asia', p. 23; Constitution of India, inserted by 42nd Amendment, 1976, Article 48A: 'The State shall endeavour to protect and improve the environment and to safeguard the forests and wild life of the country'; Article 51A(g) 'to protect and improve the natural environment including forests, lakes, rivers and wild life, and to have compassion for living creatures'.

126 The Water (Prevention and Control of Pollution) Act, 1974; The Air (Prevention and Control of Pollution) Act, 1981; The Environment (Protection) Act, 1986; The Bio Diversity Act, 2002; The Forest Conservation Act, 1980; The Wildlife Protection Act, 1972.

127 S. P. Jaswal and N. Jaswal, *Environmental Law: Environmental Protection, Sustainable Development and the Law*, India: Pioneers Publications, 2006.

128 See *Rural Litigation and Entitlement Kendra* v. *State of U.P.* (1986) Supp SCC 517; *Vellore Citizens Welfare Forum* v. *Union of India* (1996) 5 SCC 647; *M. C. Mehta* v. *Union of India* (1992) 1 SCC 358; *M.C. Mehta* v. *Kamal Nath* (2000) 6 SCC 213.

129 Jaswal, *Environmental Law*, p. 87.

130 Abraham, *Environmental Jurisprudence in India*, p. 3.

131 M. Mate, 'Globalisation, Rights and Judicial Activism in the Supreme Court of India', *Pacific Rim Law & Policy Journal*, 2016, 25: 643–671 at 647; A. Rosencranz and M. Jackson, 'The Delhi Pollution Case: The Supreme Court of India and the Limits of Judicial Power', *Columbia Journal of Environmental Law*, 2003, 28: 223–254 at 229; H. L. Banks, 'Saving the Trees One Constitutional Provision at time: Judicial Activism and Deforestation in India', *Georgia Journal of International and Comparative Law*, 2011, 40: 751–779, p. 764.

132 Rosencranz and Jackson, 'The Delhi Pollution Case', p. 229.

133 C. M. Abraham and S. Abraham, 'The Bhopal Case and the Development of Environmental Law in India', *International & Comparative Law Quarterly*, 1991, 40: 334–365 at 334.

134 *Ibid.*, p. 334.
135 S. P. Sathe, 'Judicial Activism: The Indian Experience', *Washington University Journal of Law & Policy*, 2002, 6: 29–108 at 62.
136 G. Sahu, 'Implications of Indian Supreme Court's Innovations for Environmental Jurisprudence', *Law, Environment and Development Journal*, 2008, 4(1): 375–394 at 378, www.lead–journal.org/content/08375.pdf (accessed on 3 May 2016).
137 V. N. Shukla, *Constitution of India*, Allahabad: Eastern Book Co., 2008.
138 *Virender Gaur* v. *State of Haryana* 1995 (2) SCC571; *Indian Council for Enviro-Legal Action* v. *Union of India* (Bichhri case) AIR 1996 SC 1146; *M.C Mehta* v. *Union of India* AIR 1988 SC 1037; *Rural Litigation and Entitlement Kendra, Dehradun* v. *State of Uttar Pradesh* AIR 1988 SC 2187; L. Rajamani, 'The Right to Environmental Protection in India: Many a Slip Between the Cup and the Lip?' *Review of European Community and International Environmental Law*, 2007, 16(3): 277; K. Thakur, *Environment Protection Law and Policy In India*, New Delhi: Deep and Deep Publications, 1997, p. 204.
139 M. K. Ramesh, 'Environmental Justice: Courts and Beyond', *Indian Journal of Environmental Law*, 2002, 3(1): 20; Thakur, *Environment Protection Law*, p. 204.
140 *Narmada Bachao Andolan* v. *Union of India*, 2002 10 SCC 664.
141 *Vellore Citizens' Welfare Forum* v. *Union of India*, 1996, 5 SCC 647.
142 *Vellore Citizens*, para 10.
143 *Ibid.*
144 *Vellore Citizens*, at para 11.
145 *Indian Council for Enviro-Legal Action* v. *Union of India* (1996) 3 SCC 212, p. 215.
146 *Vellore Citizens*, para 12.
147 *A.P. Pollution Control Board* v. *Prof. M.V. Nayudu*, 1999, 2 SCC 718.
148 *Narmada Bachao Andolan*, p. 735; see Philip Cullet, *The Sardar Sarovar Dam Project: Selected Documents*, London: Ashgate Publishing, Ltd, 2007, p. 185.
149 *Research Foundation for Science Technology National Resource Policy* v. *Union of India and Anr.* 2005, 10 SCC 510.
150 *M.C. Mehta* v. *Kamal Nath* 1997(1) SCC 388, p. 34;, see Philip Cullet, *Water Law for the Twenty First Century*, London: Routledge, 2009, p. 188.
151 *State of Himachal Pradesh and Ors.* v. *Ganesh Wood Products and Ors.*, AIR 1996 SC 149: (1995) 6 SCC 363.
152 *T.N. Godavarman Thirumulpad* v. *Union of India, (*1996) 9 SCR 982, (2008) 2 SCC 222.
153 A. Rosencranz and L. Sharachchandra, 'Supreme Court and Indian Forests', *Economic and Political Weekly*, February 2008, 43(5): 2–8, 2008, 11–14 at 11.
154 R. Dutta and B. Yaday, *The Supreme Court on Forest Conservation*, New Delhi: Universal Law Publishing Company, 2007.
155 A. Rosencranz, E. Boenig and B. Dutta, 'The Godavarman Case: The Indian Supreme Court's Breach of Constitutional Boundaries in Managing India's Forests', *ELR*, 2007(37): 10032–10042 at 10032, http://elr.info/sites/default/files/articles/37.10032.pdf, (accessed on 10 April 2016).
156 *T. N. Godavarman* v. *Union of India*, para 5.

157 N. Choudhary, 'From Judicial Activism to Adventurism – The Godavar-
     man Case in the Supreme Court of India', *Asia Pacific Journal of Envi-
     ronmental Law*, 2014, 17: 178–189, p. 184; L. Rajamani and S. Ghosh,
     'Avenues for Climate Change Litigation in India', in R. Lord et al. (eds.),
     Climate Change Liability: Transnational Law and Practice, Cambridge:
     Cambridge University Press, 2012, pp. 139–177.
158 Rosencranz, 'Supreme Court and Indian Forests', p. 14.
159 Rosencranz, Boenig and Dutta, '*The Godavarman Case*', p. 10032; K.
     Sivaramakrishnan, 'Environmental, Law and Democracy in India', *Jour-
     nal of Asian Studies*, 2011, 70(4): 905–928 at 905.
160 S. Jolly, 'Ninth Schedule, Basic Structure, and Constitutionalism', *Pan-
     jab. University Law Review*, 2007, 48: 51–62 at 54.
161 Rosencranz, 'Supreme Court and Indian Forests', p. 13.
162 *Ibid.*
163 'Vedanta Resources Lawsuit re Dongria Kondh in Orissa', http://busi
     ness-humanrights.org/en/vedanta-resources-lawsuit-re-dongria-
     kondh-in-orissa (accessed on 1 April 2016).
164 P. Das, 'A Mining Controversy', *Frontline*, November 20–December 3,
     2004, 21(24).
165 The CEC's mandate is to monitor and ensure compliance with orders of
     the Supreme Court concerning forests and wildlife. Report of the Four
     Member Committee for Investigation into the Proposal Submitted by the
     Orissa Mining Company for Bauxite Mining in Niyamgiri, 2010.
166 Report of the Four Member Committee for Investigation into the Pro-
     posal Submitted by the Orissa Mining Company for Bauxite Mining in
     Niyamgiri. Submitted to the Ministry of Environment & Forests Gov-
     ernment of India New Delhi, 2010, http://envfor.nic.in/sites/default/files/
     Saxena_Vedanta-1.pdf (accessed on 31 March 2016).
167 *Orissa Mining Corporation.*
168 'The Significance of Niyamgiri', *The Hindu*, May 3, 2013, www.thehindu.
     com/opinion/editorial/the-significance-of-niyamgiri/article4677438.ece
     (accessed on 27 February 2016).
169 Rio Declaration, Principle 10; Wirth, *The Rio Declaration on Environ-
     ment*, p. 165.
170 Banisar, *Moving From Principles to Rights*, p. 51.
171 N. A. Robinson, 'Ensuring Access to Justice Through Environmental
     Courts', *Pace Environmental Law Review*, 2012, 29: 363, http://digital
     commons.pace.edu/pelr/vol29/iss2/1 (accessed on 4 May 2016).
172 Writ Petition (Civil No. 337 of 1995 Supreme Court, 2013).
173 T. Ginsburg, 'International Judicial Lawmaking', *Illinois Law and Eco-
     nomics Working Papers Series*, Working Paper No. LE05-006, 2005,
     1–56, file:///Users/sau/Downloads/SSRN-id693861.pdf (accessed on 5
     May 2016).
174 [1949] ICJ Rep. 174.
175 M. Shahabuddeen, *Precedent in the World Court*, Cambridge: Cam-
     bridge University Press, 2007, p. 68.
176 R. Higgins, '*The Rule of Law: Some Sceptical Thoughts*' in *Themes and
     Theories*, Volume 11, Oxford: Oxford University Press, 2009, p. 1335.
177 R. P. Anand, *Compulsory Jurisdiction of International Court*, India:
     Hope India Publications, 2008, p. 59.

Part VI

THE WAY FORWARD

11

FEMINIST DIMENSION OF BIODIVERSITY CHALLENGES

Niharika Bahl

Women and nature are intrinsically linked. Folklore and legends across the world are replete with references testifying to the same.[1] Nature, abundance and fertility have been regarded as feminine notions in almost all cultures and beliefs.[2] Ancient Indian texts regard the primordial life-giving energy as *sakti* – female; earth itself is referred to as *aditi*; and *prithvi* – female goddess who is a nurturing mother, in whose womb seeds are germinated and who, in her infinite generosity, provides food and sustains all the flora and fauna.[3] All goddesses manifest as primal relational powers of nature, connecting all elements of life in a single biotic web.[4] The synergy between women and nature is underscored always.[5]

India is blessed with a rich treasure trove of natural resources. Embracing three major biological realms – Indo-Malayan, Eurasian and Afro-tropical – India is adorned with 10 bio-geographic zones and 26 biotic provinces.[6] With only 2.4% of the earth's land area, India accounts for 7–8% of the recorded species of the world, which includes millions of races, subspecies and local variants of species, and the ecological processes and cycles that link organisms into population, communities and all different ecosystems.[7] Demographically, it is the second largest populated country in the world[8] and a majority of its population directly depends on biological resources for livelihood.[9]

Despite this natural bounty, India is faced with a burgeoning food crisis. Its rich biodiversity is getting steadily eroded, endangering the lives of all dependent upon it, particularly women. According to the 2013 Global Hunger Index (GHI), India ranks 63rd out of the 78 hungriest countries, significantly worse than neighbouring Sri Lanka (43rd), Nepal (49th), Pakistan (57th) and Bangladesh (58th).[10] The United Nations Food and Agricultural Organization believes that 17% of Indians are too undernourished to lead a productive life. In fact, one-quarter of the world's undernourished people live in India, more than in all of Sub-Saharan Africa.[11]

The present chapter seeks to examine the challenges to biodiversity in India from a gendered feminist perspective of the rights and interests of woman food producers. It is divided into two parts. The first part examines the dependence of women upon the surrounding environment and discusses the integral role played by women in managing and sustaining biodiversity in India and the impact of its erosion on them. The second part analyses the framework of laws for protecting the country's natural wealth, ranging from the Constitution of India 1950, to the Biodiversity Act 2002, to the Forest Rights Act, 2006, and investigates whether the same protects the rights of the women food producers and whether it can be leveraged by women to safeguard their rights to grow, gather food, feed and sustain their households, and apply their conservation practices.

I. Women and biodiversity: a gendered difference

May this Earth, replete with seas, rivers and water sources, excellent food grain from agriculture, prolific and abundant living creatures, bestow upon us munificent nutrition.
Prithvi Sukta, Atharva Veda [12.1.1 to 12.1.63]

There is a lineal narrative which predominantly prevails in academic literature about women's relationship with biodiversity. It postulates a near complete dependence of women on its surrounding environment to fulfil the household needs of fuel, food and fodder which would not hold true in an urban landscape. Further, it promotes a homogenous view of women's interests, ignoring the distinctions on account of geographical location, culture and economic and social class. This is turn leads to a monolithic view of women's role as naturally inherent conservationists. Yes, rural women have a greater dependency than men on the environment. But the difference is not merely of degree, it is also qualitative. It arises on account of the gendered division of labour and limited access and control over resources. So, while men's dependency on forests is for the occasional need for timber, women's is for daily provisions of fodder and firewood, which makes their dependency not only higher but more immediate and persistent.[12]

Again, women are primarily responsible for the task of food gathering. Forests and other surrounding natural vegetation serve as supplements to daily subsistence diets and cushions during periods of scarcity. This too leads to a gendered distinction of priorities. Men prefer plant species that provide timber while women prioritize fruit bearing, fodder and firewood species. In a similar vein, there is gendered acquisition of traditional knowledge. Women are not the sole repositories of indigenous knowledge, but the wider range of products that women gather leads to gender differences of traditional know-how. Food gathering requires deep knowledge of the nutritional and medicinal properties of plants which is practiced and maintained by women in most rural communities.[13]

At the same time, though there may be commonalities of the gendered division of labour and vulnerabilities, in reality, there is a great deal of heterogeneity in women's interests and complexity in their relationship with the environment on account of class, ethnicity, region and economic capacity. For instance, while household work, child rearing and elder care largely fall within the domain of women's work, even within this domain, some women, due to their better economic standing, might not have to do the work themselves because some could afford helpers. Agarwal also points out that conflicts of interest amongst women can also be easily prevalent. Thus, while the requirement of firewood is felt by all rural women, it would be more acutely felt by women from landless households and less so by those from landed households, as they would be able to obtain firewood from their lands; therefore, landless women would be more resistant to strict forest closure rules.[14]

The foregoing discussion indicates women's response and interaction with biodiversity is intrinsically linked to the gendered access to and control over resources and the sexual division of labour. Adopting a contextual approach instead of a monolithic one is more helpful in building a better understanding of women's relationship with the environment and in developing an effective strategy to mobilize the same.

A. Women as food producers

It is in the performance of their gendered duties that rural women have over time built a vast repository of traditional knowledge, of agrarian and conservation practices. According to M. S. Swaminathan, the famous agricultural scientist, some historians believe that it was woman who first domesticated crop plants and thereby initiated the art and science of farming. While men went out hunting in search of food, women started gathering seeds from the native flora and began cultivating those of interest from the point of view of food, feed, fodder, fibre and fuel.[15]

India boasts of a rich crop diversity and agriculture forms the main-stay of the Indian economy with half of its workforce engaged in it.[16]

Unrecognized and unacknowledged, women constitute a substantial part of the agrarian workforce in India. According to the 2011 Census, there are nearly 100 million women engaged in the agricultural sector as farmers and labourers.[17] Beyond these economic figures, women perform many roles. They are simultaneously farmers, herders, for-est gatherers, drawers of water, food processors, market vendors, soil conservationists and caretakers of the natural environment – functions which may not necessarily be accounted for as an 'economic produc-tive' activity but which are integral to their existence, household food security and nutrition, and for environmental protection.[18]

A critical aspect of agrarian diversity is seed selection. Women in most rural communities in India have traditionally been the guard-ians of the seed.[19] Seeds, a symbol of the cycle of birth, life and death, are a part of India's cultural heritage. The practice of seed saving has been the cornerstone of Indian farming traditions that made agri-culture, itself, a way of life in the country. And it is women who select the seeds which set into motion the entire agrarian cycle. They identify the valuable genes and traits in these crops and have main-tained them over generations through a highly sophisticated system of crossing and selection. Selection is done on the basis of traits as diverse as disease resistance, high salt tolerance, and resistance to water logging; and traits such as drought tolerance as well as cook-ing time, taste, digestibility, milling and husking characteristics, like how much grain breaks during milling operations, are recognized and maintained.

Women are in the true sense owners of this complex seed technology and know-how.[20] It is due to their key role that India has displayed such vast crop diversity in its cereals, pulses, vegetables and fruits. Nearly 166 food/crop species and 320 wild relatives of crops have originated here. One species of rice (*Oryza sativa*) has been diversified into at least 50,000 (and perhaps up to 200,000!) distinct varieties while one species of mango (*Mangifera indica*) has yielded over 1,000 varieties, ranging from the size of a peanut to a muskmelon.[21] Studies have shown that in the proportionately small geographical areas of North East India, there is as much diversity of rice as in all of Asia. Amongst the *Garo* tribe in Meghalaya, some women are aware of over 300 indigenous varieties of rice.[22] Alongside food production, food gathering is also done. Surrounding natural vegetation is collected and utilized as food, fodder, medicine and building material.

In lean seasons or of crop failure, women are able to identify and gather 'hidden harvest' – wild fruit, tubers, roots and leaves which grow on common lands or in between crops as important sources of food supply.[23] Take, for instance, the case of 'bathua', a green leafy plant which grows as a weed in the wheat fields. When women weed the wheat field, they harvest the bathua as a rich source of nutrition for their families.[24] In a survey in Andhra Pradesh, village women identified 79 species of uncultivated leafy greens that they gathered for food, in addition to roots, tubers and fruits.[25] This wealth of 'traditional knowledge' of women has been applied and modified over the ages to maintain their household's food security and health. Thus, it is a continuing stream and it reflects an innate understanding of the contribution made by each and every living thing in maintaining the fragile ecological balance.

Women farmers, however, are a marginalized group. They have limited access to resources – financial and educational – and they lack secure land rights and have limited decision-making ability within their households and within the community of cultivators.[26] Agrarian land reforms in the country have left women in the sidelines. In spite of the progressive changes in the succession laws, ground reality indicates a dismal picture. The total number of female farmers in the country has declined 14% from 41.9 million in 2001 to 36 million in 2011.[27] In the same period, there has been an increase of 24% in the number of female agricultural labourers, from 49.5 million in 2001 to 61.6 million in 2011.[28] Tribal women in the country stand even more marginalized owing to the inherent disadvantaged status of their communities.[29]

A closer look highlights the starkness of the ground reality. A study of women's agricultural ownership in Gujarat covering 10 districts, 15 tehsils and 23 villages revealed that 4,188 men (81%) as against 561 women (11.81%) owned land.[30] Of the 403 cases of women ownership for which data could be collected, 193 women (47.89%) owned land following widowhood and 167 (41.43%) owned land to avail of some government benefit, escape the land ceiling or avoid payment of a tax or bribe to the revenue official.[31] In 18 (0.04%) cases, women obtained land from their parents who had no male heirs. And in 10 cases, women from tribal and Muslim communities received land on account of their husband marrying a second time.[32] Women not only owned less land than the male members of the household, but also land of inferior, low productive quality.[33]

B. *Shrinking biodiversity, shrinking lives*

*In daily life, on Earth, whether we are sitting, standing,
or in motion, may our activity be such as would never
cause injury or grief.*

Prithvi Sukta, Atharva Veda [12.1.1 to 12.1.63]

India's biodiversity is, however, rapidly decreasing. Rampant growth of human population and human activity has led to widespread destruction of forests and other natural habitats. Global warming and climate change have endangered many species of plants and animals. At least 10% of India's recorded wild flora and possibly more of its wild fauna are on the list of threatened species, and many are on the brink of obliteration. Of the wild fauna, 80 species of mammals, 47 of birds, 15 of reptiles, 3 of amphibians and a large number of moths, butterflies and beetles are endangered. Out of 19 species of primates, 12 are endangered. The cheetah (*Acinonyx jubatus*) and the pink-headed duck (*Rhodonessa caryophyllacea*) are among species that have become extinct.[34]

With regard to cultivated plants, too, India's diversity is eroding. This process began with the so-called 'green revolution' which displaced the indigenous seeds with those of high-yielding varieties requiring heavy use of fertilizers, pesticides and irrigation facilities.[35] Native seeds saved and selected by women were declared to be primitive and inferior despite their suitability to climatic conditions and high nutritional content. Increasing use of pesticides and herbicides has further damaged biodiversity. Further, monocultures of uniform crop cultivation again eroded crop diversity.[36] And this process continues unabated. According to the UN Food and Agriculture Organization, the planet has lost 75% of its food crop diversity in the twentieth century as farmers shifted from genetically diverse traditional food crops to high-yielding monocultures.[37]

A further threat which has loomed on the horizon is that of Genetically Modified (GM) crops. GM crops are of two variants – herbicide tolerance (HT) and insect resistance (IR). In the case of the former, a gene that makes the crop immune to a particular chemical herbicide is inserted into the plant which enables farmers to spray the herbicide without damage to crops. IR, on the other hand, involves insertion of a gene that enables the plant to produce chemicals that kill certain pests which feed on it.[38] Though no clear verdict has been made on the biosafety of GM crops, it is feared that GM crops will have an adverse impact on human health and on the surrounding biodiversity.[39] The majority of the members of the European Union have refused to permit its cultivation.[40] Multi-National Corporations (MNCs) are heavily pushing the same. In India, BT cotton is the one that has been permitted to be commercially cultivated due to claims of high insect resistance, but the results have not been satisfactory. High cost of BT cotton seeds which have not been wholly pest resistant and the heavy dependence on irrigation have led to crop failures and escalated farmer debt and distress. Thus, Vidharba, Maharashtra, where BT cotton has been grown, continues to be labelled as the farmer suicide belt of the country.[41]

All this has had far-reaching consequences on the nation's food security. From growing crops for sustenance, post-GATT (General Agreement on Tariffs and Trade) and WTO (World Trade Organization), Indian farmers have been thrown into the global market of demand–supply. Dumping of agricultural produce from subsidized countries led to a steep fall in agricultural prices and pushed them further into debt, acerbated by rising costs of production, seeds and pesticides, etc.[42] P. Sainath explains,

> it cost Rs 2500 for a typical (non-organic) farmer to cultivate one acre of cotton in Vidharbha in 1991. Today it can cost upwards of Rs 13,000. Seed (then local) could be had for nine rupees a kilogram – if you didn't produce it yourself. By 2005, a 450-gram bag of (Monsanto's) Bt. cotton seed was selling for between Rs 1650 and Rs 1800. But in 2006, cotton was fetching the farmer the same per quintal as it did in 1994 though input costs had quadrupled. In Wyanad, Kerala, a farmer raising an acre of paddy could have done it for Rs 8,000. Many, lured by marketing, urged by government, shifted to vanilla – which in 2003 cost Rs 1.45 lakh an acre to grow. The risks multiplied, the costs multiplied – income did not.[43]

Thus, one scholar concludes that 'the underlying cause of the global food crisis is a corporate-dominated, fossil fuel-dependent model of agricultural production that is ecologically unsustainable and economically unjust'.[44] From a state of food self-sufficiency, we have moved to that of hunger and starvation deaths. According to the UN Food and Agriculture Organization's report, India is home to 194.6 million undernourished people, the highest in the world.[45] This translates into 15% of its population.[46]

The cumulative impact of all this on women food growers and gatherers has been devastating and has been largely unnoticed. When one thinks of farmers' suicides, one rarely thinks in terms of women farmers. However, suicides by women farmers do occur but are not recorded as such because the land is not in their names or reported as dowry death or accident.[47] Thus, in 2013 only 126 suicides by women farmers in Maharashtra were recorded by the National Crime Records Bureau against 3,020 by male farmers.[48] Further, when male farmers die, they leave behind a crushing debt for their widows. Farm widows struggle to manage their household, feed their children and pay the debts to the extent of doing more labour than a convict sentenced to rigorous imprisonment and eating even less than the prison quota of food for the convicts![49] The impact has been not just economic in terms of poverty and indebtedness, it has translated into 'a social crisis'.[50] More and more women are forced to shoulder the responsibility of their households in the wake of farmer suicides and male migration, without land, without assets – which have gone toward paying the farm debt – and thus they stand more vulnerable to exploitation. They are increasingly pushed into trafficking and there is a rising tide of violence against women.[51] In a public hearing in 2005 conducted by the National Commission for Women on the impact of the WTO on women in agriculture, in the Bundelkhand region, which has seen a spate of farmers' suicides, women spoke about the increase in domestic violence and dowry-related ill-treatment. The rise in dowry crimes was also reported by tribal and Scheduled Caste women in West Bengal, where earlier the custom of bride price prevailed.[52]

C. Feminist voices

Feminists' views on women's role as conservationists draw on different conceptual frameworks. The most prominent feminist strand is that of Ecofeminism. Ecofeminists, notably Vandana Shiva and Maria Mies, perceive a common thread linking the exploitation of nature and women. They believe that the worldview that causes environmental degradation and injustice is the same worldview that causes a culture of male domination, exploitation and inequality for women.[53] The destruction of nature becomes the destruction of women's sources for sustenance for themselves and their families. That is the reason why women are found at the heart of every environmental struggle. The most celebrated instance is that of the *Chipko* movement in the 1970s when women in the Garhwal region hugged trees to prevent them from being felled by contractors licensed by the State.[54] This resistance was effective in restraining tree felling locally and it captured the worldwide attention.

According to Shiva and Mies, commercial agriculture displaces women as the primary seed keeper and devalues her productive role. Trade liberalization and the Trade-Related Intellectual Property Rights (TRIPS) Agreement have adversely affected women's autonomy and control over her knowledge of seed and biodiversity by transferring the same to global corporations. As globalization shifts agriculture to capital-intensive, chemical-intensive systems, women bear disproportionate costs of both displacement and health hazards. Furthermore, as their livelihoods and productive roles are displaced, and their status is devalued, women become more disempowered, triggering further violence against them. Thus, they write,

> we see the devastation of the earth and her beings by the corporate warriors, as feminist concerns. It is the same masculinist mentality which would deny us our right to our own bodies and our own sexuality, and which depends on multiple systems of dominance and state power to have its way.[55]

The process works both ways as devaluation of women's role and status leads to further erosion of biodiversity.[56] As a counterpoint to this, there is feminist environmentalism advocated by economist Bina Agarwal, who questions the belief that women are identified with nature.[57] According to Agarwal, women, particularly rural women, are likely to be affected adversely in quite specific ways by environmental degradation. At the same time, in the course of their everyday interactions with nature, they acquire a special knowledge of species varieties and the processes of natural regeneration. They could thus be seen as both victims of the destruction of nature and as repositories of knowledge about nature, in ways distinct from the men of their class. The former aspect would provide the gendered impulse for their resistance and response to environmental destruction. The latter would condition their perceptions and choices of what should be done.[58]

Agarwal writes that

> feminist environmentalism recognizes that both women and men would have an interest in forest conservation and regeneration, but their interests would stem from *different* (and at times *conflicting*) concerns, rooted in their respective responsibilities, and the nature and intensity of their dependence on these resources.[59]

As environmental activist Anil Agarwal pointed out, while men are more interested and engaged with the cash economy, women continue to deal with the non-monetized, biomass-based subsistence economy of the household. Even within the same household, one can find cases of men supplying tonnes of bamboo to paper mills, etc. and thus destroying nature to earn cash, even though it could create greater hardships for the women in collecting daily fuel, fodder, roots, tubers, etc.[60] Further, with increasing male migration to urban areas and cash economies, the importance of women's traditional knowledge has grown as women have not only retained but added to their indigenous knowledge.[61]

Both approaches underscore the higher and more continuous nature of women's dependency upon the natural environment. This moulds women's attitude and outlook towards biodiversity management in terms of their traditional knowledge in the utilization of forests and other natural resources but also with regard to conservation. Compulsions of economic distress and class disparity would also come into play. The challenge, however, lies in enabling women, as critical stakeholders, to participate in the management and conservation of biodiversity resources and safeguard their interests.

II. International legal framework: limited recognition, limited role for women

O Earth, in the villages, forest, assemblies, committees and other places on Earth, may what we express always be in accord with you.
Prithvi Sukta, Atharva Veda [12.1.1 to 12.1.63]

Law makers and policy planners have paid scarce recognition to women's vital role in biodiversity conservation – internationally and at the domestic level. The primary international instrument, i.e. the Convention on Biological Diversity, 1992, contains only a bland recognition of women's role and the need to have participation of women at all levels of policy-making and implementation for biological diversity conservation.[62] There are no concrete steps outlined to achieve this participation. In most other instruments, even this recognition is also missing. Overall, the international legal framework regarding biodiversity reflects a lack of cohesion, a need rather of balancing the conservation imperatives without compromising the regime of intellectual property rights set up under the TRIPS agreement.

The International Union of the Protection of New Plant Varieties[63] (UPOV), an intergovernmental organization established under the International Convention for the Protection of New Varieties of Plants,

1961, promotes the property rights of plant breeders through the *sui generis* form of intellectual property protection.[64] The International Treaty on Plant Genetic Resources for Food and Agriculture, 2001, which India has ratified and which came into force in 2004, is an attempt to keep in free circulation the listed agricultural genetic resources which are crucial for food security, thereby providing for their conservation and sustainable use.[65] It creates a Multilateral System (MLS) that provides for facilitated access to a negotiated list of plant genetic resources and for the fair and equitable sharing of the benefits arising from their use.[66] The treaty has been hailed as protecting the interests of farmers by permitting free access to certain germplasm.[67] However, it allows patenting of genetic material received through the MLS if it has been modified in any way,[68] and this lack of a clear ban on patents or any other Intellectual Property Rights (IPRs) on crops for food and fodder is troublesome to farmers. While lauding the farmer's contribution to biodiversity, the treaty includes no recognition of the property rights of farmers – merely a grant of permission to save, exchange and sell farm seeds.[69] Further, there is no mention of the concerns or interests of women engaged in agrarian practices or in seed conservation.

In addition, the Nagoya Protocol on Access to Genetic Resources and the Fair and Equitable Sharing of Benefits, 2010 also came into force in 2014.[70] This Protocol to the Convention on Biological Diversity again seeks to facilitate access to genetic biodiversity resources with mechanisms for equitable sharing of benefits, but the onus for the establishment of the same is upon the respective party States. The Nagoya Protocol requires parties to take into consideration indigenous and local communities' customary laws, community protocols and procedures, as applicable, with respect to traditional knowledge associated with genetic resources.[71] Parties are also mandated to support the development of community protocols, model contractual clauses for sharing of benefits of traditional knowledge by indigenous and local communities, *including women*,[72] and to take measures to enhance the capacity of women of indigenous and local communities.[73] The Nagoya Protocol marks a step forward in at least acknowledging the necessity of the involvement of women in key decision-making pertaining to access to biodiversity.

III. Domestic legal framework: more access, less conservation of biodiversity

At the national level too, there is no gendered understanding in the framework of laws on food production and biodiversity conservation.

The Constitution of India recognizes the principle of gender equality.[74] It also makes the right to life and personal liberty a fundamental right.[75] The scope of this right has been amplified by the courts to include privacy, legal aid, shelter and livelihood. However, this has not helped in checking the tide of deforestation and displacement of people from land and livelihoods.[76]

With respect to biodiversity specifically, the domestic statutory regime mirrors the same fragmented approach as found in the international plane. This is due to the fact that it has come about in compliance with India's international obligations. The foremost legislation in this regard is the Biological Diversity Act, 2002, enacted ostensibly for the sustainable use and conservation of biological resources and for the equitable sharing of the benefits arising out of the use of these resources and associated knowledge.[77] For these purposes, it establishes a three-tier structure with the National Biodiversity Authority (NBA), State Biodiversity Boards (SBBs) as well as Biodiversity Management Committees (BMCs) at the local level.[78] The main thrust of the Act, however, appears to be to regulate access to India's biodiversity and traditional knowledge for which it outlines clear procedures.[79] Conservation measures are confined to advising government and collating and publishing data on various aspects of biodiversity,[80] including documentation of traditional knowledge through the preparation of People's Biodiversity Registers by the BMCs.[81] In essence, the Act is more permissive than restrictive in its approach.

Women's participation in the process is through the mandatory one-third membership in the BMC.[82] This type of female participation is hemmed with all the obstacles that come with gender power-sharing in a patriarchal society, which raises questions about meaningful participation in any decision-making process. At present, there are 37,769 BMCs across all states and Union Territories in India,[83] and even these are not fully functional.[84] BMCs have also faced stiff opposition in certain states from the local government bodies who are unhappy about sharing power with them.[85] The low relevance of the BMCs in the working of the Act becomes apparent when even without all the BMCs operating properly, the NBA has executed, as of 2012, 100 Access and benefit-sharing (ABS) agreements[86] – though, under the Act, the NBA and SBBs are required to consult the BMCs while taking any decision relating to the use of biological resources and associated resources occurring within their territorial jurisdiction.[87]

The real challenge, according to activists and analysts, lies in ensuring that 'the monetary collections from these 100 agreements going into the National Biodiversity Fund translate into real "benefits" for

270

at least 100 local communities in India'.[88] Though there are a few reported success stories of favourable benefit sharing from the states of Andhra Pradesh and Madhya Pradesh,[89] conservation really seems to have taken a backseat under the Biodiversity Act, 2002. There is no clear roadmap on conservation ahead – no clear allocation of roles and responsibilities. The NBA too has earned a scathing criticism from the CAG for its dismal performance in this regard.[90]

The success of any conservation strategy depends upon community involvement, which is absent under the Act. The money from the Local Biodiversity Fund is to be spent on conservation and promotion of biodiversity in local areas in a manner 'as prescribed by State Government'.[91] Local participation, especially of women, at this stage would be greatly effective as it would enable employment of their strategies of conservation of biodiversity, which allow them to make sustainable use of natural resources and ensure their household's food security. Efforts are on by local women to utilize the provisions of the Act without reference to BMCs. A shining example is provided by the village women from the Zaheerabad region of the Medak district in Andhra Pradesh. These women have been practicing biodiversity conservation for generations with the use of intercropping, crop rotation practices, crop protection practices and maintaining community seed banks as well as passing the culture and practices to the next generation of women who come to the villages as daughters-in-law. They celebrate a festival of handing over seed pots to the next generation, symbolizing the handing over of the custody of the seed wealth to the next generation. Farmers in this area cultivate more than 85 types of crops, pulses, oil seeds, vegetables and medicinal plants by preserving over 80 seed varieties. In April 2010, through the intervention of a local NGO – Deccan Development Society – the farming community sent a proposal for protecting more than 50,000 acres of biodiverse farmland. This was approved by the SBB and after a considerable wait, the State government notified the area as a Bio Heritage Site (BHS) under Section 37 of the Biological Diversity Act.[92] The Medak story serves as a successful case study of women farmers successfully using the Biological Diversity Act to preserve their agrarian heritage without the BMC mechanism.

Of greater significance to women's agrarian rights is the Protection of Plant Variety and Farmers' Rights Act, 2001[93] (PPVFR). This was enacted in compliance with India's obligations under the TRIPS treaty, which required members to either opt for an Intellectual Property Rights model for plant variety protection or to develop a *sui generis* model.[94] The Act establishes plant breeders' and farmers' rights. The inclusion of farmers' rights came about as a result of a concerted campaign by farmers' advocacy groups. The Act places farmers at the same level as plant breeders and provides that they receive the same kind of protection as plant breeders. Farmers, therefore, can also apply to have a plant variety registered which will entitle them to exclusive right to produce, sell, market, distribute, import or export the variety.[95] The actual process is tedious, not free from expense and involves a lot of paperwork which is difficult for illiterate farmers to comprehend. Conservation of genetic resources of land races and their improvement entitles farmers to be rewarded from the Gene Fund set up.[96] Further, a farmer shall be entitled 'to save, use, sow, resow, exchange, share or sell his farm produce including seed of a variety protected under this Act in the same manner as he was entitled before the coming into force of this Act'.[97] This right, dubbed 'farmers' privilege', is more in the nature of exemption than a right.[98]

Critics attack the fundamental premise of importing the concept of IPRs into agriculture. Traditional farming practices in India and the related knowledge has not been the private property of anyone.[99] The Act would only ensure to the benefit of richer farmers and corporate entities who would seek to stake claims on what is otherwise collective knowledge. This would further commercialize agriculture and marginalize women and small farmers. The rising spate of farmers' suicides is a pointer to the same. In the offing is a proposed Seeds Bill to regulate the sale and quality of seeds. It also contains provisions on the registration of GM seeds, which is a matter of controversy. At present, it has been put on hold.[100]

The Scheduled Tribes and other Traditional Forest Dwellers (Recognition of Forest Rights) Act, 2006[101] or Forest Rights Act (FRA) also has bearings on women's role in ensuring food sustenance and conserving biodiversity. The Forest Rights Act of 2006 seeks to make a shift in the conservation from purely state management of forests to a more participatory–community-oriented approach through involvement of *gram sabhas* and a Forest Rights Committee (FRC) at the local level, and by vesting of community rights in the forest dwellers since they are 'integral to the very survival and sustainability of the forest ecosystem'.[102] These include cultivation rights;[103] community land rights;[104] right to reclaim any disputed land;[105] right to convert to titles leases or grants of forest lands issued by local authorities or state government;[106] rights of settlement and conservation of all forest villages;[107] user rights like collection and usage of minor forest produce[108] and to use traditional seasonal resources, such as pastures and water bodies;[109] right of access to biodiversity; and the community right to intellectual property and traditional knowledge related to biodiversity and cultural diversity.[110]

Field studies indicate a lack of awareness amongst the forest communities about the Act. Community rights and conservation provisions seem largely to be ignored in practice.[111] A study conducted amongst the tribal *bhil* women in Banswara, Rajasthan, reveals gendered discrimination against women in the implementation of the FRA.[112] The majority of the tribal women interviewed including those who were members of the *gram panchayat* and the FRC, were unaware of the nature of the rights under the Act. An overwhelming majority also reported that their access to forests and forest produce had decreased after the passage of the Act due to the restrictions imposed by the FRC, which was more interested in selling the forest produce then in self-consumption. *Bhil* women also reported a loss of decision-making authority. Prior to the coming of the Act, there was a tradition of collective forest management by both men and women. Now women's participation has become limited to the token reserved membership in the FRC and *gram panchayat* as women are prevented from ever exceeding their reserved quota. Focus is on individual property rights and not on the collective community rights. Therefore, tribal women are dependent on extra-legal or illegal means to access forest resources for subsistence. This problem has major consequences on the future of tribal communities due to their heavy dependence on the forest as a common pool resource that is now converted into private property. The study concluded that recognizing women's traditional collective access to forest land and its resources may empower women in decision-making authority, change institutional rules, and contribute to women's socio-economic and political empowerment.[113]

Similar stories emerge from other parts of India. A study conducted across three districts in Andhra Pradesh, namely Adilabad, Srikakulam and Vizianagaram district, discloses the haste with which the state government 'fast tracked' the establishment and organization of gram sabhas and FRCs within a span of a month.[114] The emphasis, as seen in other parts of the country as well, is on individual claims and not on community rights. In Vizianagaram district, only 0.02% of the original community claims were accepted.[115] The non-recognition of community rights over common lands and forests where women were accustomed to collecting food and fodder and growing vegetables and other crops, directly and adversely impacts on women's access to sources of sustenance. With regard to individual claims, the final amount of land approved to women holders was a mere fraction of the amount claimed; this is a pointer to the inherent patriarchy ingrained in state mechanisms. Even on the individual land titles, government had undertaken to heavily promote and push monocultures of plantation farming. Thus, according to the analyst, the Adivasi women, who are being 'granted' these private titles, are left with the 'free choice' of deciding between mangoes and cashew nuts.[116] In the process, the women are becoming wage labourers, carrying out the government's programmes of plantation rather than exercising their traditional knowledge to nurture forests and gain rightful livelihoods. This completely destroys the spirit of the legislation which emphasized customary traditions and indigenous livelihoods.

The above discussion indicates that the current legal regime fails to appreciate the critical role played by women in food subsistence and managing biodiversity. Women's involvement is limited to token representation without allowing for actual decision-making. Legal illiteracy, skewed gender equations and limited access to resources makes it virtually impossible for them to avail any legal redress, even where provision is there. The existing legal framework does not secure or promote women's right to grow or gather food and feed and sustain their households, as it is directed towards facilitating commercialization of agrarian and other natural resources. The problem, which is not endemic to India but prevalent worldwide, is according to a scholar, that 'the commodification of biodiversity has caused a shift in the ownership of genetic resources from communal to private. Biodiversity knowledge and resources are alienated from the original custodians and donors and become the monopoly of private industry'.[117]

Thus, India has lost its food sovereignty and is in the throes of a deepening agrarian crisis. It may be argued that recourse may be had through the constitutional guarantees of right to life and liberty, since lives of women food producers and their families are endangered but the approach of the 'globalized Indian Supreme Court', as dubbed by Prof. Baxi, since the 1990s has, barring a few stray instances, been to endorse the wave of economic liberalization reforms for the 'development of the country'.[118] Of particular significance in recent times has been the Supreme Court ruling, impelled by the starvation deaths in the country, that fulfilment of the basic necessity of food is an integral part of the right to life.[119] Thereafter, the National Food Security Act, 2013[120] was enacted. The Act seeks to guarantee entitlement to minimum food grains to people below the poverty line and to streamline the existing Public Distribution System. The thrust of both the Supreme Court orders and the Act, as well as of other food security programmes[121] implemented in the country, has been on food distribution and not on food production or on the shrinking sources of sustenance. This strategy has come under criticism.[122] The focus, according to an analyst, should be on increasing food production in a sustainable manner.[123] And the core of this problem lies in tackling 'the fundamental question of common property resources and the right of access to them'.[124]

IV. Conclusion

Women and nature share a symbiotic relationship. A shrinkage of bio-diversity entails a shrinkage of their rights to sustenance, and an erosion of women's status and decision-making ability adversely impacts the environment. The present scenario of commercialization of agri-culture and plant resources has marginalized women and depleted food and other biological reserves, endangering the health of our future generation as well as of our planet. To rectify the same requires us to throw the biodiversity resources back into the community and to restore their guardianship and management to those who know how to utilize the same in a sustainable manner. Critically, it requires a change in outlook. Noted legal thinker Dhavan, in an insightful arti-cle on the environmental protection regime in India,[125] has written 'Indian law has always been more concerned with property rights than the environment, the exploitation of natural resources than their con-servation, and the ruthless appropriation of natural spoils than with evolving a common understanding of the totality of the inheritance of nature'.

Even more fundamentally, women's voices therefore have to be not only heard but also counted. Law makers and policy planners have to ensure that women are not simply 'added' to the conservation pro-gramme, but rather, that biodiversity itself is defined in broader, more inclusive and fluid terms incorporating diverse gendered experiences of different groups. More importantly, one has to address the issue of gender differentiated access and control of resources, which in turn would lead to genuine women empowerment and decision-making. This is not a question of mere inclusion but of dismantling patriarchal institutions and biases, a recognition of the fact that women have a right to a dignified livelihood and sustenance, not as a measure of protection or welfare, but as a matter of right.

Notes

1 For instance, the Greek legend of Demeter and Persephone. F. Guirand (ed.), *New Larousse Encyclopaedia of Mythology*, New York: Prometheus

Press, 1960; see also P. Mary Vidya Porselvi, *Nature, Culture and Gender: Re-reading the Folktale*, New York: Routledge, 2016.

2 The primary deity of fertility being female in numerous mythologies, for example *Gefjun* in Nordic mythology and *Sara Mama* in Inca mythology. See Guirand (ed.), *Encyclopaedia of Mythology*.

3 *Prithvi Sukta*, (Hymn to the Earth) Atharva Veda [12.1.1 to 12.1.63].

4 In Vedic texts, the female deities include Usha – Goddess of dawn and Prithvi – Goddess of terrestrial earth. Rivers are invariably female goddesses – Ganga, Yamuna, Cauvery. See D. R. Kinsley, *Hindu Goddesses: Visions of the Divine Feminine in Hindu Religious Tradition*, Berkeley: University of California Press, 1988.

5 *Ibid.*

6 K. Venkatraman, 'India's Biodiversity Act, 2002 and its Role in Conservation', *Tropical Ecology*, 2009, 50(1): 23–30, 24.

7 Venkatraman, 'India's Biodiversity Act, 2002'; National Biodiversity Authority, *Factsheet 2011*, http://nbaindia.org/uploaded/pdf/Fact%20 Sheets.pdf (accessed on 10 September 2016).

8 Statistics Times, 'Population of India,' http://statisticstimes.com/popula tion/population–of–india.php (accessed on 10 September 2016).

9 National Biodiversity Authority, 'Defining and Explaining ABS Terminology' 2013, http://nbaindia.org/uploaded/pdf/ABS_Terminology.pdf (accessed on 10 September 2016).

10 International Food Policy Research Institute, *Global Hunger Index–The Challenge of Hunger*, 2013, www.ifpri.org/publication/2013–global–hunger–index–challenge–hunger–building–resilience–achieve–food–and–nutrition (accessed on 20 January 2016).

11 A. K. Biswas and C. Tortajada, 'India's Homemade Food Crisis', *Project-Syndicate*, August 8, 2014, www.project–syndicate.org/commentary/asit–k—biswas–and–cecilia–tortajada–attribute–shortages–and–under nourishment–to–widespread–wastage–of–output#OZSbeTjY6j7dlt6J.99 (accessed on 20 January 2016).

12 B. Agarwal, *Gender and Green Governance-The Political Economy of Women's Presence Within and Beyond Community Forestry*, New Delhi: Oxford University Press, 2010, p. 32.

13 Zweifel, 'The Gendered Nature of Biodiversity Conservation', *NWSA Journal*, Autumn 1997, 9(3) (Women, Ecology, and the Environment): 107–123; see also M. S. Swaminathan (ed.), *Gender Dimensions in Biodiversity Management*, New Delhi: Konark, 1998.

14 Agarwal, *Gender and Green Governance*, p. 22.

15 National Commission for Women (NCW), *Impact of WTO on Women in Agriculture*, New Delhi: National Commission for Women, 2005, p. 9, http://ncw.nic.in/pdfreports/impact%20of%20wto%20women%20in% 20agriculture.pdf (accessed on 4 January 2016).

16 'World Bank Data', 2013, http://data.worldbank.org/indicator/SL.AGR. EMPL.ZS?locations=IN (accessed on 10 September 2016).

17 K. Lahariya and A. S. Sethi, 'India's Quiet Women Farmers Slip into Crisis', *Business Standard*, March 10, 2015, www.business–standard.com/article/economy–policy/india–s–quiet–women–farmers–slip–into–crisis–115031000415_1.html (accessed on 4 January 2016).

18 V. Shiva, 'Women's Indigenous Knowledge and Biodiversity Conservation', *India International Centre Quarterly*, Spring–Summer 1992, 19(1/2) (Indigenous Vision: Peoples of India Attitudes to the Environment): 205–214.

19 Shiva, 'Women's Indigenous Knowledge'; see also R. Chitnis, 'Seeds and Their Keepers Are Key to Preserving India's Food Diversity', *Earth Island Journal*, June 30, 2011, www.earthisland.org/journal/index.php/elist/eListRead/seeds_and_their_keepers_are_key_to_preserving_indias_food_diversity (accessed on 6 January 2016).

20 I. Barpujari, 'A Gendered Perspective of Indigenous Knowledge', Briefing Paper 4, July 2005', New Delhi: Gene Campaign, 2005, http://genecampaign.org/ (accessed on 20 January 2016).

21 A. Kothari, 'Agro-Biodiversity: The Future of India's Agriculture', Maharashtra Council of Agricultural Education &Research (MCAER) 1999, p. 3, http://dev.mtnforum.org/sites/default/files/publication/files/47.pdf (accessed on 4 January 2016).

22 S. Krishna, 'The Gendered Price of Rice in North-Eastern India', *Economic and Political Weekly*, June 18–24, 2005, 40(25): 2555–2562.

23 Zweifel, 'The Gendered Nature of Biodiversity Conservation'; see also Chitnis, 'Seeds and Their Keepers'.

24 NCW, *Impact of WTO on Women*, p. 51.

25 Agarwal, *Gender and Green Governance*, p. 33.

26 *Ibid.*

27 Lahariya and Sethi, 'India's Quiet Women Farmers Slip into Crisis'.

28 *Ibid.*

29 According to the UNDP India Report (2007) on Human Poverty and Socially- Disadvantaged Groups in India, the Human Development Index (HDI) at all India levels for Scheduled Tribes (ST) is estimated to be 0.270, which is lower than the HDI for Scheduled Castes (SC) and non–SC/ST for the period 1980–2000, while the Human Poverty Index (HPI) for STs is estimated at 47.79, which was higher than others for the period 1990–2000, www.undp.org/content/dam/india/docs/human_poverty_socially_disadvantaged_groups_india.pdf (accessed on 15 September 2016).

30 M. Velayudhan, 'Women's Land Rights in South Asia: Struggles and Diverse Contexts', *Economic and Political Weekly*, October 31–November 6, 2009, 44(44): 74–79.

31 Velayudhan, 'Women's Land Rights in South Asia'.

32 *Ibid.*

33 *Ibid.*

34 The Wildlife (Protection) Act, 1972 (Act No. 53 of 1972) Schedule I; S, Balaji, 'Biodiversity Challenges Ahead', *The Hindu*, May 27, 2010, www.thehindu.com/opinion/lead/biodiversity–challenges–ahead/article439099.ece (accessed on 6 January 2016).

35 B. Newman, 'A Bitter Harvest: Farmer Suicide and the Unforeseen Social, Environmental and Economic Impacts of the Green Revolution in Punjab India', Food First – Institute for Food and Development Policy, Development Report No. 15, January 2007: 13–14, https://foodfirst.org/wp-content/uploads/2013/12/DR15–A–Bitter–Harvest.pdf (accessed on 15 September 2016).

36 C. G. González, 'Introduction: The Global Politics of Food', *The University of Miami Inter-American Law Review*, Fall 2011, 43(1) (LatCrit South–North Exchange: The Global Politics of Food: Sustainability and Subordination): 77–87 at 81.
37 *Ibid.*
38 K. Kuruganti, 'Transgenic Crops: A Questionable Option?' *Seminar*, March 2009: 595, www.india-seminar.com/semframe.html (accessed on 6 January 2016).
39 *Ibid.*
40 A. Nelsen, 'Half of Europe Opts Out of New GM Crop Scheme', *The Guardian*, Brussels, October 1, 2015, www.theguardian.com/environment/2015/oct/01/half-of-europe-opts-out-of-new-gm-crop-scheme (accessed on 20 January 2016).
41 A. Malone, 'The GM Genocide: Thousands of Indian Farmers are Committing Suicide After Using Genetically Modified Crops', *Daily Mail*, November 3, 2008, www.dailymail.co.uk/news/article-1082559/The-GM-genocide-Thousands-Indian-farmers-committing-suicide-using-genetically-modified-crops.html#ixzz3u053EBiF (accessed on 20 January 2016).
42 NCW, *Impact of WTO on Women.*
43 P. Sainath, 'Ways of Seeing', *Seminar*, 5 March 2009, www.india-seminar.com/semframe.html (accessed on 6 January 2016).
44 González, 'The Global Politics of Food', p. 77.
45 'India Tops World Hunger List with 194 Million People', *The Hindu*, May 29, 2015, www.thehindu.com/news/national/india-is-home-to-194-million-hungry-people-un/article7255937.ece (accessed on 20 January 2016).
46 Biswas and Tortajada, 'India's Homemade Food Crisis'.
47 P. Kakodar, 'Records May Not Show, But Women Farmers Dying Too', *The Times of India*, May 17, 2015, http://timesofindia.indiatimes.com/india/Records-may-not-show-but-women-farmers-dying- too/articleshow/47314082.cms (accessed on 12 September 2016).
48 *Ibid.*
49 P. Sainath, 'Farmer's Diet Worse than a Convict's', February 13, 2015, https://ruralindiaonline.org/articles/farmers-diet-worse-than-a-convicts/ (accessed on 12 September 2016).
50 NCW, *Impact of WTO on Women*, p. ii.
51 *Ibid.*, p. 110.
52 *Ibid.*, pp. 184 and 155.
53 R. Brinker, 'Dr. Vandana Shiva and Feminist Theory', *Women–Justice–Ecology*, July 4, 2009, https://womenjusticeecology.wordpress.com/2009/07/04/dr-vandana-shiva-and-feminist-theory/ (accessed on 4 January 2016).
54 S. Jain, 'Standing Up for Trees–Women's Role in Chipko Movement' Unasylva–Women in Forestry', 1984/4, 36(146), www.fao.org/docrep/r0465e/r0465e03.htm (accessed on 2 October 2016).
55 Maria Mies and Vandana Shiva, Ecofeminism Halifax: Fernwood Publications, 1993, p. 14
56 Brinker, 'Shiva and Feminist Theory'.
57 Agarwal, *Gender and Green Governance*, p. 41.

58 Bina Agarwal, 'The Gender and Environment Debate: Lessons from India', *Feminist Studies*, Spring, 1992, 18(1): 119–158.
59 Agarwal, *Gender and Green Governance*, p. 42.
60 Barpujari, 'A Gendered Perspective of Indigenous Knowledge'.
61 Zweifel, 'The Gendered Nature of Biodiversity Conservation'.
62 The Convention on Biological Diversity 1992, Preamble.
63 www.upov.int/portal/index.html.en (accessed on 15 September 2016); India is not yet a member but seeks to be a member.
64 International Convention for the Protection of New Varieties of Plants, 1961, as revised in 1972, 1978 and 1991, adopted in the Diplomatic Conference in Paris on 2 December 1961 and came into force on 10 August 1968 being ratified by the states of the United Kingdom, the Netherlands and Germany, Articles 6–12.
65 International Treaty on Plant Genetic Resources for Food and Agriculture, 2001, adopted at the FAO conference on 3 November 2001 and came into force on 29 June 2004 after deposit of the required fortieth instrument of ratification, Preamble.
66 International Treaty on Plant Genetic Resources for Food and Agriculture, 2001Articles 10–13.
67 Grain and Kalpavriksh, 'The International Treaty on Plant Genetic Resources: A Challenge for Asia', April 19, 2002, www.grain.org/article/entries/41–a–challenge–for–asia–the–international–treaty–on–plant–genetic–resources (accessed on 15 September 2016).
68 International Treaty on Plant Genetic Resources for Food and Agriculture, Article 12.3.
69 International Treaty on Plant Genetic Resources for Food and Agriculture, Article 9.3.
70 www.cbd.int/abs/doc/protocol/nagoya-protocol-en.pdf (accessed on 15 September 2016); India is a party to the Protocol.
71 Nagoya Protocol on Access to Genetic Resources and the Fair and Equitable Sharing of Benefits (hereinafter Nagoya Protocol), adopted on 29 October 2010 at Nagoya, came into force on the 12 October 2014 after deposit of the required fiftieth instrument of ratification, Article 12(1).
72 Nagoya Protocol, Article 12(3).
73 Nagoya Protocol, Article 22(5) (j).
74 The Constitution of India 1950, Article 14.
75 The Constitution of India 1950, Article 21.
76 See *Narmada Bachao Andolan* v. *Union of India* (2000) 10 SCC 664; *N.D. Jayal* v. *Union of India* (2004) 9 SCC 362.
77 The Biological Diversity Act (hereinafter The BD Act), 2002, (Act 18 of 2003), Preamble.
78 The BD Act, Sections 8, 22 and 41.
79 The BD Act, Sections 3–7 and Sections 19–21.
80 The BD Act, Section 41(1) read with BD Rules 2004, Rule 22.
81 BD Rules 2004, Rule 22(6).
82 BD Rules 2004, Rule 22(2).
83 As of 2 September 2015 as per the NBA website, http://nbaindia.org/content/20/35//bmc.html (accessed on 21 January 2016).
84 K. Kohli and S. Bhutani, 'Ten years of the Biological Diversity Act', *Economic and Political Weekly*, September 29, 2012, 47(39): 15–18.

85 K. Kohli and S. Bhutani, 'BMC: Lost in Numbers', *Economic and Political Weekly*, April 19, 2014, 49(16): 18–20.

86 Kohli and Bhutani, 'Ten years of the Biological Diversity Act'.

87 The BD Act, Section 41(2).

88 Kohli and Bhutani, 'Ten years of the Biological Diversity Act'.

89 Kohli and Bhutani, 'BMC: Lost in Numbers'.

90 L. Jishnu, 'Biodiversity–India's Other Scandal', *Down to Earth*, 15 January 2011, www.downtoearth.org.in/blog/biodiversityindias-other—scandal-2795 (accessed on 21 January 2016).

91 The BD Act, Section 44(1).

92 R. Awadhani, '39 Villages Fight for BHS Status', *The Hindu*, September 27, 2012, www.thehindu.com/todays-paper/tp-national/tp-andhra pradesh/39-villages-fight-for-bhs-status/article3941154.ece (accessed on 21 January 2016).

93 The Protection of Plant Variety and Farmers' Rights Act, 2001, Act 53 of 2001 (hereinafter the Plant Variety Act).

94 The Agreement on Trade-Related Aspects of Intellectual Property Rights 1994 (TRIPS), adopted on 15 April 1994 at Marrakesh and came into force on 1 January 1995, Article 27.3b.

95 The Plant Variety Act, Sections 26 and 28.

96 The Plant Variety Act, Section 39(iii).

97 The Plant Variety Act, Section 39(iv).

98 S. Bhutani and K. Kohli, 'Protection of Plant Variety and Farmers' Rights Act, 2001 – Just Leave the Seed Alone', *Business Line*, 12 March 2004, www.thehindubusinessline.com/2004/03/12/stories/2004031200100900.htm (accessed on 21 January 2016).

99 Bhutani and Kohli, 'Just Leave the Seed Alone'.

100 'Land Law Not The Only Rollback, Seeds Bill Put on Hold over GM Clause', *Indian Express*, September 3, 2015, http://indianexpress.com/article/india/india-others/land-law-not-the-only-rollback-seeds-bill-put-on-hold-over-gm-clause/ (accessed on 21 January 2016).

101 The Scheduled Tribes and other Traditional Forest Dwellers (Recognition of Forest Rights) Act, 2006, Act 2 of 2007 (hereinafter the Recognition of Forest Rights Act).

102 The Recognition of Forest Rights Act, Preamble.

103 The Recognition of Forest Rights Act, Section 3(1) (a).

104 The Recognition of Forest Rights Act, Section 3(1) (b).

105 The Recognition of Forest Rights Act, Section 3(1) (f).

106 The Recognition of Forest Rights Act, Section 3(1) (g).

107 The Recognition of Forest Rights Act, Section 3(1) (h).

108 The Recognition of Forest Rights Act, Section 3(1) (c).

109 The Recognition of Forest Rights Act.

110 The Recognition of Forest Rights Act, Section 3(1) (k).

111 J. Sathyapalan, 'Implementation of the Forest Rights Act in the Western Ghats Region of Kerala', *Economic and Political Weekly*, July 24–30, 2010, 45(30): 65–73.

112 P. Bose, 'Forest Tenure Reform: Exclusion of Tribal Women's Rights in Semi–Arid Rajasthan, India', *The International Forestry Review*, 2011, 13(2): 220–232.

113 *Ibid.*
114 S. R. Ramdas, 'Women, Forest Spaces and the Law: Transgressing the Boundaries', *Economic and Political Weekly*, October 31–November 6, 2009, 44(44): 65–73.
115 *Ibid.*
116 *Ibid.*
117 Zweifel, 'The Gendered Nature of Biodiversity Conservation'.
118 U. Baxi, 'Access to Justice in a Globalized Economy: Some Reflections', Keynote address at the Golden Jubilee Celebration of Indian Law Institute, New Delhi, 2006, http://sites.google.com/site/lawcentreone/juris workshop (accessed on 21 January 2016); See also *Balco Employees Union* v. *Union of India* (2002) 2 SCC 333.
119 WP(c) 196 of 2001, *PUCL* v. *UOI*, Important orders in this Petition include *PUCL (PDS matter)* v. *Union of India* (2013) 2 SCC 688; (2013) 14 SCC 368.
120 The National Food Security Act 2013, Act No. 20 of 2013.
121 Public Distribution Scheme, Annapurna Yojna, Mid-day meal Programme, Antyodaya Anna Yojana.
122 See S. Sahai, 'Need for a Different Food Security Law', *Seminar*, June 2012, 634, www.india–seminar.com/semframe.html (accessed on 6 February 2016).
123 *Ibid.*
124 *Ibid.*
125 R. Dhavan, 'Wealth of Nations', *Seminar*, August 2000, 492, www.india–seminar.com/semframe.html (accessed on 6 January 2016).

12

THE POSSIBILITY OF *URGENDA* IN INDIA

Shreeyash Uday Lalit

Till date, there has been very little success in holding governments responsible for fulfilling their obligations to mitigate climate change. However, the recent landmark decision in the *Urgenda Foundation case*[1] of the Netherlands Hague District Court has compelled the Government of the Netherlands to fulfil its obligations by virtue of its duty of due care to achieve its objective of 25% reduction in greenhouse gas emissions under various international obligations, including the UN Framework Convention on Climate Change (UNFCCC). This decision marks a drastic change in the approach of the courts to identify scientific results as being of crucial assistance in the determination of a sound environmental policy. It is also the first instance in international jurisprudence where a court has intervened to exercise the tort of negligence qua climate change mitigation as against the actions of a state. Furthermore, the decision raises interesting questions on the separation of powers between the legislature, the executive and the judiciary, the effect of the instant litigation on foreign jurisprudence, and the maintainability of similar suits in the future in foreign courts. With India having ratified UNFCCC, the decision and scientific conclusions arrived at in the *Urgenda case* cast a guiding light for the Indian courts to follow suit in holding the governments accountable.

Firstly, this chapter attempts to establish the link between loss of biodiversity as a direct and immediate consequence of the mitigation of climate change and provide a brief description of climate change litigation in the foreign jurisprudence; and draw out the historical development of environmental policy in the Netherlands as well as a description of the facts that subsequently led to the decision in the Urgenda case. Secondly, the chapter dissects the decision of the Hague District Court in light of the civil, constitutional and international law that informed the same, including inter alia the issue of standing, Article 21 of the Dutch Constitution, various provisions of the European

Convention on Human Rights (ECHR) as well as the international obligations of the Netherlands. Thirdly, it delves into a deeper understanding of the duty of care as well as the standard of care expected out of a government in response to a tort action. Fourthly, it analyses the decision on whether it upholds separation or balance of powers. Finally, it compares the precedent to the legal scenario in India and attempts to equate the situations that may possibly lead to the Indian courts reviewing the potential impact of importing the tort of negligence or the duty of care for the climate against the State for the State's failure to mitigate climate change. The author also attempts to address the question of the impact of the decision on foreign courts, and on the future deliberation of international climate change policy.

I. Background

The Urgenda case has had a tremendous effect on environmental law jurisprudence. However, it is imperative to first recognise the nexus between biodiversity preservation and climate change mitigation. This will help us determine if the link is remote and illusory or direct and immediate. Only if the link between the two issues has a direct and proximate relationship, would the issue of climate change mitigation assume significance in terms of biodiversity protection and preservation. Thereafter, litigation surrounding climate change will be looked into in order to appreciate the nature of the dispute in the Urgenda case. This will be followed by an analysis of the socio-political climate in the Netherlands since it is important to be apprised of the contextual policy debates that arose within the governmental hierarchy in order to truly comprehend the arguments in the climate change debate as well as in the environmental law jurisprudence.

A. Link between biodiversity and climate change

There are various gases in the atmosphere, such as water vapour, ozone, carbon dioxide and methane, which act similar to a greenhouse by barring the heat from leaving the planet, thereby warming it. Therefore, the major promoter and contributor to the growing effects of climate change are the greenhouse gases which are gradually decreasing the Earth's temperatures in the lower atmosphere to inhospitable levels, thereby slowly but surely making habitation an impossible thought.[2]

Thus, it would not be far to imagine that climate change has a direct and immediate impact on biodiversity and efforts aimed at conserving it. In the last 1.8 million years, concentrations of carbon dioxide, precipitation and temperature have undergone tremendous changes, whereas the beings on earth have been subjected to tumultuous processes of evolutionary rejection and mutation, gradually leading to a symbiosis between the living beings and the environment. However, the current rates of deforestation and greenhouse effect are only amplifying the cause for concern as the magnitude of species extinction far outweighs the normal rates. There is ample evidence to suggest that mitigation of climate change is very crucial for conserving biodiversity.[3]

In the Arctic, lessening ice coverage in the sea is inadvertently endangering the habitat of polar bears, causing them to migrate, or worse, giving them lesser time to hunt.[4] In North America, the great fluctuations in temperature have reduced the plankton populations, which are essential since they are a major source of food for the North Atlantic right whale.[5] As a result, these species of whales are reducing in number by the day, whereas the population of plankton doesn't seem to be appearing on the rise as well. In the Pacific regions, warmer temperatures are leading to a reduction in the male sea turtles available to mate.[6] Reports show that a warmer climate increases the chance of a female sea turtle being born, which may endanger the reproduction of the species.[7]

Therefore, it is imperative that the resilience of ecosystems should be strengthened and the risk of damage to natural habitats should be mitigated through effective governmental strategies. Thus, climate change mitigation and litigation are extremely important in any strategy that attempts to conserve biodiversity and restore the lost balance to our ecosystems. Also, for those strategies to be executed, it is equally important that the branches of the Government should function in tandem and not in opposition to each other.

B. Climate change litigation

Albeit there have been synchronised and simultaneous calls to action by distinguished scientists, environmental activists as well as lawyers and judges alike, countries have been largely reluctant to adopt strict targets and guidelines, fearing their inclusion would undermine the development of their state. In fact, even Pope Francis stepped up the campaign on climate change, much to the alarm of conservatives.[8] Thus, more often than not, the battle against climate change has been led by environmental non-governmental organisations (NGOs) who have used litigation as a tool to effect change and also spread awareness. Public-spirited litigation has enabled the judiciary to generate judgments that can fill the policy void left by the forbearance of the executive and the legislature.[9]

Legal action vide litigation has been attempted against both public authorities as well as private entities.[10] Against public authorities, executive or legislative action has always been tested on the foundation of judicial review, which is popularly understood as the check and balance against the *vires* of such actions. The role of this separation of powers implies that the judiciary has a pivotal role in interpreting the said legislation or executive action against constitutional principles. Judicial review has enabled the environmental agencies and authorities to take cognisance of measures aimed at reducing carbon emissions, accounting of GHG emissions as well Environmental Impact Assessment (EIA) techniques implemented by the said authorities. In *Gray* v. *Minister of Planning*, which is a landmark case in Australia, the court held that the legality of coal projects receiving validation from the Environmental Planning and Assessment Act 1979 must conduct their EIAs while also including the effect of climate change on the combustion of the already produced coal.[11] In the US, the seminal case of *Massachusetts* v. *EPA*[12] dealt with the Clean Air Act wherein the regulatory power as wielded by the Environmental Protection Agency (EPA) to contain GHG emissions was in contention. The US Supreme Court, through Justice Stevens, held that the agency could exercise its regulatory power and had an obligation to reduce such emissions. Thus, the EPA was found to be in breach of its obligations under the statutory provisions for having failed to exercise its authority.

However, in none of the above cases pertaining to climate change has a judgment ascribed to the issue of a state's duty of care towards its residents to reduce carbon emissions, outside the realm of the statutory mandate. The decision of the Dutch District Court in the *Urgenda Foundation* v. *Government of the Netherlands (Ministry of Infrastructure and the Environment)*[13] (hereinafter, '*Urgenda case*') is a landmark judgment in its own right as it is the first judgment to assess the state's duty of care and find it in breach through a tort action and not due to a failure of statutory obligations.

Against private entities, climate change litigation in the US has conventionally relied on general principles of tort law viz. negligence, public nuisance, etc.[14] Usually, these cases have been disposed of summarily owing to the argument that the cases raised political questions and not justiciable queries requiring a judicial ascertainment.[15] The doctrine of political question has been used and applied time and again to discard such assailments. However, the *Urgenda case* has attempted to reverse this trend by observing it as a facet of legal inquiry and judicial disposal.

In India as well, climate change litigation predominantly rests upon statutory duties as opposed to judicial adventurism as attempted in the *Urgenda case*. Climate change litigation connotes a multitude of actions which may include civil as well as criminal liability. However, it would be prudent to mention that this chapter adverts solely to actions brought against the Government or private entities for reducing emissions of GHG, which includes any liability that may be imposed thereunder.

C. Historical development of environmental policy in the Netherlands

As it stands today, the persisting existence of the Netherlands demonstrates the success of human determination to reign superior over the environment. With an increasing population and demands stemming from the same, the Kingdom of the Netherlands has also had to face a contest with the surrounding waters, which frequently create difficulty for its residents. With regard to the other European countries which are largely industrialised, the Kingdom of the Netherlands faces an arduous task of controlling the natural hazards created through the effects of increased pollution, which will only overburden the Dutch ecosystems, especially due to the Netherland's downstream and downwind position.

Thus, as a matter of policy, it is not surprising that the Dutch have always urged stricter standards and promoted international cooperation *vis-à-vis* environmental issues, especially those which are incapable of being disentangled through a unilateral state action. The most far-reaching of these issues is that of climate change. Between 1960 and 1970, the issue of climate change gradually dawned upon the Dutch people and slowly became an issue of significant discussion.[16] Since the late 1970s, the Netherlands has managed to project itself into environmental talks, particularly those with regard to collective action on matters of environmental policy, including those with other countries or even other NGOs and specialised interest groups.[17] In 1988, Queen Beatrix delivered a speech which is popularly remarked as the watershed moment for environmental policy action in the Kingdom of the Netherlands. She remarked that *'the earth is slowly dying and the end of life itself is becoming conceivable'*.[18] Given the trans-boundary pollution and its effects on the Kingdom of the Netherlands since time immemorial, the Dutch Government took a hardened stance at the mitigation of climate change. Over the years, the same Government has managed to exert a significant influence on the European Union (EU) to advance its agenda of mitigation of trans-boundary pollution.[19]

However, in the preceding decade, the rest of Europe to the exclusion of the Dutch managed to surpass global expectations and tirelessly expand the scope and effectiveness of global environmental policy, whereas the Dutch were not able to perform as extravagantly.[20] This tectonic shift was ascribable to a shift in perspective, as much of Europe had awoken to the alarming reduction in global temperatures. The Dutch Government, on the other hand, only denuded its resolve as could be witnessed by virtue of its opposition to the 'gold plating' of environmental policies by the EU.[21] The Dutch ban on 'gold plating' did not only preclude ambitious environmental initiatives but also concomitantly deprecated the prevailing environmental policies that went above and beyond the established European standards.[22]

D. Facts of the Urgenda case

The United Nations Framework Convention on Climate Change (UNFCCC) Conference of the Parties (COP) in Copenhagen brought forth a fruitful resolution to target that the average global temperatures should not rise beyond 2 degrees Celsius as compared to the pre-industrial estimates. The signatories to the Copenhagen Accord agree that 'climate change is one of the greatest challenges of our time. We emphasise our strong political will to urgently combat climate change in accordance with the principle of common but differentiated responsibilities and respective capabilities.'[23] Thus, they pledged to

> achieve the ultimate objective of the Convention to stabilise greenhouse gas concentration in the atmosphere at a level that would prevent dangerous anthropogenic interference with the climate system, we shall, recognizing the scientific view that the increase in global temperature should be below 2 degrees Celsius . . . enhance our long-term cooperative action to combat climate change.[24]

The same idea has been imprinted in the UNFCCC for a very long time.[25] This objective was thereafter incorporated into the Cancun Agreements[26] wherein an emission reduction target of 25 to 40% as compared to 1990 levels was agreed by Annex I parties. The Intergovernmental Panel on Climate Change (IPCC) – through its Fourth Assessment Report, 2007 as well as the Fifth Assessment Report,

2013 – noted that to achieve the 2 degree Celsius target would require the stabilisation of GHG levels at 450 ppm.[27] It further argued that

> limiting warming with a likely chance to less than 2C relative to pre-industrial levels would require substantial cuts in anthropogenic GHG emissions. . . . Scenarios that are likely to maintain warming at below 2C are characterised by a 40 to 70% reduction in GHG emissions by 2050, relative to 2010 levels, and emissions levels near zero or below in 2100.[28]

The EU commenced its emission reduction goals by fixing them at 30% relative to be achieved by 2020. However, the events of the COP resulted in the EU issuing a press release where they declared that the target would be used as an incentive and political motivator to bring in a comprehensive and cohesive global response to the issue of climate change.[29] Pursuant to the press release, the EU vide the letter to the UNFCCC committed itself to the agenda of reducing emissions by 30% provided that developed countries contribute equally and developing countries adequately according to their responsibilities.[30] The Netherlands, on the other hand, reserved for itself an initial target of a 30% emission reduction by 2020.[31] However, in 2010, the Netherlands reconsidered the targets and laid down a meek 14 to 17% reduction as opposed to the ambitious 30% earlier.[32]

The Urgenda Foundation had requested the Dutch Government to take a more ambitious and aggressive stance on climate change mitigation since it necessitates urgent intervention.[33] However, the State Secretary for Infrastructure and the Environment replied to the Urgenda Foundation by stating that overly ambitious unilateral action in the derogation of international obligations would carry a potential risk. Thus, the Urgenda Foundation in response filed a class action suit against the Government along with 886 other plaintiffs on 20 November 2013.[34]

The primary disagreement between the two factions pertained to the urgency and whether 20% is adequate or not. Pursuant to the issuance of the mandatory order to the Dutch Government, the Court defended the Urgenda Foundation's position and held that the prevailing environmental policy for 2020 was inadequate and ineffective.[35] In light of governmental negligence and forbearance from addressing the environmental obligations, the judiciary, through the recent judgment in the *Urgenda case*, compelled the Government to reconsider the mandate of the people and the supervening environmental obligation that is owed by the Government to the existing and subsequent generations. On 24 June 2015, the Dutch District Court in the Hague, after perusing multifarious constitutional and environmental issues, held that the Government owed a duty of care towards the Dutch society to mitigate the deleterious effects of climate change.[36] It also held that the present objectives of 17 to 20% reduction were inadequate, in light of scientific evidence led by the IPCC on dangerous anthropogenic climate change[37] and the causal link between emissions of GHG and tangible climate change.[38] Thus, the Court ordered the Government to implement a 25% emission reduction which would be contemporaneous with the effective discharge of the state's duty of care.[39]

By virtue of the *Urgenda case*, the Netherlands can now proudly claim to be installed back on the global map of state actors leading the efforts to propagate and advance sound environmental policy aimed at climate change mitigation. Traditionally, climate change litigation was most actively dominated by the US, Australia, and to some extent also the United Kingdom. However, the *Urgenda case* judgment is now being hailed as an unprecedented step which carries immense symbolic significance, especially for the EU.[40]

The discussion below will cull out the standing to sustain the action as brought by The Urgenda Foundation, the significance of the UNFCCC and related ancillary decisions of the COP and the relevance of the ECHR. In addition to that, there are three issues which stand out as the most important to assess while perusing the judgment of the *Urgenda case*: (1) whether the liability imposed upon the Government can have appropriate equivalents in other jurisdictions that may trigger an equivalent action; (2) in light of the separation of powers as enshrined under a Constitution, whether it is permissible to prescribe a minimum reduction goal and whether the same can withstand the test of satisfying constitutional principles; and (3) whether, as an environmental policy, the *Urgenda case* would galvanise ambitious initiatives which NGOs and specialised interest groups have been campaigning for.

Therefore, the most crucial question that requires determination is whether 17 to 20% is adequate or not, and whether the appropriate authority to decide on these matters is the executive or the judiciary as under the doctrine of separation of powers. Additionally, in the event of negligence on the part of the state for failing to discharge its duty of care, the question is whether the judiciary can issue directions to the Government for exercising an effective response or whether it should leave the question to be tested against the anvil of popular mandate and elections.

II. Principles applied in the *Urgenda* decision

A. *Standing or locus standi*

Under the Dutch Civil Code, Article 3:305a provides that any association protecting certain interests of the society as mentioned in its
articles of association or other ancillary rules with full legal representation may possess the necessary standing to protect the interests of
such similarly placed persons.[41] The succeeding provision, which is
Article 3:305b, further mandates that the association may possess the
standing only after sufficient attempts at settlement have been tried
with the defendant.

In most jurisdictions, such as the US,[42] the aspect of standing is intri-
cately linked with the doctrine of separation of powers.[43] The nature
of the question raised by the plaintiffs can severely affect the ability of
the Court to admit and hear the dispute.[44] Political questions deserv-
ing no judicial inquiry cannot be admitted by the Court as they are
directly vitiated by the doctrine of political question.

In the instant case, the Dutch Government did not contest the
standing as possessed by the Urgenda Foundation in so far as the
NGO claimed to be protecting the interests of the Dutch people with
regard to emission reductions on Dutch territory. However, it did
argue that Urgenda had no standing to protect the interests or rights
of the succeeding generations in other countries.[45] Under Article 2
of the articles of association of the Urgenda Foundation, albeit the
NGO makes it a priority to uplift the interests which are specific to
the Netherlands, it nowhere limits its mandate to only the Dutch and
not the rest of Europe. Thus, the Court found the argument tilting
towards the side of Urgenda and held that Urgenda had the necessary standing to sustain this action.[46] Furthermore, with sustainable
development as the core agenda for the Urgenda Foundation, it was
hard for the Court to resist the argument that a sustainable society is
a global concept since trans-boundary pollution is in essence a multi-
country problem.[47]

As regards the standing for future generations, the doctrine of sustainable development has in its core an inter-generational element,
and therefore any NGO attempting to protect and safeguard the sustainability of a region and protect the environment would necessarily
possess the standing to protect the current as well as the future generations. Since the by-laws and the articles of association of the Urgenda
Foundation did not limit or localise themselves to specific interests in
the Netherlands, the Dutch District Court had no qualms in holding

that the NGO had the necessary standing to bring this action against the Government.[48] In conclusion, the Court stated the following:

> Urgenda has met the requirement of Book 3, Section 305a of the Dutch Civil Code that it has made sufficient efforts to attain its claim by entering into consultations with the State, the court concludes that Urgenda's claims, in so far as it acts on its own behalf, are allowable to the fullest extent.[49]

B. Dutch constitution

Under the Dutch Constitution, Article 21 provides a mandate on the Government that it shall be their concern 'to keep the country habitable and to protect and improve the environment'.[50] As raised by The Urgenda Foundation, the issue in contention was whether the Dutch Government was fulfilling its constitutional duty of care under Article 21 or whether it was falling short in its exercise.[51]

The Dutch Court, while accepting the contentions elicited by the Urgenda Foundation, noted that although the Government possessed a *'duty of care in lieu of concern'*[52] which imposed a duty on the state to protect and improve the environment, it was not precise in its language to clearly stipulate the manner, method and form in which this duty was to be exercised, nor whether it could be subject to the discretion of the Government and therefore be engulfed by the doctrine of political question.[53] Therefore, under Article 21, the Court chose to abstain from issuing a mandamus to the Government as to how the duties and obligations inherent under Article 21 were to be fulfilled. However, pursuant to the claim for negligence under the tort law, it may be said that the Dutch District Court circumvented constitutional hurdles by indirectly deciding the tort claim in favour of Urgenda and holding that the state had breached its obligations with its current climate policy, as they did not match to the standard of care expected out of a Government.

The Court opined that the state's discretionary power under Article 21 is of a vital nature and should be considered as an important element while deciding on the standard of care and its appropriate exercise.[54] As a result, it held that the discretionary power of a Government as enjoyed under Article 21 also attracts the tests of reasonableness and adequateness of response, and that the discretion cannot be unlimited under the garb of the queries being political in nature.[55] Furthermore, the Court observed that dangerous climate change presents a high risk with grave and dangerous consequences for humanity. Thus, the corollary of the same argument leads to the inevitable conclusion that an institution mandated to protect the interests of the people and enjoying a constitutional power has the direct obligation to safeguard its residents by taking adequate and effective measures.[56]

The Court had to restrain itself from allowing the argument on Article 21 to pass muster, as it held that 'under Article 21 of the Constitution, the State has a wide discretion of power to organise the national climate policy in the manner it deems fit'.[57] Albeit the Court reserved its consideration on the implications of a breach of Article 21 of the Dutch Constitution, it nevertheless did express that 'climate change mitigation is crucial for protecting the citizens from its consequences'.[58] Thus, the Court provided an unambiguous directive to the Government to tackle the mitigation under its Constitutional obligation, although it did not explicitly state whether the Government breached Article 21 or not.

C. International law on mitigation of climate change

The Dutch Government has signed and ratified the UNFCCC and the Kyoto Protocol while also endorsing the Cancun Agreements, Doha Agreements to the Kyoto Protocol, Bali Action Plan as well as the decision of the UNFCCC in Durban, wherein states bound themselves to conclude a climate change agreement before 2015. Thus, the decisions of the COP are significant milestones for a global framework safeguarding global environmental interests. Urgenda relied on these conventions and agreements to claim that as a member of the Annex I parties, the Netherlands has adequate international obligations that may take the frontseat in mitigating climate change.[59]

The Urgenda Foundation in its action also relied on the fact that the Dutch Government has always supported the emission reduction goals of 25 to 40% by 2020 as compared to the 1990 levels. Furthermore, it also stressed the well-established 'no harm' principle, which delineates that no state has the right to cause pollution and damage to other states through the use of its own domestic territory.

It does not appear from the plain text of the *Urgenda case* judgment as to whether Urgenda wanted to argue that the damage is being caused to others or to Urgenda itself due to the state's failure to implement international obligations. Article 93 of the Dutch Constitution is also noteworthy, which provides for the direct effect of international law in Dutch territory.

It may be said that Urgenda was specifically precluded from arguing on the provisions of the UNFCCC, the Kyoto Protocol and various other decisions by the COP since they have no binding force on citizens. The same argument also applied to Urgenda's claims under EU treatises and conventions. Although it observed that the EU directives do not confer rights on the citizens *qua* the state, the Court held that the reliance on these provisions would still be necessary since 'they have a reflex effect on national law'.[60] All this does not imply that international law was immaterial in influencing the Court's decision. While agreeing that 'Netherlands is obliged to adjust its national legislation to the objectives stipulated in these directives',[61] the Court finally came to the conclusion that such 'stipulations in an EU treaty or directive can have an impact through the open standards of national law'.[62]

D. *European Convention on Human Rights (ECHR)*

The Urgenda Foundation also relied upon the articles of the ECHR to claim that it was a victim of the violations done by the Dutch Government. Under the ECHR, 'victim' has been defined and adverted to in several cases,[63] wherein the ECHR dismissed the cases as being inadmissible on account of the fact that the plaintiffs could not be construed as 'victims' under the Convention just because their religious beliefs had been offended. Religious freedom is not an attribute which the Convention appertains to, and therefore it cannot satisfactorily qualify the standard of 'victim' under the Convention. For the plaintiffs to be successful under such action, it is important to demonstrate a practical effect that is measurable and quantifiable. An abstract application to take cognisance of unspecified and vague violations of human rights would not receive any relief from the Court.[64]

In the instant case, Urgenda could not claim a violation under Article 2 or 8 of the ECHR since a legal person as opposed to a natural person cannot raise a claim regarding the violation of its physical integrity or privacy. Thus, the Court concluded that the provisions under ECHR, including inter alia, Articles 2, 8 and 34, were inapplicable to the present case and the Urgenda Foundation could not raise a claim under the same. However, it did refer to the exhaustive human rights jurisprudence as evolved by the ECHR to cast a guiding light on the objectives and the standard of care that a government ought to prescribe.[65] It also referred to the publications of the Council of Europe viz. 'Manual on Human Rights and the Environment', which is an exhaustive and comprehensive guide on the jurisprudence espoused by the ECHR.

The *Urgenda case* was a path-breaking decision in so far as it changed the climate change mitigation landscape by relying on the international consensus of the '2 degrees Celsius' emission reduction goals and thereby conclude whether the state was obliged to perform a duty of care towards its citizens, and if in the affirmative, then what was the standard of exercising that duty of care.

III. Duty of care for the climate

Climate change litigation and jurisprudence can be categorised on the basis of several factors viz. public/private nature of the nuisance or negligence,[66] proactive or reactive,[67] climate change as the motivation for the action, or as pro- or anti-regulatory.[68]

Additionally, several kinds of actions may be available to compel or prevent the Government from exercising a certain power. These actions may be premised on statutory obligations, tort claims, customary or public international law as well as arguments deriving their source from constitutional or human rights. As has already been noted, the *Urgenda case* sets itself apart from all such previous actions since it upholds a claim based on reasons outside the realm of statutes.[69]

The reason for the Court's mandatory order can be found in Section 6:162 of the Dutch Civil Code.[70] The only way liability could have been imposed on the Government would have been to first ascertain if a duty of care exists to protect the environment and the citizens from the effects of climate change. To identify the existence of the duty of care, it is important to assess whether the prevailing mitigation policies are adequate or not, given the Government's discretion as has been previously established.[71] Under Paragraph 2 of Section 6:162, the tort of negligence under the Dutch Civil Code requires a 'social standard of due care'.

There have been case laws under the Dutch Supreme Court which have laid down four parameters viz. whether the danger is apparent; whether danger is likely to manifest itself; whether the danger is serious (damage to property or bodily injury); and whether preventive measures are necessary.[72]

The Dutch District Court expanded on these parameters and increased them to six viz.[73] nature and extent of the damage; knowledge and foreseeability of the damage; likelihood of dangerous anthropogenic climate change; nature of the Government's actions or omissions; onerousness of taking precautionary measures; and the Government's discretion as exercisable under public law. The Court rejected the propositions under the Dutch constitutional as well as international law[74] while declaring that these rights are not directly enforceable by the citizens vis-à-vis the state. However, the Court did focus on the duty and standard of care explicitly provided in Section 6:162 and the six parameters outlined above.

With regard to the first three parameters viz. nature, foreseeability and likelihood of damage, the Court opined that the risk of anthropogenic damage is grave and far-reaching, thereby warranting the interference of the Court. It also stated that the measures required for implementing the said emission reduction goals were neither too costly or technologically challenging.[75] Considering the scientific evidence as adduced with regard to the IPCC Reports and the international obligations as well as the extant duty of care under the civil code, the Court held that the Government's discretion in this issue was minimal,[76] after concluding that mitigation of climate change, in light of the scientific evidence provided, was extremely essential. The argument relating to the cost or technological challenge inherent in these policies was rejected while opining that the current policies being exercised by the Netherlands were deficient in their ability to tackle the impending problem of climate change.[77]

A. Drop in the ocean argument?

To wrap up the analysis of the duty of care, the court considered the argument raised by the state which stated that a higher emission-reduction goal would result in a very minor, if not zero, reduction of GHG emissions. The response of the Court is highly meritorious:

> It is an established fact that climate change is a global problem and therefore requires global accountability. . . . The fact that the amount of the Dutch emissions is small compared to other countries does not affect the obligation to take precautionary measures in view of the State's obligation to exercise care. . . . After all, it has been established that any anthropogenic greenhouse gas emission, no matter how minor, contributes to an increase of CO^2 levels in the atmosphere and therefore to hazardous climate change. Emission reduction therefore concerns both a joint and individual responsibility of the signatories to the UN Climate Change Convention. . . . Moreover, it is beyond dispute that the Dutch per capita emissions are one of the highest in the world.[78]

Thus, the Court finally held with regard to the duty of care that

> the possibility of damages for those whose interests Urgenda represents, including current and future generations of Dutch

nationals, is so great and concrete that given its duty of care, the State must make an adequate contribution, greater than its current contribution, to prevent hazardous climate change.[79]

B. Causation

Further, the Court also addressed the arguments of causation. It was specifically averred by the Government that the responsibility of one particular state actor is minute in comparison to the pollution caused by countries worldwide.[80] Therefore, even if a stricter environmental policy was imposed upon the Government, the change in policy would hardly reduce the damage caused to the environment, since the basic facet of trans-boundary pollution is that it is a multi-country problem and not a problem which may be solved by the unilateral action of one actor.[81] However, the Court rejected these arguments and held that collective responsibility as well as minimal mitigation does not vitiate the duty of care as held by the Government or the causal link between mitigation of one actor and resulting damage.[82] For this purpose, it cited the seminal case of *Massachusetts* v. *EPA*,[83] where the US Supreme Court observed that although regulating the emissions resulting from motor vehicles will not invert the process of global warming, that does not imply that the Supreme Court lacks jurisdiction nor that the EPA could justifiably abstain from performing its duty to reduce domestic emissions.

The Urgenda Foundation also proposed a pro-rata assessment of the Government's liability, which was rejected by the Court. The Court referred to the *Kalimijnen case*,[84] in which the countries France, Germany, Luxembourg and the Netherlands caused great pollution to the river Rhine through the discharge of chloride dumps; however, individual responsibility for the actions could not be attributed, and thus all were held collectively responsible. Albeit there are various distinctions capable of being drawn between the *Kalimijnen case* and the *Urgenda case*, the Court saw ample overlap to conclude that Urgenda need not demonstrate a strict link of causation. While drawing the analogy, it also referred to the mitigation targets under the Kyoto Protocol to the UNFCCC as a reminder of the Government's obligations to reduce GHG emissions.[85]

C. Precautionary principle

Commentators Suryapratim Roy and Dr. Edwin Woerdman point out classical flaws in the judgment with regard to the procedural part of the precautionary principle and thankfully not the substantive part.[86] Akin to *Massachusetts* v. *EPA*, the Court allocates the burden of proof to the Government to prove that the measures adopted by them are adequate and effective.[87] The Commentators argue that this shifting of the burden is odd, especially when considering that the precautionary principle is usually used as a tool to either accept or reject the policy due to the scientific uncertainty of their effects on the climate. The idea as espoused in the Indian environmental jurisprudence may be construed as '*err on the safe side*'.[88] A true construction of the principle does not imply that the burden may be shifted to the Government.

Even if so, the Commentators commit a folly with regard to the *Urgenda case*, while attempting to establish an analogy with the US case of *Massachusetts* v. *EPA*, since the latter pertained only to the US Supreme Court directing the regulatory body to take action in light of absolute inaction, as opposed to the former where the Court has directed the regulatory body to *enhance* its actions on account of them not being effective.[89]

IV. Separation of powers

From the sixteenth century, separation of powers and judicial review began to enjoy a prominent part in the constitutional scheme.[90] Most modern democracies provide for a system of checks and balances between different organs of the Government viz. legislature, executive and judiciary. The judiciary is primarily charged with the responsibility to check the excesses of the legislature and the executive, and to assess the vires of their actions against constitutional principles. The responsibility of checking only the 'excesses' and not all kinds of legislative and executive actions invariably implies that the judiciary is expected to exercise a certain degree of restraint in so far as the maintainability of such actions is concerned.[91] In the Kingdom of the Netherlands, the three branches of government are referred to as the '*trias politica*' which essentially espouses the abovementioned doctrine of separation of powers.

In the *Urgenda case*, the Court emphasised that the judicial decision would be restricted only to the parties involved in the instant case. However, the mandatory order in *Urgenda* would clearly have an indirect result on third parties which were not party to the litigation, thereby bringing them within the ambit of the decision. The Dutch Government also assailed the Court's legitimacy in so far as its competence to determine the position of the Netherlands in the international arena was concerned.[92] However, this argument was also flatly rejected by the Court as it did not find any persuasive value in the contention. The possibility of international embarrassment cannot assail the rights of the citizens as are enjoyed by them.[93]

What is important to note in the judgment is that the Court has specifically observed that 'the State has failed to argue that it is actually incapable of executing the order. The State has also failed to argue here that other, fundamental interests it is expected to promote would be damaged'.[94] This is in furtherance to its observation that although the judiciary is not democratically elected, it derives its authority and legitimacy from the democratically established legislation, which assigns it the task of settling legal disputes.[95]

The major criticism that may appear with regard to the *Urgenda case* is that the Court has indulged in judicial overreach by encroaching upon the discretion available to the Government to decide upon a prudent course of action. Conventionally, the Dutch Supreme Court has always given a wide amount of discretion while being firm on the principles.[96] However, the Dutch District Court in the instant case made an extremely specific order where it ordered the Government to realign the goals to 25% emission reduction by 2020, thereby leaving minimal or even zero discretion to the Government with regard to the implementation and execution of climate change mitigation. This unequivocally impinges upon an executive function which the Court should have respected.

In the US, the two crucial doctrines governing the debates on environmental policy and its appropriateness are the doctrine of political question and the displacement doctrine.[97] While addressing the issues of public nuisance, the US Supreme Court has explicitly set aside the exception which bars the interference of the judiciary in political questions.[98] Thus, the doctrine of political question has lost a lot of weight as a useful tool in blocking climate change litigation from being entertained by the courts.

With regard to Indian jurisprudence, separation of powers is deeply entrenched through various judicial decisions. What originated with Montesquieu, which was further galvanized by Chief Justice John Marshall in *Marbury* v. *Madison*,[99] took root in the Constitutional Assembly debates. Although never explicitly prescribed with absolute rigidity in the Constitution, the Indian Supreme Court has had the occasion to imbibe the doctrine of separation of powers as being intrinsic to the basic structure of the Constitution.[100]

With the imposition of a minimum emission-reduction goal by the judiciary while circumventing the necessary position occupied by the legislature in the *trias politica*, many view this judgment as a regressive step that may dismantle the constitutional machinery and bring forth a tyranny of the unelected. Nonetheless, within Dutch academia, many have praised the judgment as it allows the state to decide on matters of extreme importance without having to conform to the considerations of re-election or political accountability.[101] Some have even put forth that the judiciary provides an alternative forum for interested and necessary parties to voice their opinions which have not been heard in the political arena.[102] The judgment aims to have a pro-regulation effect which has thus received praise from many quarters.[103] However, as was witnessed in the US, there were various judgments that were pronounced as a response to the 'pro-regulation' success achieved in the case of *Massachusetts* v. *EPA*.[104] A similar backlash may result if the *Urgenda case* survives its appeal in the Supreme Court.

V. Can the *Urgenda* decision be imported to the Indian scenario?

Firstly, in drawing a parallel between the *Urgenda case* and the Indian legal scenario, the first point of inquiry would be to determine if a tort action on similar grounds can be upheld in the Indian legal system. Prior to the promulgation of statutes, tortious actions based on nuisance or negligence were the custom and the norm.[105] In the celebrated case of *Union of India* v. *United India Insurance Co.*,[106] the Indian Supreme Court dealt with an omission to exercise a statutory power which provided that the Central Government 'may require' a railway administration to erect fences and construct gates. In the same case, a train had collided with a bus owing to non-existent crossings and gates which could have prevented the same. The Union of India was held guilty of negligence, being in breach of its common law duty. Therefore, Indian courts have since time immemorial allowed tortious actions, whether it be against a person or authority, or whether the duty stems from a statutory grant or a common law duty.

Secondly, the requirement of standing or *locus standi* in environmental matters has been considerably relaxed. India offers a highly relaxed and widened application of the principle of locus standi,[107] which is a complete departure from the 'proof of injury approach'.[108] This catalysed the formation of a new kind of litigation, known as Public Interest Litigation (PIL),[109] which is extremely effective in dealing with environmental disputes in so far as they are concerned with the rights of the community and not the individual.[110] It may entail the appointment of an *amicus curiae* and follow a non-adversarial approach.[111] Therefore, there would be no issue in importing the decision of the *Urgenda case* in so far as the standing is concerned.

Thirdly, in India, the doctrine of separation of powers has no strict position in the Indian Constitution. However, it has evolved as a part of the constitutional law developed by various jurists over time. In the Constituent Assembly debates, Dr. B. R. Ambedkar, one of the most important architects of the Indian Constitution, argued that that

> there is no dispute whatsoever that the executive should be separated from the judiciary. With regard to the separation of the executive from the Legislature, it is true that such a separation does exist in the Constitution of the United States; but many Americans themselves were quite dissatisfied with the rigid separation embodied in the American Constitution between the executive and the Legislature.[112]

Subsequently, in a catena of decisions pronounced by the Indian Supreme Court on the subject, the doctrine of separation of powers is not deeply entrenched and embossed in the constitutional framework. In the *Delhi Laws Act case*,[113] the then Hon'ble Chief Justice Kania observed that

> although in the Constitution of India, there is no express separation of powers, it is clear that a Legislature is created by the Constitution and detailed provisions are made for making that Legislature pass laws. Is it then too much to say that under the Constitution the duty to make laws, the duty to exercise its own wisdom, judgment and patriotism in making law is primarily case on the Legislature?[114]

What followed were a number of judgments which categorically upheld the separation of powers as being a salient feature of the Indian Constitution.

We have witnessed the Indian Supreme Court passing various orders in the *Vehicular Pollution case*,[115] whereby diesel-run buses were directed to be converted to CNG. In 1998, on the strength of the recommendations of the Bhure Lal Committee, the Court passed an order requiring the change to be effected within three years.[116] Thus, one could argue that there has been ample 'judicial adventurism' of late, whereby the judiciary has passed various orders to safeguard the Article 21 Fundamental Right under the Indian Constitution, against legislative or executive actions or omissions attempting to impeach or impede the effective exercise of that right, in so far as protecting the environment is concerned.

Fourthly, the precautionary principle is well established in the Indian legal system, which is the same principle incorporated into the understanding of the climate change mitigation in the *Urgenda case*. The Indian Supreme Court in the case of *Vellore Citizens' Welfare Forum* v. *Union of India*,[117] held that the precautionary principle is a prominent principle of environmental jurisprudence in India. Furthermore, in *A.P. Pollution Control Board* v. *M.V. Nayudu*,[118] the Supreme Court discussed the historical development of the principle. In the case of *Narmada Bachao Andolan* v. *Union of India*,[119] the Supreme Court observed that where a lack of data or material may disturb the ecological balance, then the benefit of doubt must go in favour of the path that is less likely to result in pollution.

A. Skepticism in importing the judgment

Thus, it may seem that the factors required for transposing the *Urgenda* judgment to India are more or less satisfied. However, there are various reservations that arise and which need extensive judicial circumspection while attempting to import the said judgment. Firstly, India is a Non-Annex I country under the UNFCCC, which thereby means that there are no direct international obligations on the Indian state as opposed to the Dutch State, which has a 'reflex effect' of international obligations on their national law, as mandated by the Dutch Constitution. Secondly, while the Kingdom of the Netherlands is a developed economy which allows it the freedom and leverage to be an Annex I party under the UNFCCC, India does not enjoy the same latitude when it comes to economic and environmental policies. Therefore, while the Dutch State failed to argue that it was incapable of following higher emissions standards,[120] the Indian State's arguments of incapability would definitely resonate more in light of the fact that India is a developing country and cannot afford to retard its economic growth and prosperity. Thirdly, while the Indian Supreme Court may seem overzealous on occasion, in light of the doctrine of separation of powers and judicial restraint exercised when it comes to policy considerations, the *Urgenda* judgment may receive an immovable roadblock in so far as its importation and application to the Indian context is concerned. The standard of care would naturally be reflective of the standard as set by the Indian State itself, and it would be improbable for the Supreme Court to intervene and posit its own discretion as to what the standard ought to be. While the task for the Dutch District Court was easier in light of the various international obligations of the Dutch State combined with the IPCC Reports, the Indian Supreme Court does not have the same leeway.

VI. Conclusion

Skepticism aside, the biggest question and the reason for academic excitement about the *Urgenda case* is because this is the first judicial adventurism attempted across foreign jurisprudence where the Court has dared to ascertain the legality of policies through their adequacy and effectiveness in the context of climate change mitigation, without the application of an available statutory mandate. Within the Netherlands itself, there have been significant improvements with a call for a parliamentary debate on the climate policies to be implemented by the Government and a subsequent expectation from academicians that the debate on mitigation would continue till it achieves a tangible outcome. In the neighbouring countries, Belgium, for example, is a matter which is *sub judice* and is on similar lines.[121] It is expected that the environment of judicial adventurism in this unexplored arena of upholding a tortious action for climate change mitigation would bear fruitful result.

Albeit the Urgenda decision is an admixture of European, international and Dutch law, it still bears a resemblance to the legal situation in India. Equally, the scholarly debate on the legal and constitutional principles inherent in this discussion would only further the progress of climate change mitigation as envisioned by various international institutions and actors. While there is reasonable scepticism and apprehension with regard to the applicability of the judgment to the Indian context, a careful and intuitive analysis would only further the scholarly debate, which would hopefully turn into a legal debate fought within the Indian courts. The academic, judicial and political discussions that will arise through this judgment would have far-reaching consequences throughout India as judges and politicians alike would be motivated, if not tempted, to apply the same principles of tort law as were espoused by the Dutch District Court in arriving at its decision.

As Lord Atkin explains in *Donoghue* v. *Stevenson*,[122] 'one must take reasonable care to avoid acts or omissions which one can reasonably foresee as to be likely to injure your neighbour'. Simply put, this is the basic standard of care expected out of the Indian Government. In today's world, in light of the insurmountable evidence including the IPCC Reports, there is a clear and unequivocal obligation on India to abide by this standard of care lest it cause injury to the very subjects that it wishes to protect. Therefore, any such deviation from this standard should invite the censure and intervention of the Indian Supreme Court, as has been successfully executed in the *Urgenda case*. After all, there is no lack of principle – only a lack of courage.

Notes

1 24th June 2015, C/09/456689, HA ZA 13–1396 (Neth.).
2 'Convention on Biological Diversity, Climate Change and Biodiversity', Introduction, www.cbd.int/climate/intro.shtml (accessed on 22 September 2016).
3 *Ibid.*
4 Derocher et al., 'Polar Bears in a Warming Climate', *Integrative and Comparative Biology*, 2004, 44(2): 163–176.
5 Woods Hole Oceanographic Institution, 'Study links swings in North Atlantic Oscillation Variability to Climate Warming', January 13, 2009, www.whoi.edu/page.do?pid=83537&tid=3622&cid=54686&c=2 (accessed on 22 September 2016).
6 Convention on Biological Diversity, Report on Biodiversity and Climate Change, www.cbd.int/doc/bioday/2007/ibd–2007–booklet–01–en.pdf (accessed on 22 September 2016).
7 C. Harvey, 'Another Weird Effect of Climate Change – Too Many Female Sea Turtles', *The Washington Post*, October 16, 2015, www.washingtonpost.com/news/energy–environment/wp/2015/10/16/another–weird–effect–of–climate–change–too–many–female–sea–turtles/?utm_term=.b56adf5e9c8f (accessed on 22 September 2016).
8 C. Davenport and L. Goodstein, 'Pope Francis Steps Up Campaign on Climate Change, to Conservatives' Alarm', *The New York Times*, April 27, 2015, www.nytimes.com/2015/04/28/world/europe/pope–francis–steps–up–campaign–on–climate–change–to–conservatives–alarm.html?_r=0 (accessed on 22 September 2016).
9 J. Peel and H. Osofsky, *Climate Change Litigation: Regulatory Pathways to Cleaner Energy*, Cambridge: Cambridge University Press, 2015.
10 J. Lin, 'Climate Change and the Courts', *Legal Studies*, 2012, 32(35): 35–57, at 37, 38.
11 (2006) NSWLEC 720 (*Gray v. Minister of Planning*).
12 (2007) 127 S. Ct. 1438 (*Massachusetts v. EPA*).
13 ECLI:NL:RBDHA:2015:7145, Rechtbank Den Haag, C/09/456689/HA ZA 13–1396 (*Urgenda case*).
14 *Native Village of Kivalina, et al. v. ExxonMobil Corporation, et al.*, 696 F.3d 849 (9th Cir. 2012); *American Electric Power Co., Inc., et al. v. Connecticut et al.*, 131 S. Ct. 2527 (2011) (*AEP v. Connecticut*).
15 *Corrie v. Caterpillar, Inc.*, 503 F. 3d. 974, 980, 9th Cir. 2007.
16 E. Tellegen, 'The Dutch National Environmental Policy Plan', *The Netherlands Journal of Housing and Environmental Research*, 1989, 4(4): 337–345.
17 C. Hey, 'EU Environmental Policies: A Short History of the Policy Strategies', in S. Scheuer (ed.), *EU Environmental Policy Handbook: A Critical Analysis of EU Environmental Legislation*, Brussels: European Environmental Bureau, 2006, pp. 17–30, at 20.
18 Netherlands Institute of Public Health and Environmental Protection, Report on 'Concern for Tomorrow', 1988.
19 M. E. Pettenger (ed.), *The Social Construction of Climate Change: Power, Knowledge, Norms, Discourses*, Hampshire, UK: Ashgate, 2007, p. 55.
20 L. H. Gulbrandsen and J. B. Skjærseth, 'Implementing the EU 2020 Climate and Energy Package in the Netherlands: Mixed Instruments, Mixed Results', *Fridtjof Nansen Institute*, 2014: 15–31.

21 J. H. Jans et al., 'Gold Plating' of European Environmental Measures?' *Journal for European Environmental and Planning Law*, 2009, 6(4): 417–435, at 429.

22 Ibid., p. 427.

23 UNFCCC, Decision 2/CP.15, Copenhagen Accord: Report of the Conference of the Parties on its fifteenth session, held in Copenhagen from 7 to 19 December 2009, http://unfccc.int/resource/docs/2009/cop15/eng/11a01.pdf (accessed on 22 September 2016).

24 UNFCCC, Decision 2/CP.15.

25 S. Randalls, 'History of the 2C Climate Target', *WIREs Climate Change*, 2010, 1(4): 598–605.

26 UNFCCC, Decision 1/CMP.6, The Cancun Agreements: Outcome of the work of the Ad Hoc Working Group on Further Commitments for Annex I Parties under the Kyoto Protocol at its fifteenth session, http://unfccc.int/resource/docs/2010/cmp6/eng/12a01.pdf#page=3 (accessed on 22 September 2016).

27 R. K. Pachauri and L. A. Meyer (eds.), *IPCC, 2014: Climate Change 2014: Synthesis Report. Contribution of Working Groups I, II and III to the Fifth Assessment Report of the Intergovernmental Panel on Climate Change, Summary Report for Policymakers*, Geneva, Switzerland: IPCC, 2014, p. 20.

28 *Ibid.*, p. 82.

29 European Commission, Press Release: 'Climate change: European Union notifies EU emission reduction targets following Copenhagen Accord', Brussels, January 28, 2010, http://europa.eu/rapid/press – release_IP – 10–97_en.htm?locale=en (accessed on 22 September 2016).

30 European Commission, Letter to the UNFCCC, 'Subject: Expression of Willingness to be Associated With the Copenhagen Accord and Submission of the Quantified Economy-wide Emissions Reduction Targets for 2020', Brussels, January 28, 2010, http://unfccc.int/files/meetings/cop_15/copenhagen_accord/application/pdf/europeanunioncphaccord_app1.pdf (accessed on 22 September 2016).

31 *Urgenda*, at para 4.31.

32 *Urgenda*, at para 5.1.

33 'Urgenda Foundation, Letter to Dutch Government', Amsterdam, November 12, 2012, www.urgenda.nl/documents/Staat–der–NL–Engels1.pdf (accessed on 22 September 2016).

34 'Urgenda Foundation, Final draft Translation of the Summons in Urgenda Foundation v. Kingdom of the Netherlands', Hague, June 25, 2014, www.urgenda.nl/documents/FINAL–DRAFT–Translation–Summons–in–case–Urgenda–v–Dutch–State–v.25.06.10.pdf (accessed on 22 September 2016).

35 *Urgenda*, para 5.1.

36 *Urgenda*, para 4.83.

37 *Urgenda*, para 4.12.

38 *Urgenda*, para 4.15.

39 *Urgenda*, para 4.84.

40 'Netherlands Ordered to Cut Greenhouse Gas Emissions', *BBC News*, June 24, 2015, www.bbc.com/news/world– europe–33253772 (accessed on 22 September 2016).

41 H. Tolsma et al., 'The Rise and Fall of Access to Justice in the Netherlands', *Journal of Environmental Law*, 2009, 21(2): 309–321, at 311–313.
42 *Massachusetts* v. *EPA*; *AEP* v. *Connecticut.*
43 Peel and Osofsky, *Regulatory Pathways*, pp. 269–278.
44 E. Kosolapova, *Interstate Liability for Climate Change – Related Damage*, The Hague: Eleven International Publishing, 2013.
45 *Urgenda*, para 4.5.
46 *Urgenda*, para 4.7.
47 *Ibid.*
48 *Ibid.*
49 *Urgenda*, para 4.9.
50 The Constitution of the Kingdom of the Netherlands 2008, Article 21, www.rijksoverheid.nl/ (accessed on 22 September 2016).
51 *Urgenda*, para 4.35.
52 *Urgenda*, para 4.36.
53 *Ibid.*
54 *Urgenda*, para 4.74.
55 *Ibid.*
56 *Ibid.*
57 *Urgenda*, para 4.55.
58 *Urgenda*, para 4.75.
59 J. Lin, 'The First Successful Climate Change Case: A Comment on Urgenda v: The State of the Netherlands', *Climate Law*, 2015, 5(1): 65–81.
60 *Urgenda*, para 4.43.
61 *Urgenda*, para 4.44.
62 *Ibid.*
63 *Ouardiri* v. *Switzerland* (Application No. 65840/09); *Ligue des Musulmans de Suisse and Others* v. *Switzerland* (Application No. 66274/09).
64 K. Boyle, 'The European Experience: The European Convention on Human Rights', *Victoria University of Wellington Law Review*, 2009, 40(1): 167–175.
65 *Urgenda*, para 4.46.
66 *Massachusetts* v. *EPA.*
67 C. J. Hilson, 'Climate Change Litigation: An Explanatory Approach (Or Bringing Grievance Back In)', in F. Fracchia and M. Occhiena (eds.), *Climate Change: la riposta del diritto*, Naples: Editoriale Scientifica, 2010, pp. 421–436.
68 D. Markell and J. B. Ruhl, 'An Empirical Assessment of Climate Change in the Courts: A New Jurisprudence or Business as Usual?' *Florida Law Review*, 2012, 64(15): 15–86.
69 Peel and Osofsky, *Regulatory Pathways.*
70 Dutch Civil Code, Title 6.3 (Tort), Section 6.3.1 (General provisions), Article 6:162 (Definition of a 'tortious act'):

(1) A person who commits a tortious act (unlawful act) against another person that can be attributed to him, must repair the damage that this other person has suffered as a result thereof;

(2) As a tortious act is regarded a violation of someone else's right (entitlement) and an act or omission in violation of a duty imposed

by law or of what according to unwritten law has to be regarded as proper social conduct, always as far as there was no justification for this behavior;

(3) A tortious act can be attributed to the tortfeasor [the person committing the tortious act] if it results from his fault or from a cause for which he is accountable by virtue of law or generally accepted principles (common opinion).

71 *Urgenda*, para 4.5.
72 Hoge Raad (Dutch Supreme Court), 6 Nov. 1965, NJ 1966/136; see also R. Cox, 'The Liability of European States for Climate Change', *Utrecht Journal of International and European Law*, 2014, 30(78): 125–135, at 129.
73 J. van Zeben, 'Establishing a Governmental Duty of Care for Climate Change Mitigation: Will Urgenda Turn the Tide?' *Transnational Environmental Law*, 2015, 4(2): 339–357, at 347.
74 *Urgenda*, paras 4.35–4.52.
75 *Urgenda*, para 4.73.
76 *Urgenda*, paras 4.74–4.77.
77 *Urgenda*, paras 4.88–4.89.
78 *Urgenda*, para 4.79.
79 *Urgenda*, para 4.89.
80 *Urgenda*, para 4.78.
81 *Ibid.*
82 *Urgenda*, paras 4.79, 4.90.
83 *Massachusetts* v. *EPA*.
84 Hoge Raad (Dutch Supreme Court), 23 Sept. 1988, NJ 1989/743.
85 *Urgenda*, para 4.79.
86 S. Roy and E. Woerdman, 'Situating *Urgenda Versus the Netherlands* within Comparative Climate Change Litigation', *Journal of Energy and Natural Resources Law*, 2016, 34(2): 165–189.
87 *Urgenda*, para 2.53.
88 *Punjab* v. *Modern Cultivators, Ladwa*, 1964 SCR (8) 273; *Rajkot Municipal Corporation* v. *Manjulben Jayantilal Nakum*, 1997 9 SCC 552.
89 Roy and Woerdman, 'Situating *Urgenda Versus the Netherlands*', p. 8.
90 J. Calvin, *Institutes of the Christian Religion (translated by H. Beveridge)*, Edinburgh: Calvin Translation Society, 1845; See also Montesquieu, *Spirit of the Laws (Translated by A. M. Cohler, B. C. Miller and H. S. Stone)*, Cambridge: Cambridge University Press, 1989.
91 M. Fordham, *Judicial Review Handbook*, Oxford: Hart Publishing, 2012; see also K. L. Hall, *Judicial Review and Judicial Power in the Supreme Court: The Supreme Court in American Society*, New York: Routledge, 2014.
92 *Urgenda*, para 4.100.
93 *Urgenda*, para 4.97.
94 *Urgenda*, para 4.101.
95 *Urgenda*, para 4.97.
96 Hoge Raad (Dutch Supreme Court), 17 Dec. 2010, NJ 2012/155 (*Wilnis*), and Hoge Raad (Dutch Supreme Court), 30 Nov. 2012, TBR 2013/72 (*Dordtse Paalrot*).
97 Zeben, 'Duty of Care for Climate Change Mitigation', p. 354.

98 *AEP* v. *Connecticut.*
99 5 US 137 (1803).
100 *Kesavananda Bharati* v. *State of Kerala* (1973) 4 SCC 225.
101 J. Spier, 'Injunctive Relief: Opportunities and Challenges: Thoughts About a Potentially Promising Legal Vehicle to Stem the Tide', in J. Spier and U. Magnus (eds.), *Climate Change Remedies: Injunctive Relief and Criminal Law Responses*, The Hague: Eleven International Publishing, 2014, pp. 1–120.
102 Hilson, 'Climate Change Litigation: An Explanatory Approach'.
103 *Ibid.*
104 Peel and Osofsky, *Regulatory Pathways*, pp. 269–278.
105 Justice B. N. Kirpal, 'Developments in India Relating to Environmental Justice', www.unep.org/delc/Portals/119/publications/Speeches/INDIA%20.pdf (accessed on 22 September 2016).
106 AIR 1998 SC 640, pp. 651, 654.
107 *Mumbai Kamgar Sabha* v. *Abdulbhai*, AIR 1976 SC 1455; *Fertilizer Corporation Kamgar Union* v. *Union of India*, AIR 1981 SC 344.
108 *Bangalore Medical Trust* v. *B.S. Muddappa* (1991) 4 SCC 54; H. Salve, 'Justice between Generations: Environment and Social Justice', *Supreme But Not Infallible: Essays in Honour of the Supreme Court of India*, New Delhi: Oxford University Press, 2000, pp. 367, 370.
109 U. Baxi, 'Taking Suffering Seriously: Social Action Litigation and the Supreme Court', *International Commission of Jurists Review*, 1982, 4(6): 107–132, http://scholar.valpo.edu/cgi/viewcontent.cgi?article=1125&context=twls (accessed by Supreme Court on 22 September 2016).
110 *Sheela Barse* v. *Union of India*, AIR 1988 SC 2211; see also G. S. Tiwari, 'Conservation of Biodiversity and Techniques of People's Activism', *Journal of the Indian Law Institute*, 2001, 43(2): 191–220.
111 *T.N. Godavarman Thirumulpad* v. *Union of India*, AIR 1997 SC 1228; *M.C. Mehta* v. *Union of India (Vehicular Pollution case)*, 1998 8 SCC 648.
112 Constituent Assembly Debates Book No. 2, Vol. No. VII, Second Print 1989, pp. 967, 968.
113 AIR 1951 SC 332.
114 *Ram Jawaya Kapur* v. *State of Punjab*, AIR 1955 SC 549; *R.K. Dalmia* v. *Justice Tendolkar*, AIR 1958 SC 538; *Chandra Mohan* v. *State of U.P.*, AIR 1966 SC 1987; *Kesavananda Bharati* v. *State of Kerala*, AIR 1973 SC 1461; *Indira Gandhi* v. *Raj Narain*, AIR 1975 SC 2299.
115 *Vehicular Pollution case.*
116 *Ibid.*
117 AIR 1996 SC 2715.
118 AIR 1999 SC 812.
119 AIR 2000 SC 3751; see also P. Leelakrishnan, 'Environmental Law', XXXVI *Annual Survey of Indian Law*, India: Indian Law Institute, 2000, pp. 252–257.
120 *Urgenda*, para 4.101.
121 'Netherlands Ordered to Cut Greenhouse Gas Emissions', *BBC News*, June 24, 2015.
122 1932 AC 562.

13

THE POSSIBILITY OF A GLOBAL ENVIRONMENTAL ORGANIZATION

Niraj Kumar

In the West, the environmental movement stemmed from a desire to protect endangered animal species and natural habitats. In India, however, it arose out of the imperative of human survival. This was *environmentalism of the poor* (emphasis supplied), which sought to promote social justice with sustainability.[1] Therefore, it appears that the Western model was more about protection of biodiversity and the Indian model was more about sustainability. At this stage, a caveat must also be issued that none of them has been immune from other influences.

For most of human history, nature has been considered something to be conquered. Only recently has the idea that nature needs to be protected and conserved received general support.[2] Peace and survival of life on earth, as we know it, are threatened by human activities which lack a commitment to humanitarian values. Destruction of nature and natural resources results from ignorance, greed and lack of respect for earth's living beings. This lack of respect extends even to earth's human descendants, the future generations who will inherit a vastly degraded planet, if world peace does not become a reality, and destruction of the natural environment continues at present rate.[3]

Rene Dubos, the French microbiologist who popularized the maxim 'think globally, act locally' in connection with the 1972 Stockholm Conference on the Human Environment convened by the United Nations (UN), came to view the relationship of 'humankind and Earth' as "a diversity of systems of symbiosis that constantly undergo adaptive changes and thus contribute to a continuous evolution process of creation."[4] His view of the future of the biosphere is an optimistic one. He posits that nature and humanity will interact and adapt, and that human creativity is an essential sustaining force whereby civilizations learn to make the most of the natural environments on Earth.[5]

There are various global associations and institutions which have taken up the task of conservation. This chapter attempts to look into the effectiveness of two of them, namely the International Union for the Conservation of Nature (IUCN) and the United Nations Environment Programme (UNEP). It first attempts to look into their functioning primarily in connection with their efforts and impact in the area of biodiversity conservation and protection. Subsequently, it tries to identify the problematic areas. Lastly, the chapter will attempt to look into the possibility of cooperation between them and any further possibility of setting up a World Environmental Organization (WEO).

I. International Union for the Conservation of Nature

The IUCN was established in 1948 with the help of 18 governments, 124 national agencies and non-governmental organizations (NGOs), and a collective of individual scientists and lawyers.[6] Part I of the IUCN Statutes declares that the International Union for the Conservation of Nature is constituted in accordance with Art.60 of the Swiss Civil Code as an international association of governmental and non-governmental members.

Over a period of time, the IUCN has become one of the world's largest global environmental organizations, with almost 1,300 governmental and NGO members and more than 15,000 volunteer experts in 185 countries.[7] Since then, it has been working tirelessly to help the nations of the world address emerging threats to the environment with the objective to influence, encourage and assist societies throughout the world to conserve the integrity and diversity of nature and to ensure that any use of natural resources is equitable and ecologically sustainable.[8]

The IUCN has entered into strategic partnership with the Global Environment Facility; United Nations Environment Programme (UNEP), United Nations Educational, Scientific, and Cultural Organization (UNESCO); World Wide Fund for Nature (WWF); and World Bank to achieve its goals. The IUCN has also been instrumental in developing the Convention on Biological Diversity (CBD)'s 'Programme of Work on Protected Area'.[9] The IUCN has also been providing secretarial assistance to many high contracting parties, e.g. in the case of the Ramsar Convention.[10] The IUCN also serves as a principal scientific advisor to the Conference of the Parties in application of Article 23, paragraph 5 under the CBD. So there is enough evidence to suggest that the IUCN interacts with many of the existing environmental instrumentalities to achieve collective goals.

The IUCN was a major factor behind the preparations for the Stockholm Conference on the Human Environment in 1972.[11] The IUCN is often characterized as a 'hybrid' international organization since it has both non-governmental and governmental members. Its three major organizational parts–sometimes referred to as the 'three pillars'–comprise its membership, its scientific and technical commissions, and its worldwide Secretariat.[12]

The IUCN was also instrumental in creation of the Convention on International Trade in Endangered Species (CITES) in 1974. However, its Secretariat was later shifted to the United Nations Environment Programme (UNEP). It also provided considerable technical and strategic support for the initial planning and for start-up of the Biodiversity Convention.[13] The IUCN was the driving force behind the influential 1982 World Charter for Nature. Both CITES and the World Charter for Nature have played an important role in bringing nature conservation to international legal attention.[14] One of the major works of the IUCN has been the bringing out of a world conservation strategy in partnership with the UNEP and World Wide Fund for Nature (WWF). The World Conservation Strategy (WCS) lays down three basic principles for conservation: that essential ecological processes and life-support systems must be maintained; that genetic diversity must be preserved; and that any use of species and ecosystem must be sustainable.[15]

The IUCN Environmental Law Programme(ELP) has its overriding rationale to assist in laying the strongest possible legal foundation for the conservation of the natural environment, thereby underpinning and supporting conservation efforts at national and international levels.[16] It has been instrumental in the preparation of draft instruments, most notably:[17]

- The African Convention on the Conservation of Nature and Natural Resources (Algiers, 1968)
- The Convention on International Trade in Endangered Species of Wild Fauna and Flora – CITES (Washington DC, 1973)
- The Convention on the Conservation of Migratory Species of Wild Animals – CMS (Bonn, 1979)
- The ASEAN Agreement on the Conservation of Nature and Natural Resources (Kuala Lumpur, 1985)
- The Convention on Biological Diversity – CBD (Rio de Janeiro, 1992)
- The International Covenant on Environment and Development, in cooperation with the International Council of Environmental Law (ICEL) and the UNEP (1995–present)

IUCN has also contributed in the form of implementation of international agreements through regular advice and technical assistance on legal issues to the secretariats of multilateral environmental treaties (in particular CITES, Migratory Species, Ramsar, and Biodiversity) and production of studies and analyses of implementation issues (e.g. for CITES, Ramsar, Biodiversity).[18]

IUCN also has been providing technical assistance to developing countries upon request ranging from review, diagnosis and preparation of inventories of existing legislation, to identifying measures for strengthening national legislation, and assisting in drafting new legal instruments. Over the years, examples of countries receiving various levels and types of technical assistance include India, Saudi Arabia, Oman, Gambia, Kenya, Sudan, Uganda, Swaziland, Indonesia, Vietnam, the Solomon Islands, Argentina, Ecuador, Chile, Brazil, the Philippines, Eritrea and Ethiopia, to cite but a few. Assistance ranged from the development of a framework of environmental laws to the improvement of specific legislation in the field of protected areas, wildlife conservation, access to genetic resources, biodiversity legislation and other specific environmental fields.[19]

With biodiversity conservation at the core of its work, the IUCN has devised a thematic approach in four key areas, namely: changing the climate forecast; naturally energizing the future; managing ecosystems for human well-being; and greening the world economy. Its purpose, *inter alia*, is to enhance the engagement of members and contribution to internationally agreed targets to reduce the rate of biodiversity loss, and to apply an environmental perspective to achieving sustainable development goals.[20]

The IUCN's unique character under international law can be evaluated from several perspectives. First, the international personality of the IUCN must be understood. Second, the IUCN's own Statutes constitute a kind of *lexspecialis* (emphasis supplied) which States acknowledge under international law. Third, State practice has been to accord the IUCN recognition as an intergovernmental organization. Fourth, and finally, the IUCN's status under municipal law facilitates its transnational mission and activities.[21]

The IUCN's legal personality has been recognized under international law. The States that established the IUCN by the Act of the UNESCO Conference in1948 and the States that have since adhered to the IUCN Statutes have established the IUCN as an international juridical body for the specifically delimited role of advancing conservation and sustainable natural resource use.[22] Thus, the IUCN serves a mission on which all States depend and which no State can accomplish alone. Beyond these constituent characteristics of the IUCN, external recognition of its functional attributes also indicates its international status. Even though the IUCN meets the generally accepted criteria for international status under public international law, it is generally acknowledged that the presence of 'international personality' under international law does not necessarily determine the organization's status under municipal law. Thus, it remains necessary for each State to also accord to the IUCN such legal capacity as the IUCN requires for its operations with and within that State.[23] But one can't rest on the laurels that just because some agreement or treaty has been established, it will ultimately work for the improvement of the environment. The above view can be appreciated by the event quoted below.

In 1997, Robert Mugabe, the President of Zimbabwe, inaugurated the 10th Conference of Parties of the Convention of International Trade in Endangered Species of Wild Flora and Fauna (CITES) at Harare. The central tenet of his speech was that every species should pay for itself and only those that paid could stay. It was a stupendous statement that held the attention of over a thousand delegates, from more than 100 countries, who had gathered to decide the fate of several endangered wild species.[24] Therefore, it has to be appreciated that once an institution or organization has been created, it attains an organic life of its own. It becomes almost impossible for institutions like the IUCN to control their course.

The IUCN invested a considerable amount of its resources in the World Summit on Sustainable Development (WSSD), both in the preparatory processes and at the Summit itself. One could ask why a conservation organization should focus on an event like the Summit. Linking the conservation and sustainable development agendas is of both pragmatic and political importance to the IUCN. International interest in the environment is essential for conservation organizations in terms of their policy influence and financial viability. At the regional and national levels, globally agreed policies provide powerful benchmarks and targets to develop regional and national policies, laws and institutional arrangements necessary for conservation.[25]

Thanks to Kofi Annan's Water, Energy, Health, Agriculture and Biodiversity (WEHAB) initiative speech shortly before the Bali PrepCom, biodiversity resurfaced as a key issue for the Summit. The WSSD recognized that biodiversity conservation and sustainable use are essential for poverty alleviation and for achieving sustainable livelihoods and cultural integrity of people. This provides an important opportunity for conservation organizations to demonstrate the relationship between the environment and sustainable development, and to provide practical examples of how conservation can reduce poverty, thus making the linkage with MDG 7.[26] The section on biodiversity includes some specific issues that are important for the conservation agenda; the IUCN advocated the inclusion of some of these at Prep-Com IV, such as invasive and sustainable use, the second objective of the CBD.[27]

II. United Nations Environment Programme (UNEP)

The United Nations Conference on the Human Environment held in Stockholm in 1972 adopted a resolution on institutional and financial arrangements that led, among other things, to the establishment of the UNEP.[28] The UNEP's purpose is to promote cooperation and coordination among nations, to recommend environmental policies and to provide general policy guidelines in the international environmental arena for all nations.[29] The UNEP Secretariat is the focal point for environmental action and coordination within the United Nations system. The Governing Council of the UNEP, the Secretariat, is headed by the Executive Director, and the Environmental Fund are all located in Nairobi, Kenya, thereby making the UNEP the first UN body to have its headquarters outside the developed world.[30]

The UNEP is actively involved in the assessment and monitoring of the global environment. Through a program called Earthwatch, information exchange, research activities, monitoring of environmental issues and a continual review and evaluation of the environment on a global scale take place periodically in order to identify new problems.[31] The UNEP's involvement has been critical in the arrangement of various protocols, conventions and other agreements. It is a relatively small UN body and is limited by personnel and financial constraints.[32] The UNEP does not have the power that one of the more specialized agencies of the United Nations has, such as the Food and Agriculture Organization (FAO), and therefore it has little influence on the environmental policies pursued by other United Nations agencies. In addition, the UNEP, financed solely by voluntary contributions to the Environmental Fund, is inadequately funded. The UNEP's major limitation is its lack of implementation and enforcement powers at the national level. Unfortunately, it must rely on the member States to implement and comply with its endeavours.[33]

Unfortunately, the UNEP does not have the ability to create binding international law. Instead it merely studies, recommends and adopts non-binding resolutions and charters; this is done with the expectation that member States will feel an obligation to abide by the provisions and cooperate in safeguarding the environment on an international scale.[34] The so-called 'soft' law that is promulgated without binding force in the environmental arena consists of periodic enforcement by the General Assembly of the United Nations of the priorities set by the UNEP.[35]

The UNEP is now the only UN body charged exclusively with international environmental matters and has played an important role in the development of international environmental law, not least through its promotion of numerous regional sea treaties, the 1985 Vienna 'Ozone' Convention, and the 1992 Biodiversity Convention. But, in general terms, the UNEP has been a weak institution, which is somewhat underfunded and of relatively low visibility.[36]

The UNEP's constituent instrument *per se* does not mention its role in international environmental law-making, but it seems that it has now come to be regarded as its 'cornerstone'. The innovative exercise of such *implied power* (emphasis supplied) gives latitude to the institution as well as facilitates in actual governance by giving effect to such a mandate. Most institutions come across a situation during their lifespan when they are required to exercise certain powers or functions so as to 'effectuate the general purposes for which they have been created, though their constitutive instrument does not expressly provide for such a specific power or function'.[37] In the specific case of the UNEP, the attitude of States, especially in the Governing Council, can act as a barometer of endorsement for the exercise of such implied powers. In this sense, as creatures of institutionalized cooperation, international environmental institutions (IEIs) evolve as dictated by the interests and expectations of the sovereign States.[38]

In the post-Stockholm era, the UNEP has generally been the centrepiece of the UN's efforts to protect the global environment. At the institutional level, the UNEP has simply been a programme, a subsidiary organ of the UN General Assembly. Though its legal status is not that of an independent international organization, it does not appear to have deprived this UN body of some measure of international legal personality as it is required to fulfil certain tasks on the international plane (such as entering into a headquarters agreement with the host country as well as other international agreements and arrangements as the host institution), as per the criteria laid down by the International Court of Justice in the *Reparations* case.[39]

The UNEP even provides secretariat services to global conventions like the Convention on International Trade in Endangered Species (CITES) and the Convention on the Conservation of Migratory Species of Wild Animals (CMS), amongst OZONE, BASEL, and CBD also.[40]

But there are many criticisms of the UNEP, such as the fact of not taking a very definitive position towards the Biodiversity Convention. For example, Melinda Chandler in the context of the CBD Convention wrote that

> the proposed texts put forth by the chairman of the Inter-governmental Negotiating Committee INC and the executive director of UNEP were even more objectionable. The text proposed in their informal note dated April 9, 1992, provided that the Biodiversity Convention would not affect the rights and obligations of any Party deriving from an existing international agreement "relating to the conservation and sustainable use of biological diversity.[41]

She further asserts that "In the press to have the Convention completed by the United Nations Conference on Environment and Development UNCED, the executive director of UNEP intervened in the negotiations to force issues to a conclusion."[42]

Part of the blame for the UNEP's weak mandate can be attributed to the relatively recent arrival of international environmental principles onto the international agenda. Since international environmental problems were not recognized when the UN was being developed, environmental protection was not included in the UN Charter. Authority for establishing environmental programs within the UN has come from the Charter's determination to promote better standards of life. However, environmental protection and the idea of sustainable development were not given the validity that specific mention in the UN Charter would have given.[43]

Moreover, lack of an environmental directive to a specific agency has resulted in the development of many separate environment programs within the UN specialized agencies and other organizations. For example, specialized agencies that deal directly with environmental protection issues include the International Labour Organization (ILO); the Food and Agriculture Organization; the International Maritime Organization; the United Nations Educational, Scientific and Cultural Organization; the World Health Organization; the World Meteorological Organization; and the World Bank.[44]

III. Cooperation between the UNEP and the IUCN

Twice a Memorandum of Understanding (MoU) was entered into between the United Nations Environment Programme and the

IUCN-the World Conservation Union for Cooperation on Biodiversity Conservation and Protected Areas. So for all practical purposes, it has been recognized that Biodiversity Conservation requires a unified and concerted effort on the part of both the UNEP and the IUCN. If one looks into the internal affairs of both entities, it will become quite apparent that both of them are facing a funding crunch. It can also be noticed that in many cases, the funding agencies are the same. It makes a sheer financial sense that at least in the area of common interest, they must work together to optimize utilization of resources.

The Mediterranean marine fauna and vegetation have evolved over millions of years and the Mediterranean has a high biodiversity, with 7.5% of the recorded global marine fauna and 18% of the marine flora (out of them, 28% endemic), and can be considered a marine species diversity hotspot.[45] The continuous demand for nature and natural resources by human activities puts the Mediterranean at risk. There is severe impact on the marine environment due to an increasing population living on the coasts of the riparian countries. The threats are complex and the biodiversity loss calls for well-coordinated responses (UNEP/MAP-RAC/SPA, 2003).[46]

There are currently several organizations addressing the management and conservation of the Mediterranean. Among them, the Regional Activity Centre for Specially Protected Areas (RAC/SPA), based in Tunis, works under the framework of the UNEP-Mediterranean Action Plan (MAP) for the Barcelona Convention and its Protocols, which aims to protect the Mediterranean marine and coastal environment and mitigate pollution.[47] The RAC/SPA assists the Secretariat of the Barcelona Convention and its Parties in the implementation of their commitments under the Protocol Concerning Specially Protected Areas and Biological Diversity in the Mediterranean (SPA/BD Protocol), which came into force in 1999. It also assists them in the implementation of the 'Strategic Action Programme for the Conservation of Biological Diversity in the Mediterranean Region' (SAP BIO), adopted by the Contracting Parties in 2003.[48] ACCOBAMS, GFCM, IUCN-Med and RAC/SPA have been collaborating since 2001 and have established several cooperation agreements (MoUs, MoCs) in order to harmonize efforts and strengthen synergies in supporting conservation and sustainable use of marine living resources in the Mediterranean.[49]

In addition to a MOC between IUCN-Med and UNEP/MAP, IUCN-Med and RAC/SPA have an MoU on collaboration for the Med Open Seas project. They have collaborated on the open seas subject in several thematic reviews, including on canyons and sea mounts, and on definition files for Mediterranean EBSAs; and collaborated with experts of UNEP/MAP and the CBD, as well as on the UNEP/MAP priority conservation areas embracing the open seas, including deep seas. The IUCN–Med is cooperating with RAC/SPA on its efforts to identify and make spatial mapping of areas fulfilling criteria to be candidate sites for inclusion on the list of Specially Protected Areas of Mediterranean Importance (SPAMIs). In addition, scientific and technical reports have been produced in close collaboration and presented in consultation processes and several fora.[50]

The threats on the marine environment are increasing, and biodiversity loss is occurring every day. The financial, human and logistical resources to prevent them are limited. The above-mentioned bodies are following diverse mandates that in certain cases request them to act through different approaches on common fields of interest of their respective parties or members, accordingly.[51]

Therefore, there is an urgent need for collaboration among all of the above bodies in a much more structured way, considering at the same time their different perspectives. It needs to be considered as to how to use their different areas of expertise to pursue common tasks and make compatible and stronger eventual overlapping ones. Despite the active engagement, there is no common strategy amongst the organizations mentioned above to act simultaneously. The goal pursued is to strengthen regional ocean governance, by increasing the synchronous collaboration amongst the GFCM, ACCOBAMS, IUCN–Med and RAC/SPA in cooperation with MedPAN, with a specific attention, as a first concrete step, on how to address the issues of common interest in the Alboran Sea, the Adriatic Sea and the Sicily Channel.[52]

The IUCN and UNEP have also come up with detailed elements for a strategic framework in biodiversity conservation and natural resources management. They are, *inter alia*:

i   Monitor marine biodiversity and natural resources in line with principles, indicators under the Ecosystem Approach,
ii Pursue multilateral meetings and ad-hoc experts workshops and exchange to address priority areas of interest linked to specific species groups, areas or sub-regions,
iii Mitigate impacts of fisheries and aquaculture on marine habitats and species,
iv Promote ecosystem approach and sustainable use of natural resources in Mediterranean region,
v   Identify key areas regarding biodiversity and/or marine life resources in need of appropriate spatial management measures

and pursue synergic ways to plan/institutionally assist their management.[53]

The International Union for Conservation of Nature and Natural Resources (IUCN), the United Nations Environment Programme (UNEP), and the WWF (World Wide Fund for Nature) recently signed a MoU pledging their support to the IUCN Conservation Monitoring Centre.[54] Signing on behalf of the three organizations were Dr Martin W. Holdgate, the IUCN's Director-General; Dr Mostafa K. Tolba, the UNEP's Executive Director; and Mr Charles de Haes, the WWF's Director-General.[55] Each of the three partners has pledged £200,000 a year towards establishing the Centre as a focal point for documentation and distribution of information on the state of the planet's threatened species, habitats and living resources.[56] To be renamed the World Conservation Monitoring Centre (WCMC), the Centre will aim for a global overview of conservation data while at the same time making data available for those carrying out conservation and natural resources assessments at regional and national levels.[57] The WCMC will also develop a network of national databases as a means of supporting conservation action at the local level in the less-developed countries.[58] "It will no longer be good enough for officials to make bad decisions and claim they didn't have enough information," said Dr Robin Pellew, Director of the WCMC.[59]

The above discussion was intended to have a sense of the cooperation which is already happening. It also reflects the realization of the importance of communication. At the same time, it also establishes the possibilities of cooperation. But the more important issue in the context of the theme of this chapter is to explore the possibilities of the creation of one Global/World Environmental Organization.

IV. An enquiry into the feasibility of a World Environmental Organization

There have been 40 years of debate over a World Environmental Organization (WEO), starting with calls from US foreign policy strategists for an International Environment Agency. Instead of creating an agency, following the 1972 Stockholm Conference on the Human Environment, the United Nations Environment Programme (UNEP) was created the following year. Despite this positive step, this was a weaker reform than many proposed, and effectively curtailed further debate over the need for a specialized agency.[60]

The logic of collective action in the international environmental arena is clear. Absent cooperation, the spillover of pollution from one country onto its neighbours or into the shared space of the global commons, as well as the over-exploitation of shared natural resources, promise not just environmental degradation but also economic inefficiency, political instability and diminished social welfare. Ecological interdependence, expanded economic inter-linkages, and tensions at the trade–environment interface have also made environmental cooperation an important element in the process of establishing the terms of engagement for international commerce.[61]

The discussion is also motivated both by recent calls for such an organization in light of WTO trade and environment conflicts and the relative absence of internalization of global externalities.[62] Just as the GATT/WTO tries to liberalize trade in goods and services by removing border impediments on trade through negotiated exchanges of trade policy concessions, so it can be argued that a World Environmental Organization (WEO) should focus on removing impediments to bargaining (and trades) on the global environment.[63] If bargains were struck, the result would be improved environmental quality and transfers of resources for developmental purposes to poorer countries who, in the main, are custodians of these assets.[64] Evaluation, both the rationale for and possible organizational form of such a WEO, in light of experience with the WTO, and its implications (both positive and negative) for the developing countries, is required to be done.[65] Those from a trade and environment background may tend to see a WEO as modelled on a WTO, but as the preceding discussion makes clear, the problems each seeks to solve are different, and hence the two are at best only loosely related.[66] Both are agencies which in varying ways seek to promote bargaining – one over trade and the other over bargaining – but one seeks gains from trade while the other seeks internalization.[67] To the degree that the WTO would like to see a parallel agency as the focal point for environmental debate to take trade and environment pressure off the WTO, institutional support from the WTO for the creation of a WEO may be quickly forthcoming.[68]

Another rationale for a global organization can be found in the following statement: "The globalization of environmental problems means that environmentalists – and economists, labour leaders, and other citizens – need to embrace a concept increasingly being touted, albeit in rarefied policy circles: a single global group with real power, a World Environment Organization."[69] Global problems require international strategies and coordination, particularly for such goals as maintaining biodiversity, keeping the oceans clean and lowering carbon emissions. But all too often, national sovereignty divides the world into untenable slices. The European Union, for instance, has developed a coordinated approach to tackling greenhouse gases through conservation measures and technological development. Yet, absent cooperation from the United States, China and other major powers, such a task might come to seem quixotic. The formation of a World Environment Organization would provide global environmental standards with real teeth. It would also provide an umbrella for environmental organizations, and counteract the problem of single-issue solutions for linked problems.[70]

The existing system promotes international environmental policy through small, uncoordinated steps that must each go through the international political process before being ratified and implemented. This method of governing does not keep up with the pace of environmental problems. The formidable task of creating environmental policies that can be agreed upon both by industrialized and developing countries has led to treaties that address individual issues and that are filled with ambiguities.[71]

The idea of an international agency for the environment is by no means new. The attention to the environment in the early 1970s led some analysts to propose the establishment of an international agency. In a lead article in *Foreign Affairs* in April 1970, George Kennan proposed an "International Environmental Agency" as a first step towards the establishment of an "International Environmental Authority." One of the most comprehensive proposals in that era was developed by Lawrence David Levien, who called for a "World Environmental Organization" modelled on the practice of the International Labour Organization (ILO), which had been created in 1919. The establishment of the UNEP by the UN General Assembly in 1972 settled the organizational question, although some observers at the time viewed the answer as unsatisfactory.[72]

The most important suggestion came from Sir Geoffrey Palmer, the former Prime Minister of New Zealand, who advocated new methods of making environmental law and called for action at the Rio Conference to establish a specialized UN agency for the environment. Palmer proposed the creation of an "International Environment Organization" borrowing loosely from the mechanisms of the ILO. Palmer saw an opportunity for a "beneficial restructuring" of the world's environmental institutions, that "would involve cutting away existing overlaps in international agencies".[73]

Within a couple of years, new support for institutional change came from a different direction–the international debate on 'trade and the environment', which had been rekindled in 1990and was in full swing by 1993. Both camps in this debate saw the weak state of the environmental regime as a fundamental concern.[74] With one foot in both camps, Daniel C. Esty became a champion of a new international environmental organization. His article 'GATTing the Greens' contended that solving the trade and environment conflict would necessitate not only greening of trade rules but also a stronger organization of environmental governance. In 1994, Esty optimistically named his proposed institution the Global Environmental Organization (GEO) and, in a series of studies, he strengthened the environmental arguments for institutional change by showing how the level of concerted action needs to match the level of the externality.[75]

Ford Runge was another early advocate of institutional reforms. In 1994, he called for a World Environmental Organization (WEO)to give a stronger "voice" to environmental concerns. Runge suggested that a new organization could serve as a "chapeau" to the growing number of international environmental treaties. In his most recent study, Runge argues that a GEO could alleviate environmental pressure on the World Trade Organization. The ranks of academic advocates for a World Environment Organization (WEO) have expanded in recent years. Rudolf Dolzer, for example, has proposed a global environmental authority "with the mandate and means to articulate the international interest in an audible, credible and effective manner". Frank Biermann, a research fellow at the Potsdam Institute for Climate Impact Research, has provided the most systematic analysis of what a WEO would do. Economists John Whalley and Ben Zissimos have defined an economic role for a WEO. Peter Haas has advocated a Global Environmental Organization (GEO) to centralize support functions like research, technology databases, and training for the various environmental regimes. Jeffrey Frankel argues that the UNEP "is so weak an institution that it should be replaced from scratch". The German Advisory Council on Global Change has recommended that the UNEP be upgraded into an International Environmental Organization as a separate entity or a specialized agency within the UN system.[76]

Many environmental problems cannot be combated at the national level. Professor Daniel Esty presents a three-pronged argument supporting the necessity of a GEO. A GEO is needed "to address global environmental harm; to facilitate cooperative approaches to common environmental problems; and to reduce competitiveness tensions and the stresses such tensions place on efforts to achieve rigorous environmental policies on a national basis".[77] Esty's first argument recognizes the inherent global nature of many environmental problems. Pollution does not respect boundaries. Although a polluting activity may have taken place in one country, the harm may be felt in another. For example, emissions pollutants from one nation can easily move into the air space of other nations. Therefore, nations must create a collective plan of action to repair damage that has already been done to the environment and to prevent future damage.

Esty's second argument focuses on local problems that are commonly experienced in many nations. An International Environmental Organization could coordinate research and gather and disseminate information on these types of problems. The current structure is such that "national governments as well as other international organizations are duplicating each other's work and squandering precious environmental resources". A Global Environmental Organization could coordinate such national and international efforts and further the development of technology to preserve and protect environmental resources.[78]

Esty's third argument confronts concerns that environmental protection measures would hamper free trade:

> Governments recognize that burdening their own domestic industry with cleanup costs, which would largely benefit others around the world, might disadvantage their own producers competing in the global marketplace against companies whose governments do not require similar spending on pollution abatement. Thus, governments choose not to adopt stringent environmental standards.[79]

A GEO could facilitate the development of common environmental standards for implementation by all countries. By holding all producers to the same standard, no one country would have an economic advantage due to less stringent environmental regulations. Despite the increasing need for collaboration among nations, collaboration alone is not enough; "there must be a recognized authority to enforce property rights and to regulate behaviour" in order to make the market work on behalf of the environment.[80]

The central objective for such an organization would thus be to facilitate cross-country deals on environmental issues with the aim of raising environmental quality.[81] These deals are in our view unachievable with present institutional arrangements, and these arrangements have proved themselves inadequate to the task. For a WEO to do this, a variety of impediments to such deals have to be addressed, including property rights ambiguities, free riding, and verification and enforcement of contracts.[82] We would also see a series of spin-off benefits that can also be realized from meeting this central objective, such as underpinning domestic environmental policies, and particularly so in developing countries.[83]

The international health problems and programs that led to the establishment of WHO are similar to the environmental problems and programs now being faced by the international community. WHO has earned tremendous success over the last 50 years. Accordingly, WHO can be used as a model for establishing a Global Environmental Organization. It is generally accepted that the contemporary world can be better understood, and future trends more clearly perceived and assessed, if one possesses some familiarity with the past. In particular, the future of an international organization may indeed be more successfully perceived if the experiences of the last century, in particular, are kept in mind.[84]

A politically viable first step could be upgrading the UNEP to a UN specialized agency. Although some may say that proposing a WEO denies political realities, WFM would point out that historic progress is occurring in other sectors.[85] The recently established International Criminal Court (ICC) could have wide-ranging implications for international sustainable development and environmental law.[86] While the ICC will not be a court for environmental disputes, the establishment of the ICC, and its coming into force only four years after the adoption of the Rome Statute, demonstrates that international justice may soon be employed to address social, environmental and economic rights as it is now being done for human rights.[87]

Notably, a number of governments have come forward with proposals for establishing a global agency for environmental protection, among them Brazil, France, Germany, New Zealand, Singapore and South Africa. The French government has now taken the lead by announcing its intention to use its presidency of the European Union in the second half of 2000 for an initiative to replace the UNEP with an "organisation mondiale de l'environnement".[88] Yet most actors in this debate mean different things when talking about a new organization, and no consensus on its optimal design has yet emerged. In essence, proposals can be grouped into three different models for a World Environment Organization: the "cooperation," the "centralization," and the "hierarchization" models.[89] In the cooperation model, the UNEP would merely be upgraded into a specialized UN agency, such as the World Health Organization (WHO) or the International Labour Organization (ILO), and no other agencies or regimes would be disbanded.[90] Advocates of a centralization, or 'streamlining', model call for wider reform. They want to integrate various existing agencies, programs and regimes into a World Environment Organization, which they expect to result in efficiency gains and improved environmental policy coordination. The integration of environmental regimes could follow the model of the WTO, which has integrated various multilateral trade agreements under its umbrella.[91] A third model is the hierarchization model. This model calls for a quasi-supranational agency on environmental issues that would have decision-making and enforcement powers vis-à-vis a minority of non-consenting states if global commons are at stake. This would help, it is argued, to overcome the free-rider problem in global environmental governance. The only example for such a body so far is the UN Security Council with its far-reaching powers under Chapter VII of the UN Charter.[92]

Of course, for both North and South, a powerful World Environment Organization would hardly be acceptable if decision-making procedures did not grant them sufficient control over the outcome of negotiations and the organization's future evolution.[93] Thus, a strong organization seems feasible only with a double-weighted majority system comparable to that of the Montreal Protocol as amended in 1990 or of the Global Environment Facility as reformed in 1994.[94] In both institutions, decisions require the assent of two-thirds of members that must include the simple majority of both developing and developed countries.[95] Such decision-making procedures based on North–South parity – that is, veto rights for both South and North as a group – could ensure that the World Environment Organization would not evolve into a conduit of eco-colonialism as some Southern actors suspect.[96]

One of the important structural issues which has been explored is the role of non-governmental organizations in this kind of organization. Generally, international organizations allow representation by State players only. Even if one fundamentally disagrees with the membership of non-governmental organizations, one can argue for some status, like observer status, for them, given the role played by them in environmental issues historically.[97] The barriers to achieving these goals are also many folds. There is a general sense of unease among proponents of free trade in accepting the environment as a trade barrier. However, these concerns are not well founded and interaction between a centralized environmental organization and other issue-specific international organizations will result in fewer conflicts.[98]

Establishing a World Environment Organization would also create a number of welfare gains by reducing bureaucratic overlap and by increasing the overall efficiency in the system. For instance, the sometimes-minuscule secretariats of multilateral environmental agreements could be integrated into the new organization.[99]

V. Conclusion

There has been a general trend in the post-World War I and II era towards creation of global organizations, for example, the WTO, World Intellectual Property Organization (WIPO)and ILO, to name a few. Some of them proved to be very potent like the WTO, and others like the ILO were not so effective. So if one is inclined to generalize on the basis of success or failure of the existing organization, one may find evidences inconclusive on either side.

The lack of adequate international environmental governance (IEG) is a result of a fundamental injustice in the current state of global governance: tremendous power and resources have been concentrated in international finance and trade without a corresponding legal and institutional authority for the environment, social concerns and human rights.[100] The lack of a WEO remains a clear sign that the environment is not given the same global status as trade, health and labour, or even maritime affairs, intellectual property and tourism – all of which are represented by UN agencies.[101] The reform of the UNEP is a step in the right direction, but the job is far from done. Will there be renewed calls for more cohesive global governance following the latest report from the Intergovernmental Panel on Climate Change (IPCC)? The UN Economic and Social Council does have the mandate to create a specialized agency, a move that is strongly supported by 35 nations but not backed by the US, Russia or China. Forty years on from Stockholm and 20 years on from Rio, a World Environment Organization still seems a long way away.[102] Some critics in the environmental and development communities claim that a WEO would reflect the kind of centralized structure that has failed in the past, and that only local solutions can work. These critics make the mistake of applying to globalization-as-we-know-it the lesson that globalization is, by its very nature, bad. Suspicion of a WEO may stem from fear of the World Trade Organization (WTO), which looms as a model of monolithic decision-making largely in the interests of corporations tied to the global North. Globalization as currently practiced, for instance, often leads to a reallocation of resources, labour and waste to countries with weak environmental-protection laws. But these environmental injustices, too, must be tackled on a global scale.[103]

Equally flawed is the critics' argument that environmental protection is too complex an issue to be dealt with by one agency. That almost all countries have established a distinct ministry for the environment reveals that environmental policy can indeed be dealt with by one focal point within an administrative system.[104]

Judged against the weak UNEP, a World Environment Organization would be in a much better position to embark on a new global capacity-building and technology-transfer initiative. It could also host the clean development mechanism and the clearinghouse for the future emissions trading scheme.[105]

Hence, it seems that a World Environment Organization would be in the interests especially of developing countries. It could provide, for example, for a more efficient and more effective transfer of technology and financial assistance, and it could establish a more efficient negotiating system that would increase opportunities for southern nations to raise their voice in global FORA.[106]

It goes without saying that the critics treating it as a behemoth bureaucracy make use of the prevailing clichés of the United Nations as a gargantuan bureaucracy. As usual, this argument loses some of its power once UN agencies are compared to national bureaucracies.[107]

But an enquiry in the light of cost-benefit analysis is the demand of the day. This chapter holds the view that the current international environmental structure is having many fault-lines. It also argues that creation of a Global Environmental Organization must not be a monolith with a great centralizing tendency. The major arguments in favour of a Global Environmental Organization are *inter alia*; it will lead to coherent emphasis on environmental issues both internally, i.e. between various environmental organizations like the IUCN, UNEP, etc., and externally *vis-à-vis* other issue-based international organizations like the WIPO, WTO and WHO, and maybe also with international agencies like the IAEA. Lack of coordination has resulted in a scenario where, after the collapse of the Kyoto Protocol, there doesn't exist even a single binding climate change treaty. Most of the world entities agree on the consequences of climate change – for example, WHO will say climate change will affect human health and the WTO will agree that there will be economic costs of climate change – but in the absence of coordination, there is no unified stand.

The existence of powerful international trade and financial regimes without comparable legal and institutional structures for social and environmental standards allows the WTO to act as the de facto arbiter on environmental issues.[108] However, the WTO is an institution that not only lacks a core competency on environmental issues and policy, but views the environment as a commodity to be exploited rather than a resource requiring management and conservation.[109] The result is that environmental, social and human rights issues, treaties and commitments are trumped by finance and trade interests.[110] Rather, it should be the case that these considerations get prioritized ahead of finance and trade.[111]

I'd like to make the same submissions as made by John Whalley and Ben Zissimos,[112]

> No wide ranging proposal such as ours will be adopted overnight, and we recognize many impediments to it which we discuss in the text. But if global environmental quality worsens in the years ahead, as many fear, perhaps these ideas will play a role in providing intellectual underpinnings, to institutional change in this area.

Achieving the above goals will not be very problematic because of the fact that most other organizations and agencies have in-house environmental branches. It has been accepted by almost all players that there can't be even benign neglect of the environment.

Notes

1 R. Guha, *Environmentalism: A Global History*, India: Penguin Allen Lane, 2014, pp. xi–xii.
2 L. E. Christoffersen, 'IUCN: A Bridge-Builder for Nature Conservation', *Green Globe Yearbook*, 1997: 59, www.fni.no/ybiced/97_04_christoffersen .pdf (accessed on 6 October 2016).
3 V. Menon and M. Sakamoto (eds.), *Heaven and Earth and I*, New Delhi: Penguin, 2002, p. 20.
4 N. A. Robinson, 'IUCN as Catalyst for a Law of the Biosphere: Acting Globally and Locally', *Environmental Law*, 2005, 35: 250.
5 *Ibid.*
6 *Ibid.*
7 See www.iucn.org/secretariat/about (accessed on 13 September 2016).
8 Statute of the IUCN, Section 2.
9 See www.cbd.int/history/ (accessed on 13 September 2016).
10 See www.ramsar.org/about/the–ramsar–convention–secretariat (accessed on 13 September 2016).
11 Christoffersen, 'IUCN: A Bridge-Builder'.
12 *Ibid.*
13 *Ibid.*, p. 63.
14 S. Bell and D. McGillivray, *Environmental Law*, New York: Oxford University Press, 2008, p. 149.
15 IUCN Bulletin, 17: 7–12, www.fni.no/ybiced/97_04_christoffersen.pdf (accessed on 2 October 2016).
16 www.iucn.org/about/work/programmes/environmental_law/elp_about_ achieve (accessed on 29 January 2016).
17 www.iucn.org/about/work/programmes/environmental_law/elp_about_achieve.
18 www.iucn.org/about/work/programmes/environmental_law/elp_about_ achieve, p. 2.
19 www.iucn.org/about/work/programmes/environmental_law/elp_about_ achieve.

20 IUCN/WCC-4, 'At the Crossroads of Conservation', *Environmental Law & Policy Journal*, 2009, 59: 39.
21 Robinson, 'IUCN as Catalyst', p. 298.
22 *Ibid.*, p. 300.
23 *Ibid.*, pp. 298–305.
24 Menon and Sakamoto, *Heaven and Earth and I*, p. 3.
25 'Scaling the Summit: IUCN at the World Summit on Sustainable Development', www.iucn.org (accessed on 3 February 2016).
26 Scaling the Summit: IUCN.
27 *Ibid.*
28 Bell and McGillivray, *Environmental Law*, p. 146.
29 See www.unep.org/about/ (accessed on 13 September 2016).
30 A. W. Samaan, 'Enforcement of International Environmental Treaties: An Analysis', *Fordham Environmental Law Journal*, 1993–1994, 5: 263.
31 *Ibid.*
32 *Ibid.*
33 *Ibid.*
34 *Ibid.*, p. 264.
35 *Ibid.*, p. 269.
36 Bell and McGillivray, *Environmental Law*, p. 149.
37 B. H. Desai, 'UNEP: A Global Environmental Authority?' *Environmental Law & Policy Journal*, 2006, 36: 137.
38 *Ibid.*, p. 138.
39 *Ibid.*
40 *Ibid.*, p. 140
41 M. Chandler, 'The Biodiversity Convention: Selected Issues of Interest to the International Lawyer', *Colorado Journal of International Environmental Law & Policy*, 1993, 4: 149.
42 *Ibid.*, p. 175.
43 K. Tyler Farr, 'A New Global Environmental Organization', *Georgia Journal of International and Comparative Law*, 1999–2000, 28: 502.
44 *Ibid.*
45 Available at www.unepmap.org/ (accessed on 2 February 2016).
46 www.unepmap.org/.
47 www.unepmap.org/.
48 www.unepmap.org/.
49 www.unepmap.org/.
50 www.unepmap.org/.
51 www.unepmap.org/.
52 www.unepmap.org/.
53 UNEP(DEPI)/MED WG.408/17, pp. 5–6, www.unepmap.org/.
54 International Cooperation on World Conservation Information Centre: Joint Announcement by UNEP, IUCN, and WWF, www.cambridge.org/core/services/aop–cambridge–core/content/view/9FE988500CE2D3E9638DEA22996C7A47/S037689290002991Xa.pdf/international–cooperation–on–world–conservation–information–centre–joint–announcement–by–unep–iucn–and–wwf.pdf (accessed on 10 September 2016).
55 *Ibid.*
56 *Ibid.*
57 *Ibid.*

58 *Ibid.*
59 *Ibid.*
60 See http://theconversation.com/why–is–there–still–no–world–environment-organisation–25792 (accessed on 20 September 2016).
61 *Daniel C. Esty*, 'Climate Change and Global Environmental Governance', *Global Governance*, 2008, 14: 111–112.
62 See J. Whalley and B. Zissimos, 'A World Environmental Organization?' www2.warwick.ac.uk/fac/soc/pais/research/researchcentres/csgr/papers/workingpapers/2000/wp6300.pdf (accessed on 20 September 2016).
63 *Ibid.*
64 *Ibid.*
65 *Ibid.*
66 *Ibid.*
67 *Ibid.*
68 *Ibid.*
69 E. Goffman, 'Why We Need a World Environment Organization', http://grist.org/article/goffman/ (accessed on 20 September 2016).
70 *Ibid.*
71 Farr, 'A New Global Environmental Organization', p. 506.
72 S. Charnovitz, 'A World Environment Organization', *Columbia Journal of Environmental Law*, 2002, 27: 325.
73 *Ibid.*
74 *Ibid*, p. 326.
75 *Ibid.*
76 *Ibid.*, pp. 327–328.
77 Farr, 'A new global Environmental Organization', p. 495.
78 *Ibid.*
79 *Ibid.*
80 *Ibid.*, pp. 495–496.
81 Whalley and Zissimos, 'A World Environmental Organization?'.
82 *Ibid.*
83 *Ibid.*
84 Farr, 'A New Global Environmental Organization', p. 510.
85 W. R. Pace and V. Clarke, 'The Case for a World Environment Organization', www.federalist–debate.org/index.php/current/item/575–the–case–for–a–world–environment–organization (accessed on 19 September 2016).
86 *Ibid.*
87 *Ibid.*
88 F. Biermann, 'The Emerging Debate on the Need for a World Environment Organization: A Commentary', p. 44, http://glogov.net/images/doc/BiermannWEOGEP2001.pdf (accessed on 19 September 2016).
89 *Ibid.*, p. 45.
90 *Ibid.*, p. 46.
91 *Ibid.*
92 *Ibid.*, p. 47.
93 *Ibid.*
94 *Ibid.*
95 *Ibid.*
96 *Ibid.*
97 For example, Greenpeace, WWF, etc.

98 The author has taken inspiration from writings of Steve Charnovitz, Daniel Esty, Lawrence David Levien, C. Ford Runge and Paul C. Szasz in the formation of his opinion on the issue.

99 Biermann, 'The Emerging Debate'.

100 Pace and Clarke, 'The Case for a World Environment Organization'.

101 See http://theconversation.com/why–is–there–still–no–world–environment–organisation–25792 (accessed on 20 September 2016).

102 Pace and Clarke, 'The Case for a World Environment Organization'.

103 Goffman, 'Why We Need a World Environment Organization'.

104 Biermann, 'The Emerging Debate'.

105 *Ibid.*

106 *Ibid.*

107 See *Ibid.*

108 Pace and Clarke, 'The Case for a World Environment Organization'.

109 *Ibid.*

110 *Ibid.*

111 *Ibid.*

112 Whalley and Zissimos, 'A World Environmental Organization?'.

TABLE OF CASES

331

INDEX